117 Advances in Polymer Science

High Performance Polymers

Editor: P. M. Hergenrother

With contributions by
F. E. Arnold, Jr., F. E. Arnold, K. J. Bruza,
P. Chen, E.-W. Choe, T. S. Chung, J. W. Cornell,
J. E. Economy, K. Goranov, J. L. Hedrick,
P. M. Hergenrother, M. Jaffe, R. A. Kirchhoff,
J. W. Labadie, S. Makhija, H. D. Stenzenberger,
W. Volksen

With 170 Figures and 89 Tables

Springer-Verlag
Berlin Heidelberg New York
London Paris Tokyo
Hong Kong Barcelona Budapest

Volume Editor:

Paul M. Hergenrother
NASA Langley Research Center
Hampton, VA 23681-0001, USA

ISBN 3-540-58238-X Springer-Verlag Berlin Heidelberg New York
ISBN 0-387-58238-X Springer-Verlag New York Berlin Heidelberg

© Springer-Verlag Berlin Heidelberg 1994
Library of Congress Catalog Card Number 61-642
Printed in Germany

Typesetting: Macmillan India Ltd., Bangalore-25
Printing: Saladruck, Berlin; Bookbinding: Lüderitz & Bauer, Berlin
SPIN: 10126816   02/3020 - 5 4 3 2 1 0 - Printed on acid-free paper

Preface

High performance polymers can be broadly defined as materials that exhibit properties superior to those of state-of-the-art materials. However, many scientist and engineers refer to high performance polymers as materials that exhibit not only a unique combination of properties superior to those of state-of-the-art materials but also significantly better elevated temperature behavior. Thus high performance polymers are generally considered to be materials containing high aromatic content and/or heteorcyclic units with relatively rigid segments that impart high glass temperatures, good thermooxidative stability, and high mechanical properties. Representative well-known families of high performance polymers are poly(arylene ether)s, polyimides, polybenzimidazoles and rigid rod or extended chain heterocyclic polymers. This book discusses these types of polymers.

It is well-recognized that many polymers containing little or no aromatic content having low or moderate glass transition temperatures and more conventional thermal stability can also be classified as high performance polymers. Representative examples are certain silicones, organofluoro polymers, phosphorus containing polymers, aramids, modified high impact resistant polymers and high modulus, extended chain fibrous polymers. These materials, however, are not discussed in this book.

High performance polymers probably reached the peak of their popularity in the late 1980' s. At that time, high performance or specialty polymers were under development in many organizations particularly aerospace, chemical, computer and electronic companies as well as in academic and goverment laboratories. With the end of the cold war and the downturn of the economy worldwide in the early 1990' s, many companies abandoned or significantly reduced their work on high performance polymers in an attempt to retrench and concentrate on existing product lines. This also had an adverse effect upon research funding at the goverment and university level. By 1993, at the time this book was being organized, activity in high performance polymers was still on the wane and is not expected to increase substantially until the middle or latter part of this decade.

Perhaps the biggest thrust for the development of high performance polymers over the next 10 years will be in the aerospace industry where materials will be required for a fleet of high speed civil transports (supersonic transports). At a speed of Mach 2.4, an aircraft surface temperature of about 150 to 180°C will be generated. The life requirement of materials at these temperatures will be about 60000 hours. Many different types of materials such as adhesives, composite matrices, fuel tank sealants, finishes and windows will be needed. These materials must exhibit a favorable combination of processability, performance and price. The potential market for these materials total several billions of US dollars.

This book contains information on high performance polymers which has been generated primarily during the last 10 years. The chemistry and the mechanical and physical properties of select families of high performance polymers are presented. The book is written for material scientists, polymer chemists and plastic engineers particularly those needing information on polymers for high performance applications.

A special thanks goes to the authors of the various chapters for their excellent contributions, to Dr. Marion Hertel of the chemistry editorial section of Springer-Verlag for initiating this project and to Springer-Verlag for undertaking this effort.

Paul M. Hergenrother

Editors

Table of Contents

Polymers from Benzocyclobutenes

R.A. Kirchhoff and K.J. Bruza
The Dow Chemical Company, Central Research & Development,
Materials Science & Development Laboratory,
Midland Michigan, 48674, USA

Benzocyclobutene, a thermally activated precursor to the highly reactive intermediate orthoquinodimethane, has been under extensive investigation by researchers at the Dow Chemical Co. and by
many others. Orthoquinodimethane is quite reactive and will react either with a second molecule of
othoquinodimethane or with a dienophile in a Diels Alder fashion. The efficient reactivity of
benzocyclobutene via orthoquinodimethane has been used to prepare thermoset, thermoplastic and
in some case elastomeric resins.

Advances in Polymer Science, Vol. 117
© Springer-Verlag Berlin Heidelberg 1994

In this review, the term benzocyclobutene is used to refer to structure **1**, with the numbering system shown below. This was done to be as consistent as possible with the majority of the references used in the preparation of this review. It should be noted however, that according to CAS, the name benzocyclobutene actually refers to benzocyclobutadiene **130** and what is commonly called benzocyclobutene is 1,2-dihydrobenzocyclobutadiene. The prefered CAS name for what is commonly called benzocyclobutene **1** is bicyclo[4.2.0]octa-1,3,5-triene with the numbering system shown below in **130**. Benzocyclobutene **1** has also been referred to as benzocyclobutane, cardene, benzocyclobutene-1,2-dihydro, cyclobutabenzene and generically as a cyclobutarene.

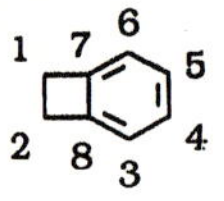

Benzocyclobutene as used in this review and in common usage in the literature.

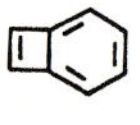

Benzocyclobutadiene; according to CAS Benzocyclobutene has this structure and therefore what is commonly called benzocyclobutene **1** is actually 1,2-dihydrobenzocyclobutadiene.

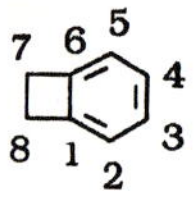

Bicyclo [4.2.0]octa- 1,3,5-triene **1** is the prefered CAS name

1 Introduction and Historical

Benzocyclobutenes constitute the basis of a broad new family of polymer forming technologies which in various expressions have been used to prepare both thermosetting and thermoplastic materials [1–14]. The first documented synthesis of a benzocyclobutene derivative was reported by Finkelstein in a 1910 paper dealing with the displacement reactions of aliphatic chlorides and bromides with iodide ion [15]. During the course of this work, it was found that α, α, α′, α′ tetrabromo-o-xylene reacted with sodium iodide in ethanol to yield 1,2-dibromobenzocyclobutene. Following this initial report, there were apparently no further research efforts in benzocyclobutene chemistry until 1956 when Cava and co-workers repeated Finkelstein's work and subsequently prepared benzocyclobutene hydrocarbon for the first time [16]. These workers also demonstrated that o-quinodimethanes were intermediates in the synthesis of benzocyclobutenes by Finkelstein's method. In 1959, Jensen and Coleman prepared 1,2-diphenylbenzocyclobutene and discovered that it readily reacted with maleic anhydride at room temperature to yield 1,4-diphenyl-1,2,3,4-tetrahydronapthalene-1,2-dicarboxylic-anhydride [17]. This product was proposed to arise from the thermal conversion of the benzocyclobutene to an o-quinodimethane which then entered into a Diels–Alder reaction with the maleic anhydride. As a result of these reports, other researchers entered the field and a steady growth in the understanding and applications of benzocyclobutene chemistry ensued. As these efforts progressed, many workers reported on the various cycloaddition and dimerization reactions of benzocyclobutene. For the most part, these research efforts were of a theoretical nature or involved the synthesis of novel small molecules and diverse natural products [18–27].

In the late 1970's, Kirchhoff at The Dow Chemical Company initiated a research program on the use of benzocyclobutenes in polymer synthesis and modification. These efforts culminated in 1985 with the issuance of the first patent on the use of benzocyclobutenes in the synthesis of high molecular weight polymers [3]. Similar work was reported separately and independently by Tan and Arnold working at the Air Force Wright Laboratories [11–14]. Since these initial discoveries, the field of benzocyclobutene polymers has rapidly expanded to include currently 75 issued patents and numerous publications by a variety of researchers. These numbers are expected to increase considerably since benzocyclobutenes constitute the basis of a new and versatile approach to the synthesis of high performance polymers for applications in the electronics and aerospace industries.

The basic benzocyclobutne technology involves a family of thermally polymerizable monomers which contain one or more benzocyclobutene groups per molecule. Depending on the degree and type of additional functionality, these monomers can be polymerized to yield either thermosetting or thermoplastic polymers. The polymerization is believed to proceed through the thermally initiated ring opening of a benzocyclobutene to give an o-quinodimethane (*o*-xylylene) intermediate. The subsequent fate of this intermediate depends largely on the number and type of other functional groups present in the monomer molecule. For the class of monomers which contains only benzocyclobutene moieties as reactive groups (i.e., containing no reactive sites of unsaturation), the *o*-quinodimethane groups react rapidly with one another to give what are believed to be linear or cyclic polymeric structures which, for a bisbenzocyclobutene monomer, quickly leads to a highly crosslinked polymer network [1].

Bisbenzocyclobutenes readily react with molecules which contain sites of reactive unsaturation such as bismaleimides [10, 13, 31, 32]. This is in essence, a novel type of Diels-Alder polymerization in which the bis-diene is latently embodied within two benzocyclobutene moieties. The properties of these polymers depends strongly on the mole ratio of the monomers and when it is equimolar, can result in some exceptionally tough high Tg resins [33, 34].

Another class of benzocyclobutene monomers is the one in which the monomers contain sites of unsaturation which are themselves capable of reacting with an *o*-quinodimethane to give cycloaddition products [4, 5, 28, 29]. Typically, alkenes and alkynes have been used as the *o*-quinodimethane reactive moieties. In the particular case where the monomer molecule contains only one benzocyclobutene group and only one site of unsaturation then, providing that the unsaturated moiety is not a vinyl group, the cycloaddition polymerization goes on to produce what is essentially a linear thermoplastic polymer. Monomers of this sort are often termed AB monomers to represent their dual and mutually interactive functional groups. Some of the benzocyclobutene – maleimide AB type monomers polymerize to yield exceptionally tough high Tg resins [36].

2 Monomer Syntheses

The parent benzocyclobutene hydrocarbon 1 is the starting material for almost all of the monomers and polymers that are described in this paper [35]. Treatment of the hydrocarbon with bromine provides an excellent yield of 4-bromobenzocyclobutene, 2 (Fig. 1) [36, 37].

Bromobenzocyclobutene 2 is converted to 4-benzocyclobutenyl carboxylic acid 3 by either of two routes (Fig. 2). The first method of preparation proceeds from the corresponding Grignard reagent of 2 followed by reaction with carbon dioxide [38]. The acid 3 is obtained in a yield of 60–70%. The second method for the formation of 3 again starts with 4-bromobenzocyclobutene, but in this case 2 is reacted with a palladium zero catalyst in the presence of carbon monoxide and methanol to provide 4-carbomethoxybenzocyclobutene, 4 in a yield of > 95% [39, 40]. Ester 4 is hydrolyzed under standard conditions to provide 3 in an overall yield of 90% for the two steps [36].

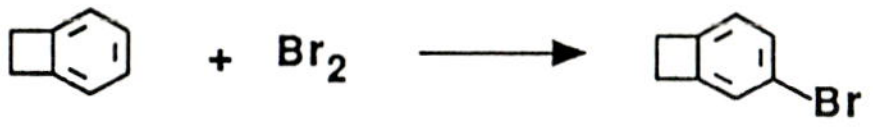

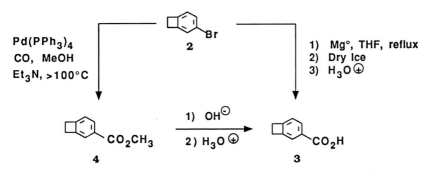

Fig. 1. Formation of 4-bromobenzocyclobutene, **2** from benzocyclobutene, **1**

Fig. 2. Preparation of carboxylic acid **3** and methyl ester **4** from 4-bromobenzocyclobutene **2**

After acid **3** is converted to the acid chloride **5** (thionyl chloride), it can be reacted with either a diamine in the presence of a base to give a bisamide **6**, or with a diol, under similar reaction conditions to afford a bisester **7** (Fig. 3) [36, 41].

The R group between the diamine or diol functionality can be widely varied to include straight chain and branched alkyl groups as well as additional functionality such as either olefin or acetylenic groups. R can also be an aromatic group which may or may not contain additional functionality. In general, all of the monomers are crystalline solids with melting points lower than 200 °C which is the minimal temperature required for the onset of efficient homopolymerization.

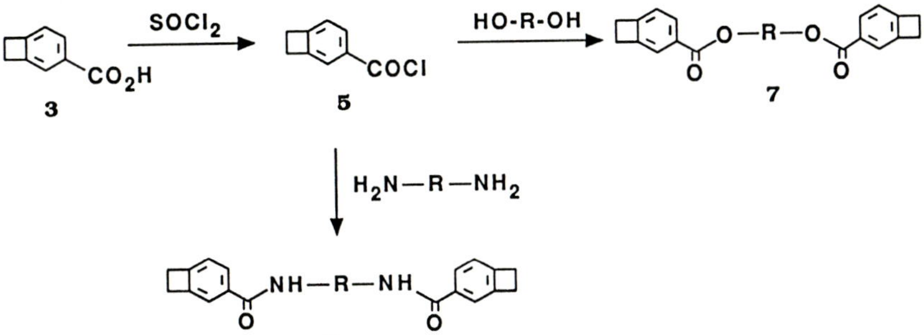

Fig. 3. Formation of bisbenzocyclobutenyl amides and esters from 4-benzocyclobutenyol chloride, **5**

The bisamides and bisesters provide two different families of benzocyclobutene monomers and polymers derived from bromobenzocyclobutene **2**. Heck and coworkers have demonstrated that aromatic bromides and iodides react with olefins in the presence of a palladium catalyst to afford products where the vinyl group is directly bonded to the aromatic ring [40, 42, 43]. This technology has been used with 4-bromobenzocyclobutene **2** as the starting aromatic halide, in order to prepare more highly functional bis- and monobenzocyclobutenes (Fig. 4)

When the Heck vinylic arylation reaction is carried out with an olefin which is either a gas or a low boiling liquid, it must be run in a pressure vessel. Depending upon the product desired, the olefin is usually present in an excess concentration in order to insure formation of vinyl substituted products. In the case of the reaction of ethylene with **2** the stoichiometry of the reactants is critical (Fig. 5). If 4-bromobenzocyclobutene **2** is present in higher concentrations than ethylene, a different monomer, 1,2-di(4-benzocyclobutenyl)-ethylene **11** is obtained.

The reaction of 4-bromobenzocyclobutene **2** with monosubstituted olefins such as styrenes and alkyl substituted olefins is quite general and provides a

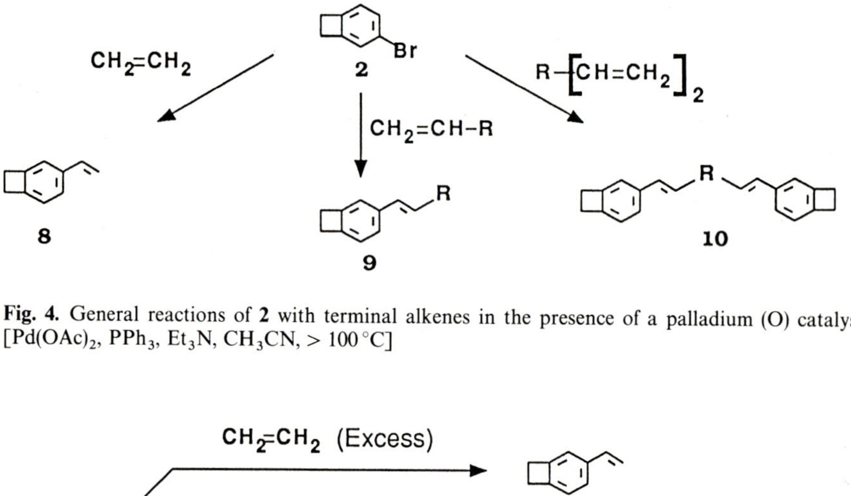

Fig. 4. General reactions of **2** with terminal alkenes in the presence of a palladium (O) catalyst [$Pd(OAc)_2$, PPh_3, Et_3N, CH_3CN, > 100 °C]

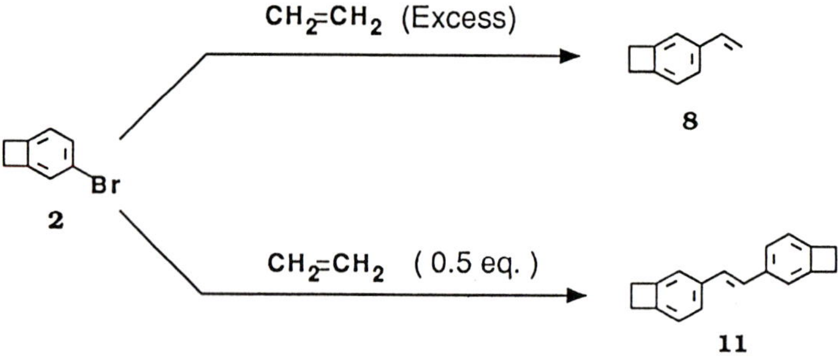

Fig. 5. Effect of concentration of ethylene upon product formation

wide variety of different monomers containing at least one benzocyclobutene group. This same chemistry has been used to prepare bisbenzocyclobutenes where the olefin containing reactant has two separate olefins capable of undergoing the vinylic arylation reaction with **2**. Two important molecules prepared via his reaction are shown in Fig. 6.

Monomer **12** is a crystalline solid which, when homopolymerized, affords a high Tg, thermally stable polymer that has potential application in both the microelectronic and aerospace areas. Monomer **13** (mixed isomers) is a liquid at room temperature and, when homopolymerized, also provides a high Tg, thermally stable polymer that has a low moisture uptake and a low dielectric constant. This polymer has been targeted into the microelectronics area because of this interesting set of properties, combined with the prepolymer's unique ability to planarize over underlying topography.

The monomers described so far have all been prepared by starting with 4-bromobenzocyclobutene, **2**. A different approach to the preparation of monomers begins with the parent hydrocarbon benzocyclobutene **1** by carrying out electrophilic aromatic substitution reactions [36]. Benzocyclobutene readily undergoes a Friedel–Crafts benzoylation reaction with a variety of substituted acid chlorides (Fig. 7).

Many of the common Lewis acids which have been used for the Friedel-Crafts reaction such as $AlCl_3$, $TiCl_4$, $SnCl_4$, $SbCl_5$, and $FeCl_3$ also function effectively with benzocyclobutene **1**. In general, the catalyst must be present in slightly greater than one equivalent for every equivalent of acid chloride functionality present. The reaction may be run with or without solvent and can

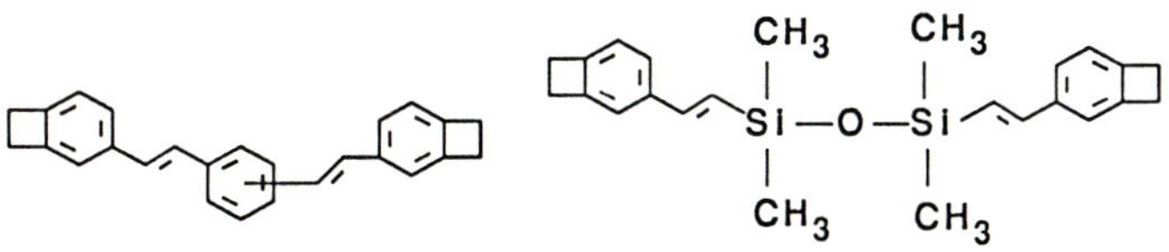

Fig. 6. Bisbenzocyclobutene-bisalkene momomers **12** and **13** prepared by using a palladium (O) catalyst

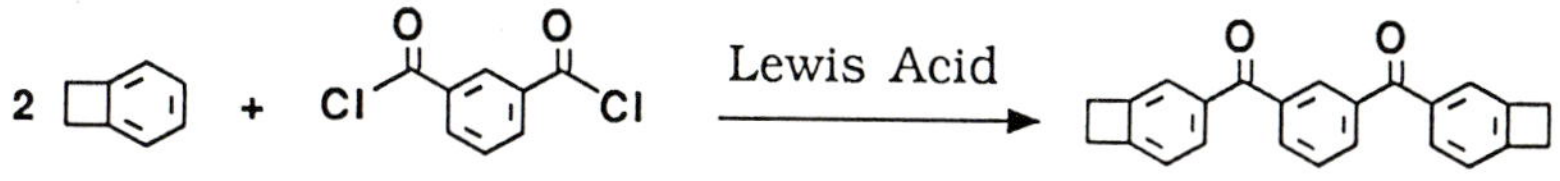

Fig. 7. Reaction of isophthaloyl chloride with **1** to prepare the bisketone-bisbenzocyclobutene monomer **14**

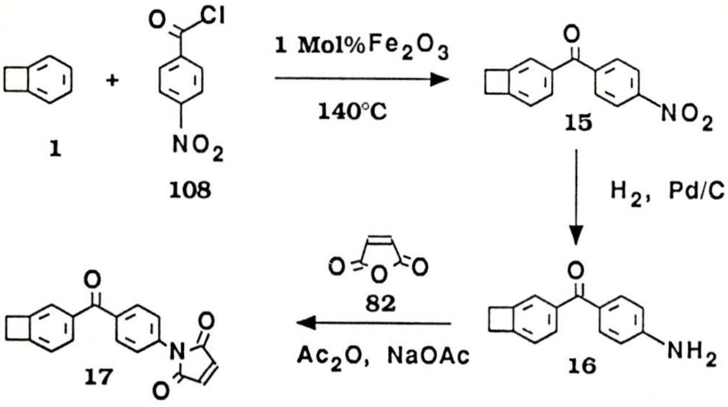

Fig. 8. Formation of AB benzocyclobutene maleimide **17** from hydrocarbon **1** by Friedel-Crafts benzoylation

be carried out at subambient to elevated temperatures depending upon the nature of the catalyst. Some care must be exercised when running this type of reaction in order to minimize the formation of by-products. The principal by-product is 2-phenylethyl chloride, which presumably arises from small concentrations of HCl that are often generated in the presence of the Lewis acid and can then react with four membered ring of benzocyclobutene. Monomer **14** (MP = 152 °C) is an example of a material prepared in this fashion. The homopolymer from **14** has been investigated as to its use as a matrix resin in high performance composite applications.

The Friedel–Crafts type of technology can also be used for the preparation of monomers that contain one benzocyclobutene and a second functional group which can react with the benzocyclobutene. These types of molecules are commonly called AB monomers. An example of this class of monomer is shown in Fig. 8.

Reaction of 4-nitro-1-benzoyl chloride with benzocyclobutene **1** provided the benzoylated product **15** [36]. The nitro group of **15** was reduced with hydrogen in the presence of palladium on charcoal to afford he amine product **16** [44]. Reaction of the amine with maleic anhydride provided the amic acid which was converted to the maleimide **17** by cyclodehydration with acetic anhydride and sodium acetate at 95 °C [45–47]. This monomer and its homopolymer will be discussed in greater detail in a later section.

3 Polymerization Mechanism

There seems to be a general agreement in the literature that the first step in the polymeriztaion of a benzocyclobutene monomer is a thermally induced ring opening reaction to yield an *o*-quinodimethane intermediate [1, 48]. Following this, the subsequent details of the polymerization process depends largely on the presence or absence of reactive unsaturation in the monomer molecule. Reactive unsaturation in the present context refers to doubly or triply bonded functional groups that can readily enter into a cycloaddition (Diels–Alder) reaction with an *o*-quinodimethane. When such unsaturation is lacking, the polymerization mechanism of benzocyclobutenes has largely been discussed in terms of the thermolysis reactions of benzocyclobutene and the self oligomerization products of *o*-quinodimethane. For those benzocyclobutene monomers which do contain reactive sites of unsaturation, there appears to be available another and in some instances predominating reaction path characterized by *o*-quinodimethane 4 + 2 (Diels–Alder) cycloaddition reactions. Overall, the mechanism of the polymerization of benzocyclobutene monomers and the structure of the resulting polymers is with a few exceptions rather poorly understood. What can be said however is that the process appears to involve *o*-quinodimethane as the key reactive intermediate and that bis or polybenzocyclobutene monomers with or without sites of unsaturation, all polymerize to yield highly crosslinked structures.

Benzocyclobutene undergoes a thermally induced electrocyclic ring opening to yield an intermediate which has been formulated as being either an *o*-quinodimethane 18 or a diradical 19 as in Fig. 9. While the nature of the ring opened species has been considerably debated in the literature, the evidence both from experiment and calculations tends to support the conclusion that *o*-quinodimethane is the product that is formed [49–53, 26, 54–68]. The thermal ring opening reaction is, as expected, first order in benzocyclobutene moieties and is exothermic by approximately 104.6 kJ per mole of benzocyclobutene. The activation energy for the ring opening of benzocyclobutene itself has been variously reported to be around 163.2 kJ per mole. Roth has discussed the activation parameters for this reaction and from his findings the half lives of benzocyclobutene relative to its conversion to *o*-quinodimethane have been calculated (Table 1) [69, 54].

From this data it is evident that benzocyclobutene should be relatively stable to cycloreversion up to about 200 °C and that it should have an almost indefinite shelf life at room temperature. The data in Table 1 is only true however for the parent hydrocarbon with all of the substituents on the four membered ring being hydrogen atoms. The presence of either electron donating or withdrawing groups in these positions significantly lowers the temperature at which the ring opening occurs.

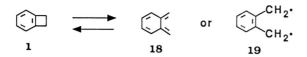

Fig. 9. Proposed thermolysis intermediates from benzocyclobutene 1

Table 1. Benzocyclobutene — — ⟶ o-quinodimethane

1	**18**

T (°C)	K(s^{-1})	T 1/2 (h)
25	2.5×10^{15}	7.6×10^{10}
100	1.7×10^{-9}	1.1×10^{5}
150	9.6×10^{-7}	2.0×10^{2}
200	1.4×10^{-4}	1.4
250	7.8×10^{-3}	2.5×10^{-2}

Table 2 shows a summation by Oppolzer of the effects that various substituents on the four membered ring have on the ring opening temperature of benzocyclobutene [20].

Any study of the polymerization kinetics of a bisbenzocyclobutene monomer is complicated by the lack of understanding of the resulting polymer's structure and the fact that as the polymerization proceeds, the reaction mixture crosslinks and vitrifies. This vitrification limits somewhat the number of quantitative methods which can be used to study the bisbenzocyclobutene polymerization kinetics. Some techniques are however useful under these constraints and good kinetic results have been obtained by both infrared and thermal analysis methods.

The polymerization kinetics of the bisbenzocyclobutene diketone monomer **14** (Fig. 10) were studied in the melt at various temperatures by infrared spectroscopy [48]. This technique has the advantage that it is relatively insensitive to the physical state of the system as it proceeds from monomer melt through the gel point and into the vitreous state. In addition, quantitative

Table 2. Ring opening temperature of 1-substituted benzocyclobutenes

Substituent (R)	Temperature (°C)[a]
H	200
$-CH_2-$	180
$C=O$	150
$-NH-CO-R$	110
$-OH$	80
$-NH_2$	25

[a] Reaction time: 18 h.

14   Fig. 10. Bisbenzocyclobutene diketone monomer **14**

measurements can be made with infrared spectroscopy without the need for physically isolating either startinlg materials or products. In the study on the diketone monomer **14** the polymerization was easily followed by monitoring changes in the infrared band at 1067 cm^{-1} which has been assigned as a combination bending mode of the cyclobutene ring. Figure 11 shows the changes in the absorption of this band over time when the monomer was held at 190 °C. A series of twelve other infrared bands were studied in this way at temperatures ranging from 170–210 °C. From these measurements, the authors concluded that the polymerization of the bisbenzocyclobutene diketone **14** was first order in benzocyclobutene moieties with a rate constant of 0.3 hr^{-1}. The activation energy arrived at from this study was 167.4 kJ per mole which agrees well with the literature reference value of 166.9 kJ per mole.

A study of benzocyclobutene polymerization kinetics and thermodynamics by differential scanning calorimetry (DSC) methods has also been reported in the literature [1]. This study examined a series of benzocyclobutene monomers containing one or two benzocyclobutene groups per molecule, both with and without reactive unsaturation. The study provided a measurement of the thermodynamics of the reaction between two benzocyclobutene groups and compared it with the thermodynamics of the reaction of a benzocyclobutene with a reactive double bond (Diels–Alder reaction). Differential scanning calorimetry was chosen for this work since it allowed for the study of the reaction mixture throughout its entire polymerization and not just prior to or after its gel point. The monomers used in this study are shown in Table 3. The polymerization exotherms were analyzed by the method of Borchardt and Daniels to obtain the reaction order n, the Arrhenius activation energy Ea and the pre-exponential factor log Z. Tables 4 and 5 show the results of these measurements and related calculations.

From this work, the authors concluded that the reaction of two benzocyclobutene groups with each other is exothermic to the extent of 221.7 ± 12.5 kJ. By comparison, the reaction of one mole of benzocyclobutene moieties with one mole of double bonds in a Diels–Alder fashion is exothermic by 184.1 ± 12.5 kJ. Thus, the thermal dimerization of benzocyclobutenes is thermodynamically favored over the cycloaddition reaction of a benzocyclobutene with a double bond by about 37–38 kJ. By contrast, the authors found that for monomers that could react either by a benzocyclobutene dimerization or by a benzocyclobutene - double bond cycloaddition (e.g. monomer **13** in Table 3) the reaction products appeared to be dominated by the cycloaddition pathway. From this data, it was concluded that the Diels–Alder reaction was kinetically

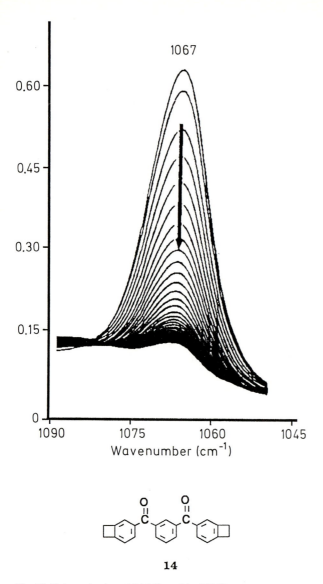

Fig. 11. Polymerization of **14** followed by FT-IR

preferred. While this conclusion is consistent with the data in the study, the generalization of the conclusions to monomers significantly dissimilar to those in Table 3 should be done with caution.

The kinetic results of the differential scanning calorimetry studies on the monomers in Table 3 are presented in Tables 4 and 5. The reaction order n with the exception of example **78a** in Table 3 was approximately unity for all of the

Table 3. Physical properties of selected BCB monomers

Monomer Structure	Sample appearance at 25 °C	Molecular weight (g/mol)	Melting point (°C)	ΔH_{melt} (kJ/mol)
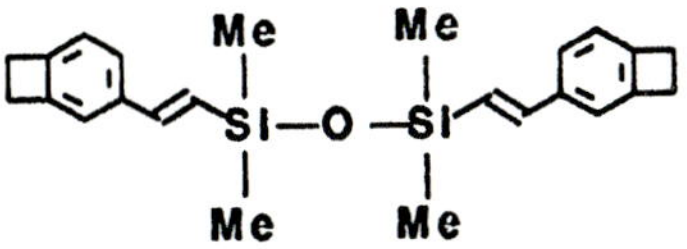 (mixture of isomers) **13**	Clear yellow liquid	390.7	Liquid	–
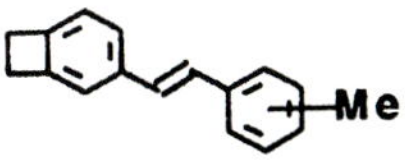 **78c**	White crystalline solid	220.8	36–38	–
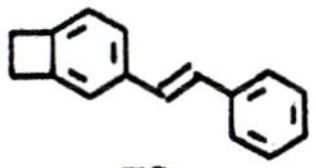 **78a**	White crystalline solid	206.3	80–83	22.09 ± 0.21
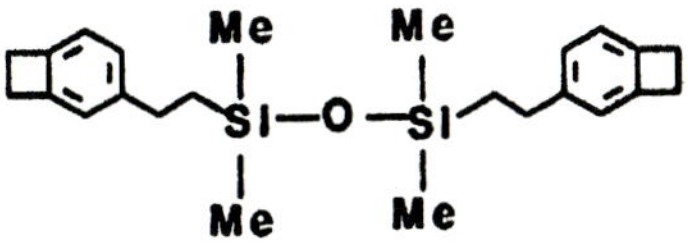 **120** (mixed isomers)	Amber liquid	394.7	Liquid	–
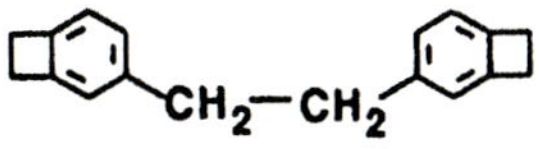 **36**	White crystalline solid	234.3	82–86	35.15 ± 0.71
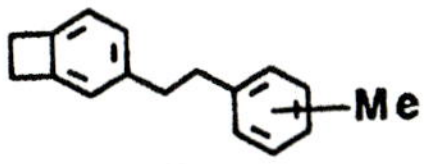 **121**	Clear liquid	222.3	–	–

monomers studied. The pre-exponential factor log Z was found to be between 11 and 17 min^{-1} and the activation energy (again with the exception of entry **78a**) for all of the monomers was approximately 175.7 kJ per mole. This latter result is in good agreement with the value (167.4 ± 12.5 kJ) obtained in the previously discussed study of the diketone monomer **14** by infrared spectroscopy, and with the value (166.9 kJ) obtained by Roth and coworkers in their study of the

Table 4. Kinetic parameters for selected BCB monomers

Monomer structure	Reaction order	Log Z (1/min)	Ea (kJ/mol)
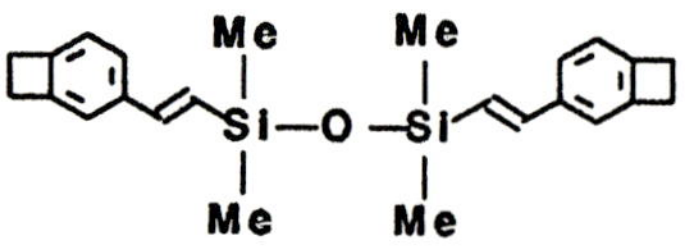 (mixture of isomers) **13**	1.03 ± 0.01	15.7 ± 0.1	161.1 ± 0.8
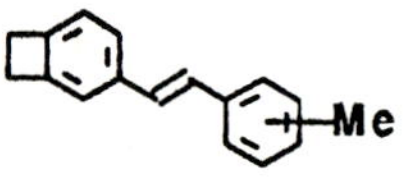 **78c**	0.86 ± 0.13	14.6 ± 1.0	149.8 ± 10
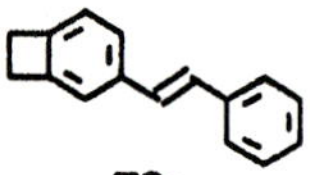 **78a**	0.43 ± 0.27	11.3 ± 2.1	118.8 ± 20
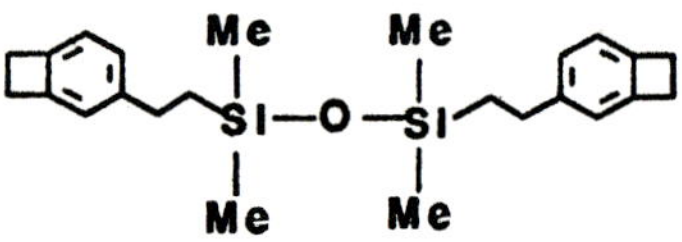 **120** (mixed isomers)	1.24 ± 0.3	17.0 ± 1.2	174.9 ± 13
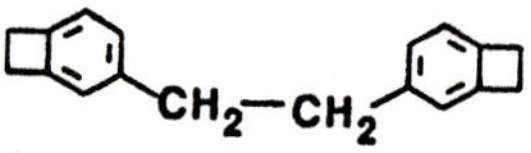 **36**	1.07 ± 0.03	16.6 ± 0.3	170.7 ± 2.9
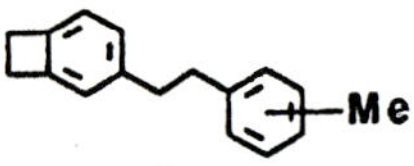 **121**	1.02 ± 0.18	16.3 ± 0.3	170.7 ± 2.9

thermal opening of a benzocyclobutene ring. The reaction order n and activation energy Ea for the polymerization of monomer **78a** were anomalously low and not explained by the authors.

Overall, then in summary, the kinetic and thermodynamic evidence available thus far suggests that the polymerization of a benzocyclobutene monomer is first order in benzocyclobutene moieties with an activation energy of approximately 167.4 kJ per mole. Further, the thermal reaction between two benzocyclobutenes is thermodynamically preferred over that of a benzocyclobutene

Table 5. Reaction enthalpies for selected BCB monomers

Monomer structure	Molecular weight (g/mol)	Dominant reaction	ΔH_{rxn} (kJ/mol)	ΔH_{rxn} (kJ/mol-BCB)
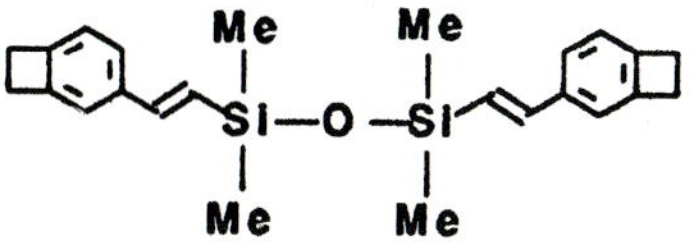 Me, Me, Si–O–Si, Me, Me (mixture of isomers) **13**	390.68	BCB + Olefin	-404.2 ± 26.4	-202 ± 13.4
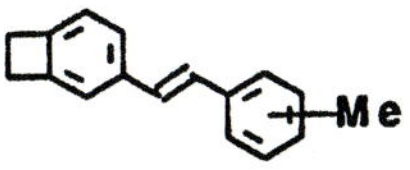 —Me **78c**	220.81	BCB + Olefin	-180.7 ± 10.9	-180.3 ± 10.9
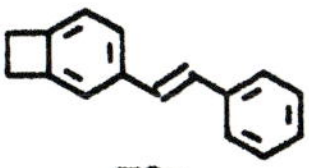 **78a**	206.29	BCB + Olefin	-176.6 ± 4.2	-176.6 ± 4.2
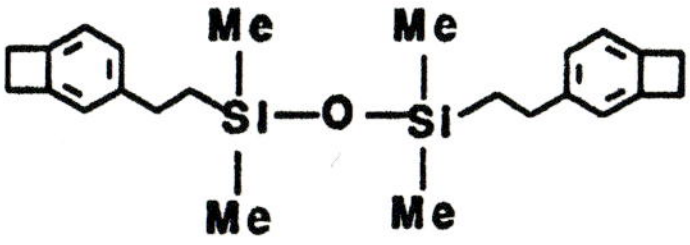 Me, Me, Si–O–Si, Me, Me **120** (mixed isomers)	394.71	BCB + BCB	-205.4 ± 11.3	-102.5 ± 4.6
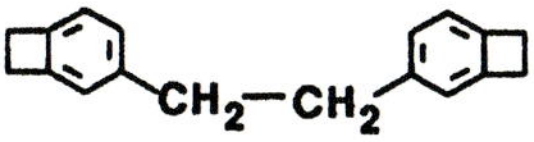 CH_2–CH_2 **36**	234.34	BCB + BCB	-222.6 ± 8.4	-111.3 ± 4.2
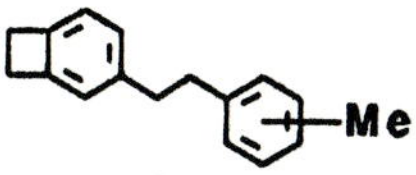 —Me **121**	222.33	BCB + BCB	-120.9 ± 12.6	-120.9 ± 12.6

with a double bond. The latter reaction however appears to be kinetically preferred.

4 Polymer Structure

The structure of the polymer derived from a polybenzocyclobutene monomer depends strongly on the presence or absence of reactive unsaturation in the monomer molecule. For the most basic case of a bisbenzocyclobutene connected by a nonreactive linking group, the structure of the polymer will be closely related to the structure of the thermolysis products of benzocyclobutene itself.

The thermolysis of benzocyclobutene either in the condensed or vapor phase has been shown by a variety of techniques to initially yield an o-quinodimethane (Fig. 9). The fate of the o-quinodimethane in the absence of other coreactive species appears to depend to a large extent upon the conditions under which it is generated. Thus for example, the thermolysis of benzocyclobutene hydrocarbon itself in a sealed flask at 200 °C gave a 24% yield of 1,2,5,6-dibenzocyclooctadiene **20** along with a mixture of unidentified oligomeric materials (Fig. 12) [70].

Cava and Deanna carried out a study on the pyrolysis of the sulfone **21** and found that the major components in the reaction mixture depended strongly on the conditions of the pyrolysis as shown in Fig. 13 [71]. Thus for example, when the sulfone **21** was heated as a melt at 280 °C, there was obtained a 16% yield of an 80:20 mixture of 1,2,5,6-dibenzocyclooctadiene **20** and benzocyclobutene **1** along with a 3.2% yield of o-xylene. This the authors interpreted as being due to the sulfone expelling sulfur dioxide to form o-quinodimethane which then either underwent ring closure to benzocyclobutene or else abstracted a hydrogen atom to yield o-xylene. This latter reaction was explained as being the result of o-quinodimethane having considerable biradical (**19**) character. The benzocyclobutene dimer **20** was proposed to arise in a similar fashion from the coupling of two o-quinodimethanes to give a bibenzyl biradical which then underwent ring closure to **20**. When the sulfone **21** was heated as a solution in diethylphthalate at 300 °C, only the benzocyclobutene cyclooctadiene dimer **20** was obtained. These conditions were regarded by the authors as being unfavorable

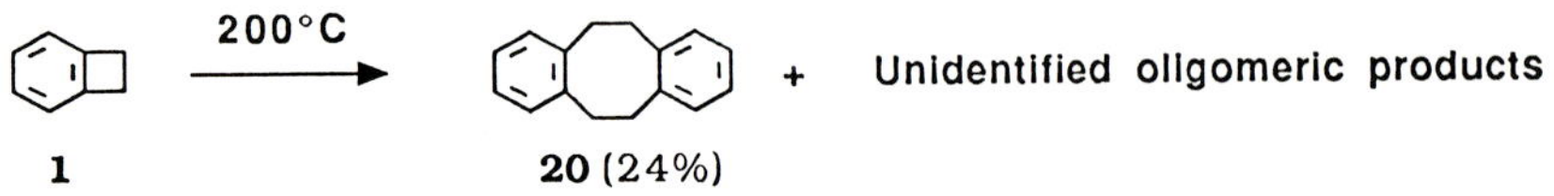

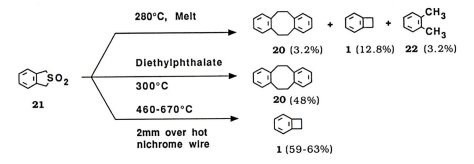

Fig. 12. Condensed phase thermolysis products of benzocyclobutene hydrocarbon **1**

Fig. 13. Thermolysis products of 1.3-dihydro-isothianapthalene-2.2-dioxide **21**

for hydrogen atom abstraction and so the o-quinodimethane underwent dimerization and ring closure to **20**. When the sulfone was pyrolyzed at low pressure in the vapor phase at 460–470 °C, there was obtained a 59–63% yield of benzocyclobutene **1**. Presumably, under these dilute conditions, bimolecular reactions such as dimerization and hydrogen atom abstractions are disfavored and therefore intramolecular ring closure predominates.

In another series of related experiments Errede and coworkers prepared o-quinodimethane itself by the flash pyrolysis of o-methylbenzyltrimethylammonium hydroxide **23** [72, 73]. The conditions of this experiment were such that the o-quinodimethane was quenched soon after it was formed by cooling to − 78 °C. The product trapped out under these conditions was an approximately 25:75 mixture of 1,2,5,6-dibenzocyclooctadiene **20** and the spiro o-quinodimethane dimer **24** (Fig. 14). Dimer **24** can readily be seen to be the result of the Diels-Alder reaction of one o-quinodimethane bis-exo-methylene diene unit across one of the exo-methylene groups of another o-quinodimethane. The spirodimer

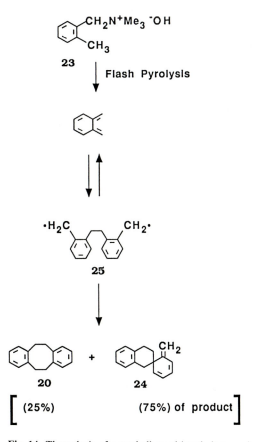

Fig. 14. Themolysis of o-methylbenzyltimethylammonium hydroxide **23**

24 was stable at low temperature but upon warming up, polymerized to poly(o-xylylene) **25** [74–76]. There was no experimental evidence in this or other related papers of any 1,2,5,6-dibenzocyclooctadiene **20** being formed by heating the spirodimer **24**. Thus, it appears that the polymerization of the spirodimer and the formation of 1,2,5,6-dibenzocyclooctadiene **20** occur through different intermediates. The mechanism proposed by Errede to account for these transformations is similar to that discussed by Cava to account for the sulfone **21** pyrolysis products in that both transformations involve the initial formation of an o-quinodimethane **18**. When this is rapidly quenched, it can form a bibenzyl

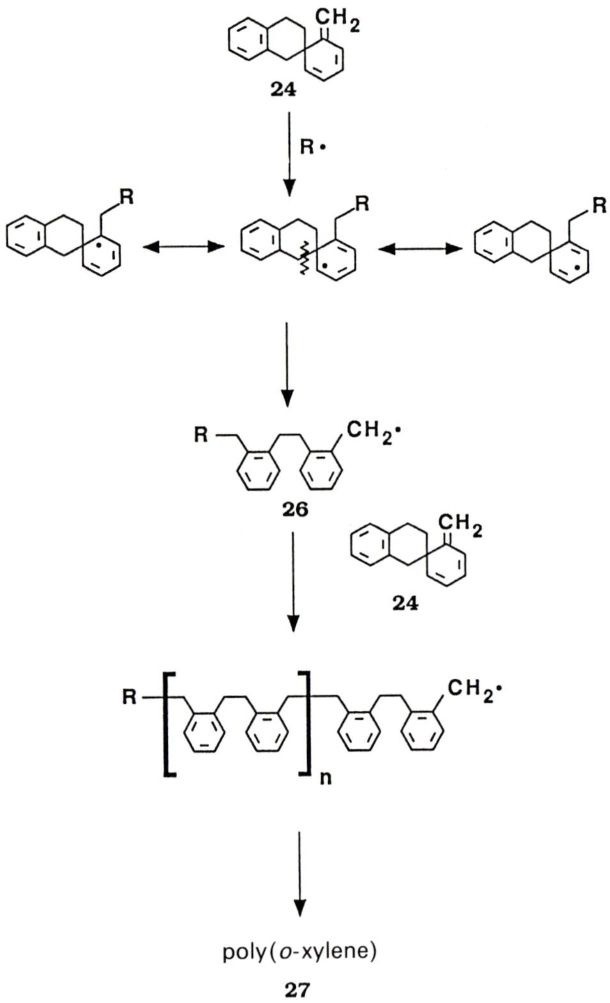

Fig. 15. Proposed mechanism for the formation of poly(o-xylylene) from the thermolysis of o-quinodimethane spirodimer **24**

biradical **25** which can close to either the spirodimer **24** or 1,2,5,6-dibenzocyclooctadiene **20**. While the 1,2,5,6-dibenzocyclooctadiene **20** is thermally stable, the spirodimer **24** is not. Upon warming, it can undergo some unspecified radical initiation to yield a mono bibenzyl radical **26** which then attacks another molecule of spirodimer **24** to begin a chain type polymerization to yield poly(*o*-xylylene) **27** (Fig. 15).

Overall then, the literature indicates that the pyrolysis products of benzocyclobutene arise from the initial formation of *o*-quinodimethane which can, depending upon the reaction conditions, react to form either cyclic products or linear oligomers. Relating all of this to the structure of the polymers obtained from bisbenzocyclobutenes is somewhat difficult since the chemical systems and reaction conditions described in the various papers are rather different to those encountered in a typical bisbenzocyclobutene polymerization. Nevertheless, certain speculations have been made which are at least to a first approximation consistent with the experimental evidence available thus far.

Simple bisbenzocyclobutene monomers which do not contain reactive sites of unsaturation appear to have available to them two possible polymerization paths that can lead to different structures for the resulting polymers. In both paths, the initial step is the thermally induced opening of the benzocyclobutene ring to give an *o*-quinodimethane. If this intermediate reacts predominantly with another *o*-quinodimethane by forming 1,2,5,6-dibenzocyclooctadiene linkages, then the resulting polymer would be a substantially linear, uncrosslinked poly(1,2,5,6-dibenzocyclooctadiene) **28** as shown in Fig. 16. There is a considerable amount of mechanical and rheological data however, which suggests that the polybisbenzocyclobutenes are in fact highly crosslinked networks [1]. In some instances the molecular weight between crosslinks is that of one monomer unit. Further, structure **28** is a linear polymer and would be expected

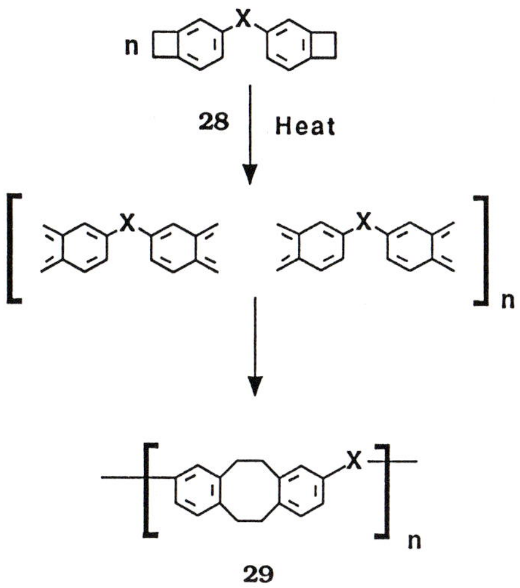

Fig. 16. Proposed linear polymerization of a bisbenzocyclobutene to give a poly(1,2,5,6-dibenzocyclooctadiene **29**)

to be soluble to some extent. In general however, the poly(bisbenzocyclo-butenes) have been found to be very solvent insensitive which also suggests a highly crosslinked structure.

An alternative polymerization mechanism and polymer architecture has been proposed by Kirchhoff [1, 2, 3], Tan and Arnold [77]. By this mechanism, polybenzocyclobutenes which do not contain reactive sites of unsaturation are proposed to polymerize by the 1,4 addition of the *o*-quinodimethane inter-mediates to give a substantially linear poly(*o*-xylylene) structure. Since the monomers all contain at least two benzocyclobutene units the net result of this reaction will to a first approximation be a ladder type polymer as shown in Fig. 17. The formation of a true ladder polymer however would require that all

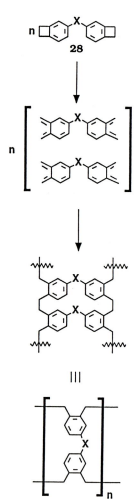

Fig. 17. Proposed polymerization of a bisbenzocyclobutene to give a ladder type poly(*o*-xylylene) polymer **30**

of the bisbenzocyclobutene monomers line up in the proper fashion to form two parallel chains of poly(o-xylylene). This seems somewhat unlikely and, more probably, the monomer units add together sequentially with a random orientation to give a complex three dimensional network (Fig. 18). Such a network would be consistent with the observed mechanical properties of the polymers which are

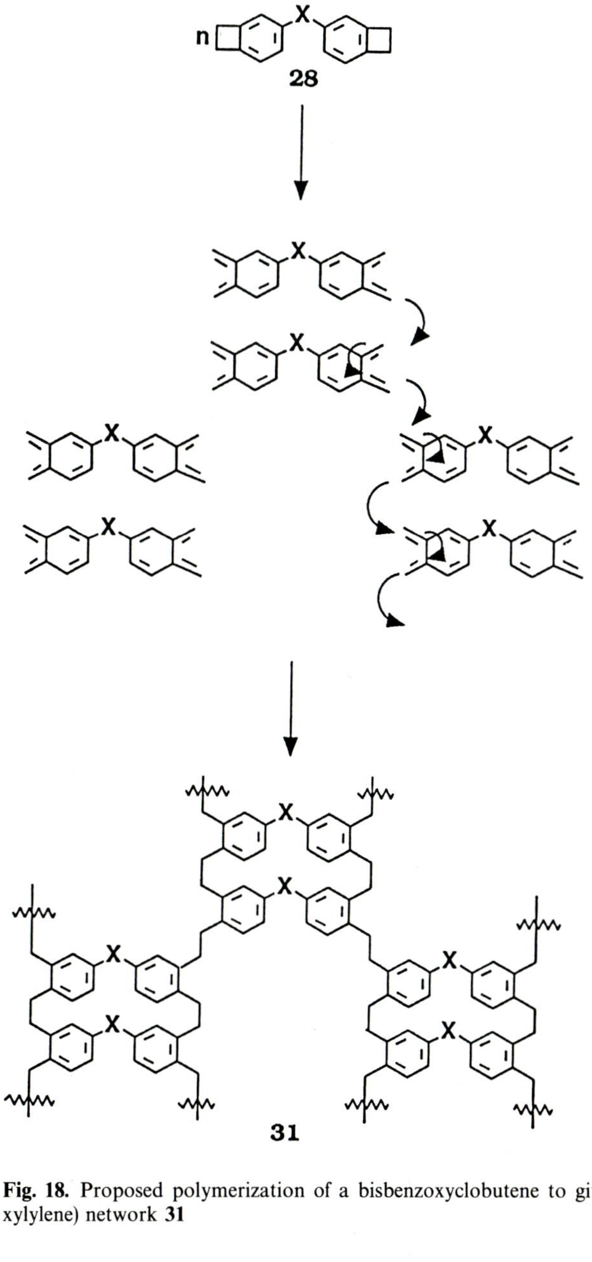

Fig. 18. Proposed polymerization of a bisbenzoxyclobutene to give a three dimensional poly(o-xylylene) network 31

those of a highly crosslinked material [1]. Such a network structure is also consistent with the finding that the crosslink density of some poly(bisbenzo-cyclobutenes) is very high, with the molecular weight between crosslinks being near that of one monomer unit. The insolubility and low solvent pick up of poly(bisbenzocyclobutenes) is also consistent with the network structure proposed in Fig. 18.

Polybenzocyclobutene monomers which contain one or more reactive sites of unsaturation have also been shown by rheological methods to be highly crosslinked structures. The situation is more complicated because, relative to monomers without unsaturation, when such sites are available, the polymerization can proceed by more than one reasonable reaction path. Specifically, the o-quinodimethane intermediate can either react with another o-quinodimethane to give poly(o-xylylene) moieties or else it can undergo cycloaddition to a double bond to give tetralin linking groups. A combination of both paths is also of course possible. A study of the polymerization of the bisbenzocyclobutene divinyldisiloxane monomer 13 was consistent with the proposal that the polymerization was predominantly the result of the cycloaddition of the benzocyclobutene to the 1-aryl-2-siloxy substituted double bonds [1]. This conclusion is also consistent with the finding that the cycloaddition of an o-quinodimethane to a double bond is kinetically preferred over the dimerization of two o-quinodimethanes. In essence then, the polymerization of the monomer 13 is a Diels–Alder polymerization between a bis-diene and a bis-olefin wherein both components are contained within the same molecule. The net result therefore is a highly crosslinked structure.

5 Polymer Properties

A general characteristic of the benzocyclobutene derived polymers is that of high thermal stability [2, 3, 33, 69–71, 78–80]. The thermal stability for the purpose of this review is defined as the maximum temperature at which the polymer showed zero weight loss by thermogravimetric analysis (TGA) at a ramp rate of at 10 °C/min under flowing nitrogen [2]. This experimental approach provides a measure of short term thermal stability and is at best a first approximation as to what the long term thermal stability of the polymer will be. In practice, this latter temperature will always be lower than the temperature observed by the procedure adopted herein. Table 6 shows the short term zero weight loss temperatures for some representative bisbenzocyclobutene polymers. Of particular note is the good thermal stability of all of the polymers, including those such as the ester 33 which contain relatively large proportions of aliphatic linkages. The polymers derived from the hydrocarbon monomers 32–34, 36 are of particular note since they exhibit no appreciable weight loss up to 365–425 °C (689–797 °F).

Table 6. Thermal stability of polybisbenzocyclobutenes

Polymer from 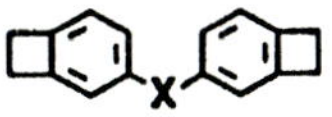 X	Maximum temperature showing zero percent weight loss in N_2 by TGA: $°C (°F)$ (10 °C/min)
Direct bond **37**	400 (752)
	385 (725)
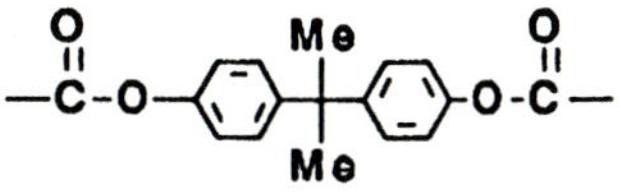	400 (752)
	370 (698)
$-CH_2-CH_2-$ **36**	425 (797)
$-CH=CH-$ **11**	475 (887)
$-C\equiv C-$ **38**	410 (770)
	365 (689)
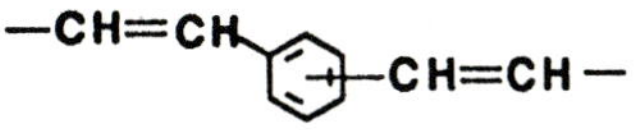 Mixture of *m*- and *p*-isomers	425 (797)
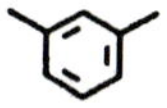 **55**	410 (770)

To a first approximation, the thermal stabilities of the polymers follow the conventional perception of functional group thermal stability. Entry **37** with a direct bond between the aryl groups of the benzocyclobutene moieties would be expected to give some indication of the ultimate thermal stability of a bisbenzo-cyclobutene polymer, since in this example there are no intervening functional groups. That is, the polymer from **37** should exhibit a thermal stability that is a strong reflection of the stability of the structures obtained by reacting two or more benzocyclobutenes together via the proposed o-quinodimethane polym-erization intermediates. That this assertion is not necessarily true is indicated by the data which shows that the polymer from **37** is not, by this test, significantly more stable than a variety of other bisbenzocyclobutene polymers containing diverse aryl linking groups. Further, the data in Table 6 shows that for some structures such as the ethane, ethylene and ethynyl-linked bisbenzocyclobutenes **36, 11** and **38** the presence of non-aryl functional groups in fact enhances thermal stability. The data suggests that in these instances, the linking groups are not in fact inert carriers of the benzocyclobutene moieties but rather take an active role in the polymerization process to modify the final polymer architec-ture. Independent evidence from rheological studies previously mentioned strongly supports the idea of the linking groups' involvement in the polymeriz-ation process. Specifically, these studies along with others suggest that the polymerization of monomer **11** first involves almost exclusively the cycloaddi-tion of a benzocyclobutene to the double bond followed by subsequent extensive crosslinking through the other benzocyclobutene groups at final higher func-tional group conversion. Such a highly crosslinked network would, on intuitive grounds at least, be expected to exhibit exceptional thermal stability. In fact, this notion is supported by the experimental facts in Table 6, using the TGA thermal stability method referenced above.

Thermal stability as measured by these ramped TGA experiments of the sort previously described are not the definitive test of a polymer's utility at elevated temperature. Rather, for a polymer to be useful at elevated temperatures, it must exhibit some significant retention of useful mechanical properties over a pre-determined lifetime at the maximum temperature that will be encountered in its final end use application. While many of the bisbenzocyclobutene polymers have been reported in the literature, only a few have been studied in detail with regards to their thermal and mechanical performance at both room and elevated temperatures. Tables 7–10 show some of the preliminary mechanical data as well as some other physical properties of molded samples of polymers derived from amide monomer **32**, ester monomer **40**, diketone monomer **14** and polysiloxane monomer **13**. The use of the term polyamide, ester etc. with these materials is not meant to imply that they are to be regarded as merely modified linear thermoplastics. Rather, these polymers are for the most part highly crosslinked thermosets.

As the data in Tables 7 and 8 indicate, these polymers exhibit an excellent retention of their room temperature mechanical properties out at least to 200–250 °C. Note in particular the very high Tg (> 350 °C) of the tetramethyl-

Table 7. Properties of poly

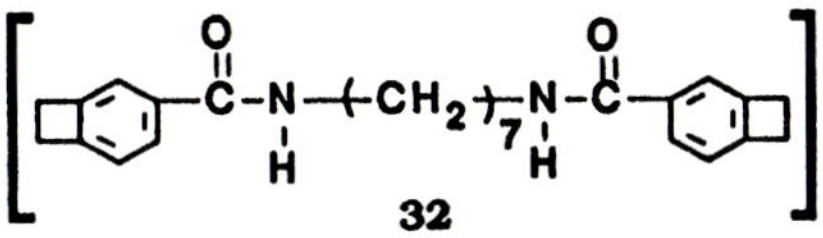

32

Tensile modulus (GPA)	
(25 °C)	2.48
(250 °C)	1.32
Shear modulus (GPa)	
(25 °C)	0.95
(250 °C)	0.50
Tensile strength (MPa)	
(25 °C)	82.8
(150 °C)	41.4
Adhesion (Tensile lap shear)	
(MPa on steel)	
(25 °C)	13.5
(200 °C)	11.4
Coefficient of thermal expansion	
(μm/m °C)	
(25–250 °C)	4.1×10^{-5}
(250–325 °C)	9.8×10^{-5}
Water pick-up (% at 48 h)	
(100 °C)	1.4
Tg	> 270 °C
0% Weight loss temperature	385 °C

Table 8. Properties of poly

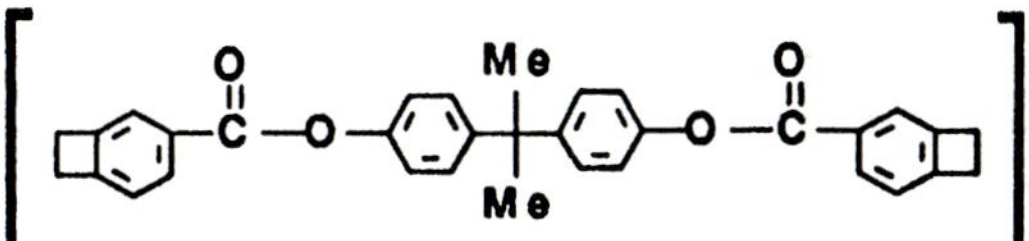

40

Tensile modulus (GPa)	
(25 °C)	2.66
(250 °C)	1.86
Shear modulus (GPa)	
(25 °C)	1.00
(250 °C)	0.70
Tensile strength (MPa)	
(25 °C)	27.6
Coefficient of thermal expansion	
(μm/m °C)	
(25–250 °C)	8×10^{-5}
Water pick-up (% at 48 h)	
(100 °C)	0.9
Tg	> 270 °C
0% weight loss temperature	400 °C

Table 9. Properties of poly

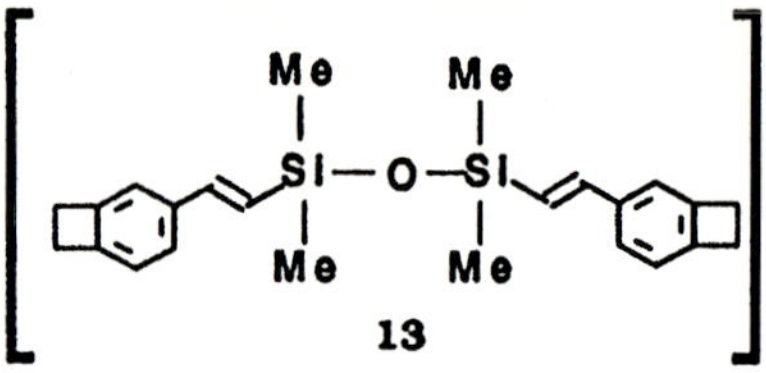

13

(mixed isomers)

Tg (TMA, DMA)	> 350 °C
Flexural modulus (DMA; 25 °C)	2.0 GPa
Tensile strength	85 MPa
Elongation	6.0%
CTE, (TMA)	52 µm/m °C
Water absorbtion	0.2%
Dielectric constant	
(1 MHz)	2.7
Dissipation factor	
(1 MHz)	0.0008

[a] Properties reported are those of the polymer derived from Cyclotene 3022[b] resin which is a formulated version of the monomer **13** from the Dow Chemical Company.
[b] Trademark of the Dow Chemical Company.

Table 10. Properties of poly

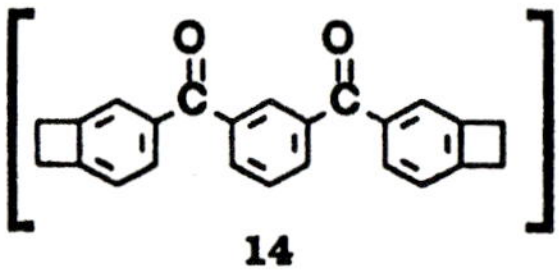

14

Tg (TMA)	310 °C
Flexural strength (MPa)	151.7
Flexural modulus (GPa)	4.25 (30 °C)
(DMA)	2.94 (300 °C)
Coefficient thermal expansion	
Below Tg	35 µm/m °C
Above Tg	150 µm/m °C
G_{1c}	80 J/m^2
Elongation (at break)	4.1%
Thermal stability	
5% weight loss temperature	445 °C (N$_2$)
(TGA: 10 °C/min)	

[a] Prepared directly from monomer.

divinyldisiloxane bisbenzocyclobutene polymer made from monomer **13**. Overall, the absolute values of some of these mechanical properties are quite comparable to those of other high temperature thermosets and suggests that the bisbenzocyclobutene polymers would be useful as matrix resins in this temperature range. Long term thermal aging studies of the polyester from **40** and the

Table 11. Solvent resistance of poly

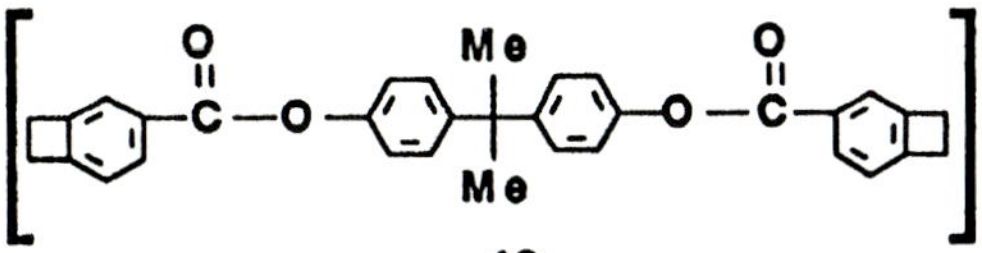

40

Solvent	Weight increase (%) Four week exposure at 70 °C
N,N-Dimethylformamide	17.0
Heptane	0.0
Isopropyl alcohol	0.6
Methyl ethyl ketone	18.0
Propylene glycol	0.0
5% Aqueous sodium hydroxide	0.3
Toluene	2.9
Trichloroethylene	47.0
Water	1.2

polydiketone from **14** at 250 °C suggest that this is the maximum continuous use temperature for these polymers [118].

Many of the bisbenzocyclobutene polymers are relatively unaffected by organic solvents and aqueous media. In Table 11 are shown some of the results which were obtained in a solvent pick-up study carried out on the bisbenzocyclobutene polyester **40** [2]. Of all of the solvents, only trichloroethylene and methyl ethyl ketone were absorbed to any significant extent at 70 °C over the four-week course of the experiment. None of the polymer samples dissolved in the solvents that were tested and only a slight swelling was observed with those liquids which were significantly absorbed. The small effect of aqueous sodium hydroxide on the bisbenzocyclobutene polyester is deserving of note since a control sample of a commercial polyimide (Vespel®) dissolved completely in two days under the conditions of this test.

Water absorption is a uniquely important property owing to water's ubiquitous nature and its ability to strongly influence a polymer's mechanical and electrical properties. In general, high water absorption is undesirable in a polymer, although in some cases such as with the nylons and many biopolymers, absorbed water is critical to the proper functioning of the material. To a first approximation, the water absorption of bisbenzocyclobutene polymers follows in an intuitive way the hydrophilic or hydrophobic character of the linking group x which connects the benzocyclobutene moieties in the parent monomer. It should be noted that all the water absorption data shown in Tables 7, 8, 9 and 10 are rather small and for the most part consistent with what one would expect of a crosslinked polymer.

6 Diels–Alder Polymerizations

The use of benzocyclobutenes in Diels–Alder polymerizations is a fairly recent discovery and has provided a new and valuable way of employing this reaction in polymer synthesis [10, 11, 13, 78, 81]. Diels–Alder polymerizations have been reported in many varieties and forms but with few exceptions have seldom led to the production of high molecular weight or commercially attractive polymers [82]. Several distinct problems encumber the use of the Diels–Alder reaction in polymer synthesis. First and foremost perhaps is the scarcity of suitably reactive and readily accessible bisdienes. A major problem is that bisdienes are usually just as prone to reacting with themselves as they are to reacting with some bisdienophile. Side reactions of this sort often terminate the growing polymer chain and limit molecular weight development. Thus for example, the homo Diels–Alder polymerization of cyclopentadiene only goes on to produce low molecular weight oligomers [83–87]. The attempted homopolymerization of biscyclopentadienes also only produced materials with a low degree of polymerization [88, 89].

By contrast bisbenzocyclobutenes are readily prepared and are thermally equivalent to a highly reactive bis-o-quinodimethane. o-Quinodimethane readily undergoes a Diels–Alder reaction with dieneophiles to yield cycloadducts which cannot by an energetically favorable process revert to starting materials. The reason for this is that the conversion of an o-quinodimethane to a cycloadduct results in the formation of a very stable benzene ring. Reversing this cycloaddition reaction would require a significant disruption of the aromatic moiety and is energetically unfavorable. As a result, the reaction of bisbenzocyclobutenes via o-quinodimethane intermediates with bisdieneophiles easily goes to completion and yields high molecular weight polymers. Further, the conversion of a bisbenzocyclobutene to a bis-o-quinodimethane and its subsequent cycloaddition reaction proceeds without the evolution of volatile by products. This is a significant advantage in the use of bisbenzocyclobutene compositions in closed mold manufacturing operations since there is little tendency for void formation due to the generation of gaseous by-products. Thus, the reaction of bisbenzocyclobutenes with bisdienophiles appears to be potentially the most favorable approach discovered thus far for the successful execution of Diels–Alder type polymerizations.

Early reports by Kirchhoff, Hahn, Tan and Arnold describe the use of polybenzocyclobutene monomers in Diels–Alder type polymerizations [10, 13]. For the most part, these early studies focused on the thermal characterization of mixtures of polybenzocyclobutenes and bismaleimides. Several reports describe the thermal analysis of mixtures of the bisbenzocyclobutene **41** with either the bismaleimide **44** or the diacetylenes **42** and **43** as shown in Fig. 19 [13, 79, 81, 90–93].

Equimolar mixtures of **41** separately and independently with each of the dieneophiles **42, 43,** and **44** were examined by differential scanning calorimetry (DSC). The thermograms obtained from these experiments are shown in

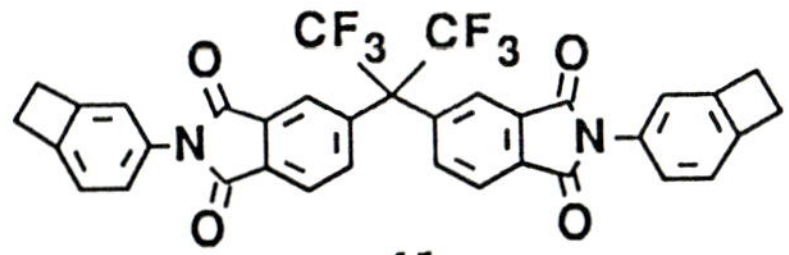

Fig. 19. Structures of the bisbenzocyclobutene **41**, the diacetylenes **42, 43** and the bismaleimide **44** used in the DSC and isothermal weight loss experiments described in Figs. 20–24

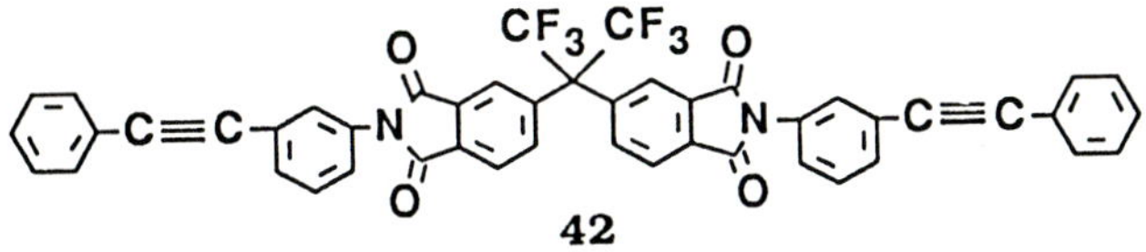

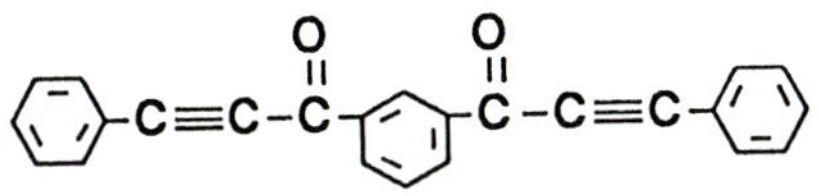

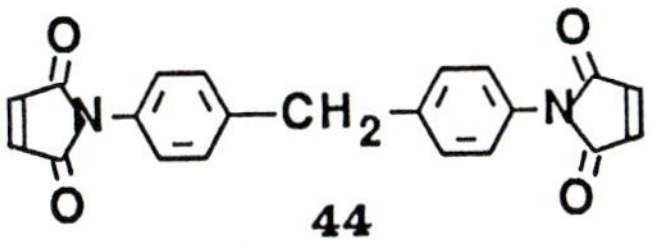

Figs. 20–24. As a control experiment, the pure bisbenzocyclobutene **41** and the pure bismaleimide **44** were also examined by DSC techniques and found to exhibit polymerization exotherm peak temperatures at 257.6 and 234 °C respectively. The DSC scan of the mixture of the bisbenzocyclobutene **41** with the diacetylene **42** showed two significant although energetically unequal exotherm peaks at 267.5 °C and 366.5 °C. The value of the high temperature (366.5 °C) exotherm peak is fairly characteristic of what is observed for the homopolymerization of unactivated diarylacetylene. By contrast, the DSC of the mixture of the bisbenzocyclobutene **41** with the activated diacetylene **43** exhibited a major exotherm peak at 263.7 °C with a much smaller one at 307 °C. Unfortunately, a DSC of the pure diacetylene **43** was not presented. Nevertheless. this data does suggest that there is either a coreaction of the bisbenzocyclobutene **41** with the diacetylene **43** or else they both undergo an exothermic transition at the same temperature. In the case of the reaction of the bisbenzocyclobutene **41** with the unactivated diacetylene **42** however, there appears to be very little, if any, coreaction taking place between the two monomers.

The reaction of the bisbenzocyclobutene **41** with the bismaleimide **44** is similar to the reaction of **41** with the activated diacetylene **43**. From the DSC thermogram in Fig. 24 it is apparent that there is one major exotherm peak at 258.8 °C associated with this reaction. This could be interpreted as either being the polymerization exotherm peak for a benzocyclobutene-maleimide Diels–Alder reaction or else it is the coincidental overlap of the exotherm peaks associated with each pure component. This latter case is a distinct possibility since both the pure bisbenzocyclobutene **41** and the pure bismaleimide **44** have

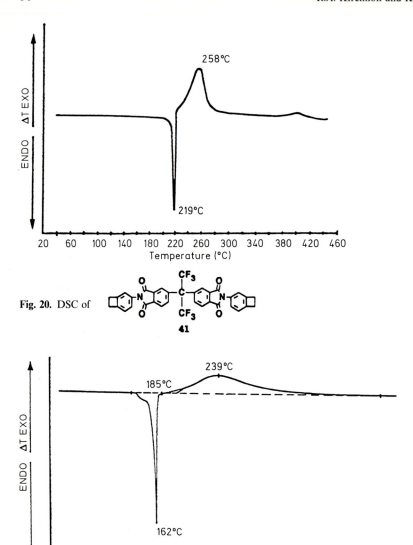

Fig. 20. DSC of **41**

Fig. 21. DSC of **BMI 44** monomer

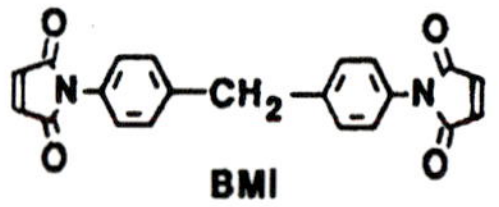

exotherm peaks at nearly the same temperature. It seems most likely however, that this exotherm peak is associated with a Diels–Alder reaction since it has been shown that the benzocyclobutene **45** reacts with *N*-phenylmaleimide **46** to give a good yield of the cycloaddition product **47** (Fig. 25) [90].

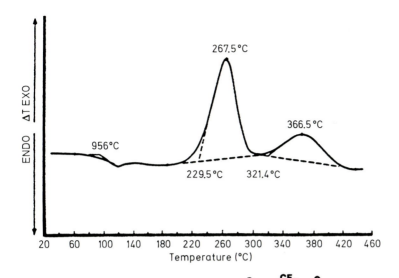

Fig. 22. DSC of an equimolar mixture of ⟨structure⟩ **41** and

⟨structure⟩ **42**

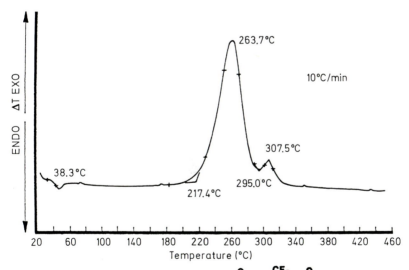

Fig. 23. DSC of an equimolar mixture of ⟨structure⟩ **41** and

⟨structure⟩ **43**

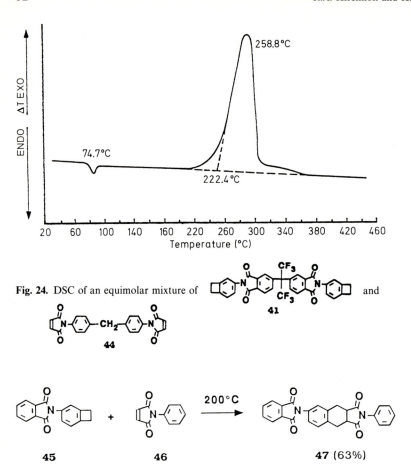

Fig. 24. DSC of an equimolar mixture of and

Fig. 25. Formation of the cycloaddition product from N-phthalimido-4-benzocyclobutene **45** and N-phenylmaleimide **46**

Further evidence supporting the proposal of there being a Diels–Alder reaction between the bisbenzocyclobutene **41** and the bismaleimide **44** is provided by a study of the thermooxidative stabilities of copolymers of various mixtures of **41** and **44**. In this work, the monomers **41** and **44** were blended in several different molar ratios of **41** to **44** and then copolymerized by means of a cure cycle that had a maximum temperature of 250 °C. The resulting cured copolymer samples were then isothermally aged in air at 650 °F for 200 h. The weight loss of the samples over time was measured during the course of the experiment. The compositions used in this experiment are shown in Table 12 and the results of the isothermal aging studies are depicted graphically in Figure 26. As indicated by the weight loss versus time plot, the pure polybismaleimide was very susceptible to oxidation under these conditions. It rapidly lost weight during the course of the experiment and was nearly completely vaporized at its

Table 12. Thermal properties of blends of 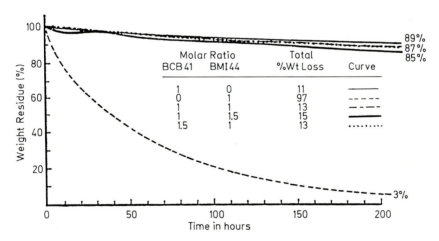 and

Mole rated		Tg Initial	Tm	Tpoly		Tg Cure	T10%[a] (air)
BCB	BMI			Onset	Max		
1	0	11[b]	219	232	258	281[b]	496
0	1	–	162	185	239	–[c]	492
1	1	61[d]	–	224	259	293	500
1	1.5	68[d]	–	221	257	298	520
1	3	70[d]	–	225	257	–[c]	492
1.5	1	68[d]	–	222	257	298	515

[a] Temperature at which 10% weight loss seen by TGA.
[b] By TMA.
[c] Not observed by DSC.
[d] Previously dried in vacuo 40 °C, 18 h.
[e] Temperature in °C.

Molar Ratio		Total	
BCB 41	BMI 44	%Wt Loss	Curve
1	0	11	————
0	1	97	– – – –
1	1	13	– · – · –
1	1.5	15	▬▬▬▬
1.5	1	13	··········

Fig. 26. Isothermal aging study on various mole ratio copolymers of

and

in air at 343°C (650 °F) for 200 h

termination. By contrast, all of the bismaleimide/bisbenzocyclobutene copolymers retained most of their weight over the 200 h that the experiment was run. This weight retention was even observed for those copolymer samples wherein the bismaleimide component was in considerable excess over the bisbenzocyclobutene. Also shown in Fig. 26 is the isothermal weight loss profile of the pure polybisbenzocyclobutene **41**. This was found to be nearly identical to the weight loss curves obtained on the various copolymers. These results were interpreted as evidence of some type of interaction, probably a Diels–Alder reaction, between the bisbenzocyclobutene and the bismalemide [13, 79, 81, 90, 91].

A similar isothermal aging study on the copolymers of the bisbenzocyclobutene **41** with K-353®, a eutectic bismaleimide, has also been reported in the literature [81]. The composition of the K-353® maleimide formulation is given in Fig. 27 and the mole ratio compositions of the copolymers studied is presented in Table 12. These copolymers were also isothermally aged in air at 650 °F to give the weight loss versus time curves shown in Fig. 28. Here again, it was found that the pure polybismaleimide was nearly completely vaporized after 200 h at 650 °F. Under these same conditions, however, all of the copolymers retained a very significant fraction of their initial weight. Particularly noteworthy was the high weight retention observed for the copolymer (Table 13,

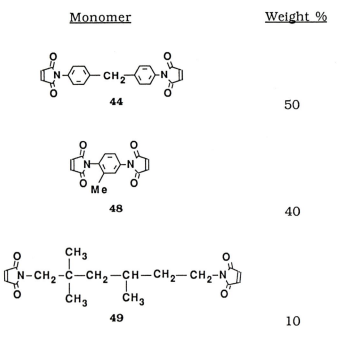

Monomer	Weight %
44	50
48	40
49	10

Fig. 27. Weight % composition of K-353 bismaleimide monomer mixture

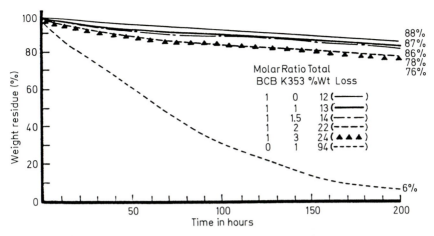

Fig. 28. Isothermal aging study on various mole ratio copolymers of

41

and

K-353 bismaleimide composition in air at 343 °C (650 °F) for 200 h

Table 13. Composition of K-353®/ [structure **41**] copolymers

Composition	Molar ratio	
	Bis-BCD. (41)	K-353®
1	1.0	0.0
2	1.0	1.0
3	1.0	1.5
4	1.0	2.0
5	1.0	3.0
6	0.0	1.0

entry No. 5; Fig. 28) wherein the molar ratio of bismaleimide to bisbenzocyclobutene was three to one.

The enhancement of thermal stability as measured by these isothermal aging
experiments by the addition of bisbenzocyclobutenes to dienophilic thermosetting monomers has also been observed with bis activated acetylenes and
dicyanates. The isothermal weight loss results obtained on a one to one mole

ratio copolymer of the bisacetylene **43** and the bisbenzocyclobutene **41** are shown graphically in Fig. 29.

As shown in Fig. 29, the pure polymerized bisacetylene began to lose weight at a relatively low temperature while the pure polybisbenzocyclobutene was much more resistant to thermooxidative weight loss. The one to one mole ratio copolymer of **41** and **43** however exhibited a much better weight retention than either pure polymer alone. In this example and in the previous maleimide cases, it might be expected that if these copolymerized compositions were just mixtures of the pure individual homopolymers, then the percent weight loss of the mixture would be proportional to the weight loss of each pure component factored by its weight percent in the initial composition. That is, the copolymerized compositions which were predominantly maleimide (e.g. No. 5 in Table 13) would have been expected to lose far more weight than they did. The fact that the presence of a bisbenzocyclobutene comonomer always resulted in the copolymers having enhanced thermal stability has been interpreted as being evidence of a Diels–Alder type polymerization having taken place.

Further support for the occurrence of a Diels–Alder polymerization in these somewhat intractable copolymer systems is contained in a study of some alkynyl substituted monobenzocyclobutenes [91, 94, 95]. The monomers used in this work had either primary or secondary alkynyl groups connected to a benzocyclobutene moiety by a bridging member which was either activating or deactivating of the triple bond to Diels–Alder type reactions. The pure monomers were examined by DSC and then homopolymerized and subjected to an isothermal aging study as above. The authors considered the DSC profile as being indicative of the degree to which each monomer polymerized by a

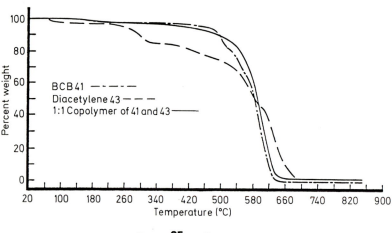

Fig. 29. TGA trace of

and a 1:1 mole ratio copolymer of **41** and **43**

Diels–Alder mechanism. It was found that those polymers which had the least weight loss in the isothermal aging study were derived from those monomers which by DSC polymerized to the greatest extent by a Diels–Alder mechanism. It was inferred therefore that if similar considerations apply to the copolymers of bismaleimides and bisbenzocyclobutenes then these materials too might be the result of predominantly Diels-Alder type polymerizations. To the extent to which these conclusions are correct, the enhanced thermal stability of the bisbenzocyclobutene/bismaleimide copolymers is likely to be due to their being Diels–Alder polymers of some sort.

Tan and Arnold have suggested that the enhanced thermal stability of these Diels–Alder polymers might be due to their ability to be easily oxidized to structures far more thermooxidatively stable than the original polymer itself [90]. The origin of this enhanced stability was proposed to arise from a series of oxidations as depicted in Fig. 30 and 31. In these reaction sequences, the Diels–Alder polymer resulting from the cycloaddition of a bisbenzocyclobutene and a bismaleimide proposed to have the tetralin imide structure 50 (Figs. 30 and 31). The benzylic hydrogens in 50 would be readily susceptible to oxidation with the proposed products being either the benzoquinone 51 or the naphthalene 52. The authors maintain that either of these structures would be more thermooxidatively stable than the original polymer. By way of debate however, it should be noted that even though these processes could convert the cure sites into seemingly more oxidatively stable moieties, the polymers still contain bisbenzylic methylene groups in the backbone of the molecule. Oxidation at these sites might be equally likely and potentially capable of cleaving the polymer into smaller fragments. Speculation aside, the data from all of these experiments strongly suggest that benzocyclobutenes can enter into Diels–Alder

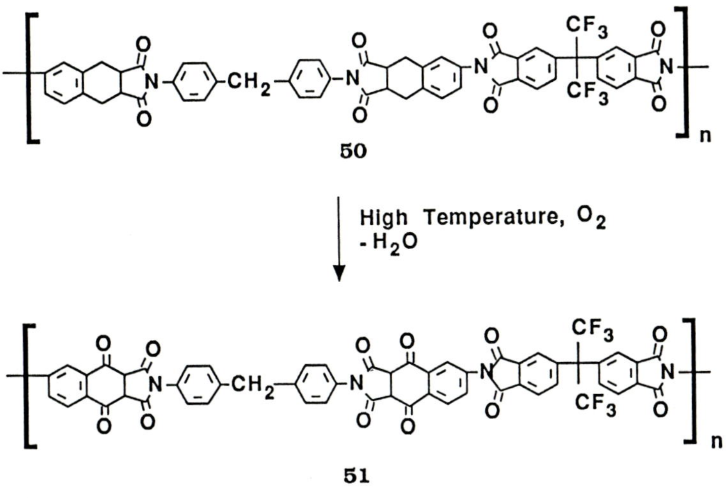

Fig. 30. Proposed air oxidation of the polyimide 50

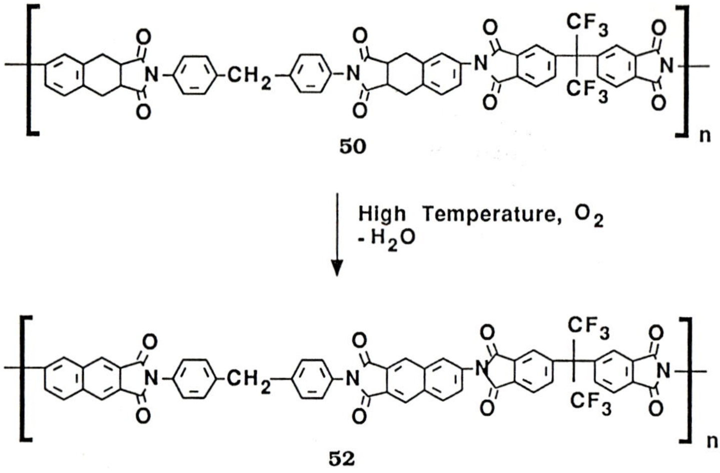

Fig. 31. Second proposed air oxidation product from polyimide **50**

type polymerizations with selected dienophiles to give polymers exhibiting fairly significant thermooxidative stability.

An interesting example of the use of bisbenzocyclobutenes in the synthesis of what appears to be linear polymers is contained in a patent by Bartmann [78]. This patent describes the copolymerization of benzodicyclobutene **53** with a variety of bismaleimides (Fig. 32). The polymerizations were all carried out in a high boiling solvent such as diphenylether or sulfolane. Further, the reactions were all run in the presence of a free radical inhibitor which was claimed to suppress the homopolymerization of the bismalemide. The polyimides prepared in this way were all reported to be soluble in their respective reaction media and were isolated by pouring into a nonsolvent such as methanol. Some of the polymers were described as being thermally stable with weight losses of less than 5% being observed at 400 °C. One of the polymers **54** from 3,6-dimethylbenzo-(1,2,4,5)-dicyclobutene **53** and the bismaleimide **44** was found to have molecular weight of 90,000 by GPC. Little else in the way of physical properties was reported for these materials. Nevertheless, the work did demonstrate that the Diels–Alder polymerization of a bisbenzocyclobutene and a bismaleimide can give rise to a seemingly linear polymers.

The Diels–Alder polymerization of a bisbenzocyclobutene with a bismaleimide to give linear polymers of the sort described above is potentially capable of being a general phenomena. That is, it should be possible to utilize the cycloaddition reaction of a bisbenzocyclobutene with a bismaleimide to prepare high molecular weight linear polymer. o-Quinodimethane is the key intermediate in this process and the course of the reaction to high molecular weight materials is dependent upon minimizing the reaction of the o-quinodimethane with itself. It has previously been shown that benzocyclobutenes, via the intermediacy of o-quinodimethanes, react faster with an activated double bond than they do with themselves [1]. Support for this assertion derives from a study

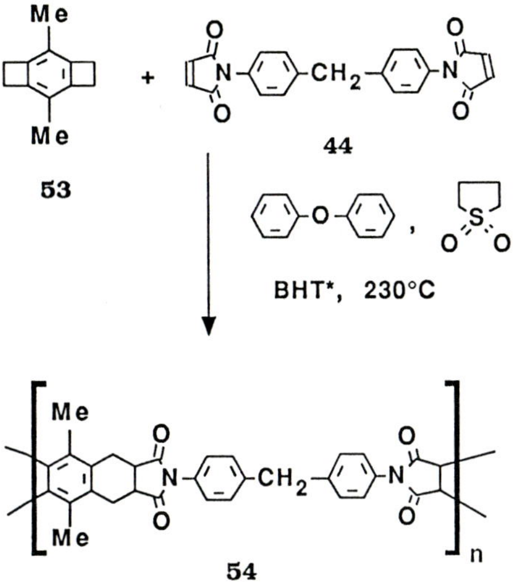

* BHT = 2,6-Di-tertbutyl-4-methylphenol

Fig. 32. Copolymerization of the benzodicyclobutene 53 with the bismaleimide 44

of a series of mono and bisbenzocyclobutene monomers containing either stilbene or 1-aryl-2-dialkylsiloxy double bonds (Table 3). These materials were examined by DSC techniques and the conclusion reached that the reaction of a benzocyclobutene with a double bond was kinetically preferred over the reaction of a benzocyclobutene with itself. The latter reaction however was found to be thermodynamically preferred. Maleimides are exceptionally good dienophiles and should be even more reactive to o-quinodimethane than the olefins used in the aforementioned study. The fact that soluble and presumably linear polymers were obtained from the reaction of benzodicyclobutene with bismaleimides strongly supports the conclusion that the reaction of an o-quinodimethane with a maleimide is much faster than the homopolymerization of an o-quinodimethane.

The advantage of using free radical inhibitors to facilitate the copolymerization of a bisbenzocyclobutene with a bismaleimide was first noted in a patent to Bartmann [78]. Subsequent to this, Corley in a series of patents described some detailed experiments on the copolymerization of bisbenzocyclobutenes with bismaleimides both with and without the addition of a free radical inhibitor [33, 34]. The structures of the bisbenzocyclobutenes used in this study are shown in Fig. 33. The bismaleimide component that was used was a mixture of three different bismaleimides in the molar ratio shown in Fig. 34. The individual bisbenzocyclobutenes were blended at elevated temperature with varying amounts of the bismaleimide composition. In some of the experiments, the free radical inhibitor phenothiazine was added at a 0.5 mole % level. The various monomer mixtures were then copolymerized using one of the cure schedules described in Table 14. The copolymers were then physically characterized using a variety of techniques. Table 14 shows the results obtained from copolymers derived from equimolar mixtures of the various bisbenzocyclobutenes in Fig. 33 with the bismaleimide composition in Fig. 34. As indicated, all of the copolymers are relatively high Tg materials which retain a significant fraction of the flexural strength and modulus even when wet [96]. The data also show that the fracture toughness of the copolymers depended upon whether or nor they had been prepared in the presence of phenothiazine. Overall, the data shows that those copolymers prepared in the presence of low levels (~ 0.5 mole %) of phenothiazine exhibited significantly higher fracture toughness.

55

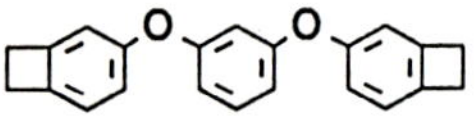

56

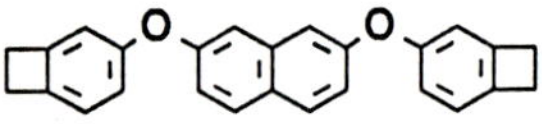

57

Fig. 33. Structures of the bisbenzocyclobutenes **55, 56, 57** used in the copolymerization studies summarized in Tables 15, 16, and 17

Bismaleimide Composition

Component	mol % in Mixture
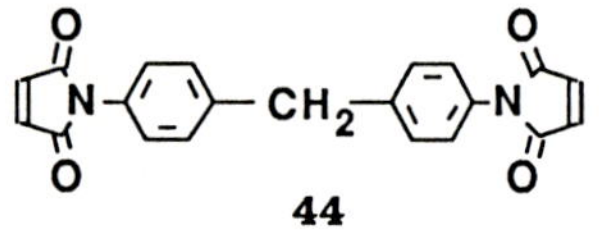 **44**	70
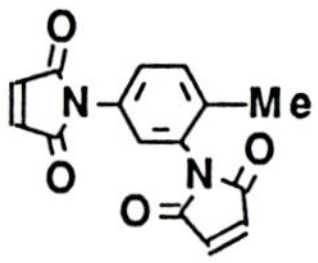 **58**	15
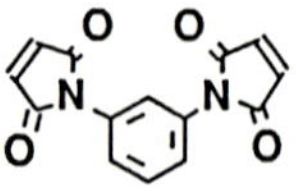 **59**	15

Fig. 34. Molar compositon of the bismaleimide mixture used in the copolymerization studies summarized in Tables 15, 16 and 17

Table 14. Preparation of characterization of bisbenzocyclobutene-bismaleimide copolymers

BCB used[a]	56	56	57	57	55
Phenothiazine used	No	Yes	No	Yes	Yes
Cure cycle	A	A	A	A	B
Tg (Dry) Rheo. Tan δ Peak	242	235	273	270	310
(Dry) Flex:					
Strength (MPa)	136	138	140	142	134
Modulus (GPa)	3.93	3.71	3.44	3.30	3.7
Elong (%)	3.7	4.0	4.5	4.8	3.9
93°C (Wet) Flex					
Strength (MPa)	109	90	109		10
Modulus (GPa)	3.49	3.38	3.37		3.16
Elong (%)	5.8	9.8	7.8		4.0
Compact tension Fracture toughness Kq (MPa) $\times$ m$^{0.5}$)	2.63 ± 0.04	4.01 ± 0.16	3.03 ± 0.03	3.42 ± 0.08	3.13 ± 0.07

Resins:

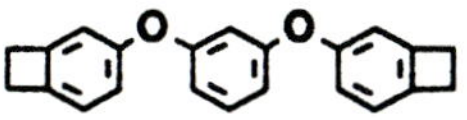

56 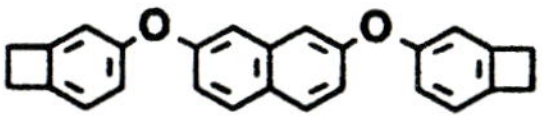   **57** 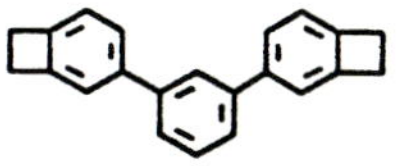   **55**

[a] All mixtures contained an equimolar amount of the BMI formulation described in Fig. 72.
[b] When phenothazine was used, it was incorporated at a level of 0.53 to 0.54 mole% based on total BMI.
[c] Cure cycles:
A. 3 h 210 °C, 15 min 220 °C, 15 min 230 °C, 15 min 240 °C, 2 h 250 °C.
B. 2 h 210 °C, 15 min 230 °C, 15 min 250 °C, 15 min 270 °C, 1 h 290 °C.

In another series of experiments, the bisbenzocyclobutene resorcinol ether monomer (Fig. 33; **56**) was blended with varying amounts of the bismaleimide composition in Table 15. The amounts of reactants used were chosen so as to provide a range of mixtures wherein the mole ratio of the bisbenzocyclobutene to bismaleimide range from about five to one down to one to one. In addition, the mixtures were all blended with phenothiazine ($\sim$ 0.5 mole%) as before and then polymerized using one of the cure cycles in Table 15. Following this initial cure cycle, the copolymers were all post cured at 250 °C for 15 min. The exact mole ratios of the monomer mixtures and details of the cure cycles are presented in Table 15. This table also shows the values of some of the physical properties of the copolymers. Note that the last entry in Table 15 is a control experiment wherein the pure bisbenzocyclobutene **56** was polymerized by itself. The data in Table 15 clearly shows that the copolymerization of even relatively low levels of the bismaleimide comonomer composition from Fig. 34 with the bisbenzocyclobutene **56** resulted in a significant increase in the latters fracture toughness. Concomitant with this was a rather dramatic initial drop ($\sim$ 50 °C) in the glass transition temperature, relative to that of the pure poly(bisbenzocyclobutene) **56** (Tg > 300 °C). As the maleimide content increased, the changes in the glass transition temperature became much smaller and in fact, appeared to level off at about 227 °C. The fracture toughness was found to reach a maximum at a mole ratio of bismaleimide to bisbenzocyclobutene of one to one. This corresponds to the exact stoichiometry required for the linear Diels–Alder polymerization of a

Table 15. Preparation and characterization of copolymers of and a bismaleimide blend[e]

56

Samples post cured at 250 °C for 15 m (molded with phenothiazine)

BCB/BMI (Mol. ratio)	BCB Resin (Mol. fract.)[a]	Cure cycle[b]	Tg[d] (°C)	Rubbery plateau (Mod. MPa)	Compact tension fract. tough (MPa × m$^{0.5}$)	% H20 pick up (2 Weeks/93 °C)
1.002	0.500	A	227	2.0	3.40 ± 0.14	1.47
1.504	0.601	B	227	2.0	2.17 ± 0.13	1.26
2.006	0.667	C	231	6.0	2.29 ± 0.72	1.10
3.007	0.750	B	238	10.0	1.34 ± 0.05	
5.018	0.834	B	250	17.0	1.09 ± 0.14	1.50
	1.00	D	> 300	–	0.74[c]	0.51

[a] BCB resin mole fract = $\dfrac{\text{Moles BCB resin}}{\text{Moles BCB + Moles BMI}}$.

[b] Cure cycles
  A. 2 hr 10 min 210 °C, 20 min 220 °C, 20 min 230 °C, 25 min 240 °C, 15 min 250 °C.
  B. 2 hr 30 min 210 °C, 15 min 220 °C, 15 min 230 °C, 15 min 240 °C, 15 min 250 °C.
  C. 2 hr 15 min 210 °C, 15 min 220 °C, 15 min 230 °C, 15 min 240 °C, 15 min 250 °.
  D. 2 hr 210 °C, 15 min 220 °C, 15 min 230 °C, 15 min 240 °C, 15 min 250 °C.
[c] Only one sample tested. Others too brittle.
[d] Tan δ peak.
[e] See Fig. 34 for composition of BMI blend.

bisbenzocyclobutene with a bismaleimide. In the previous set of experiments (Table 15) all of the bisbenzocyclobutene to bismaleimide ratios were either nearly unity or greater than one. A second series of similar experiments with the same bisbenzocyclobutene monomer has also been reported. In this example however, the mole ratios of the monomers were quite different. Specially, this series of experiments examined the effect of leaving the bismaleimide component in molar excess over the bisbenzocyclobutene. The samples that were used started out with bisbenzocyclobutene to bismaleimide mole ratios of less than one and then by the successive incremental addition of the bisbenzocyclobutene component approached unity. Phenothiazine was added to some of the monomer mixtures at a 0.4 to 0.6 mole % level.

The monomer compositions were copolymerized using several different cure cycles. All of the cure cycles in this series included an additional post cure at 290 °C for one hour. The exact monomer compositions used, their detailed cure cycles and the physical properties of the resulting copolymers are shown in Table 16. As in the previous examples, here too the glass transition temperature went down as the fracture toughness increased. As before the fracture toughness rose as the mole ratio of bisbenzocyclobutene to bismaleimide approached unity. The presence of phenothiazine appeared to increase the fracture toughness in all of the examples although, its effect appeared most pronounced when

Table 16. Preparation and characterization of copolymers of and a bismaleimide blend[c]

56

(Samples molded[b] with and without phenothiazine and post cured at 290 °C for 1 h)

BCB/BMI (Mod. ratio)	BCB Resin (Mol. fract.)	Phenothiazine	Tg (°C)	Rubbery plateau (Mod. MPa)	Compact tension fract. tough (MPa × m^{0.5})[a]
0.429	0.300	Yes	302	–	0.76 ± 0.04
0.429	0.300	No	304	–	0.90 ± 0.01
0.666	0.400	Yes	265	6.0	1.95 ± 0.29
0.666	0.400	No	274	9.0	1.70 ± 0.06
0.745	0.427	Yes	258	5.0	2.29 ± 0.09
0.764	0.427	No	246	5.0	2.38 ± 0.10
0.828	0.453	Yes	248	4.0	3.01 ± 0.15
0.873	0.466	No	248	3.0	2.72 ± 0.13
0.927	0.481	Yes	237		4.10 ± 0.40
1.002	0.501	Yes	239	2.0	3.71 ± 0.14

[a] ASTM E-399-83.
[b] Cure cycle 3 h 210 °C, 15 min 230 °C, 15 min 250 °C 15 min 270 °C, 1 h 290 °C.
[c] See Fig 34 for composition of BMI blend.

the monomer mole ratio approached one. Overall, the work described in these reports demonstrates the importance of monomer mole ratio and the effect of free radical inhibitors on the properties of copolymers of bisbenzocyclobutenes and bismaleimides. In terms of fracture toughness, the best properties were obtained for copolymers prepared from an equimolar mixture of bismaleimide and bisbenzocyclobutene and cured in the presence of phenothiazine. The copolymerization and limited physical characterization of an equimolar copolymer of the fluorinated bisbenzocyclobutene imide monomer **41** and the bismaleimide **44** has also been reported in the literature [93, 97, 98]. The glass transition temperature of the copolymer was reported as being 358–372 °C based upon a TICA (Torsional Impregnated Cloth Analysis) of the monomer mixture [99]. An equimolar blend of the monomers had a minimum viscosity of 400 poise at 185 °C. The monomer blend was polymerized under pressure in a closed mold using a cure cycle ranging from 185 to 260 °C. The resulting copolymer had a room temperature tensile strength of 77.2 MPa.

7 AB Benzocyclobutene Monomers and Polymers

7.1 Introduction

In general, the predominant methods used to prepare polymers derived from benzocyclobutene monomers have focused upon the reaction of a bisbenzocyclobutene with either a second molecule of bisbenzocyclobutene or with a molecule containing additional reactive sites of unsaturation. These approaches can be further delineated by using the AB nomenclature, wherein A and B are structurally distinct, but mutually reactive functional groups. Classically, these types of systems can be viewed as A–A + B–B reaction pathways and can lead to the formation of high molecular weight polymers. However, polymers prepared from monomers where both a benzocyclobutene and reactive site are contained within the same molecule have received less attention. The AB type benzocyclobutene monomers and polymers were first described in a series of patents issued to R.A. Kirchhoff and K.J. Bruza of the Dow Chemical Company [3–7, 46, 47, 100–102] and in a sequence of papers by Tan and Arnold at the Air Force Wright Laboratories [11, 12, 14, 28, 91, 95, 103]. These types of molecules are the AB monomers and have the inherent advantage of containing an exact 1:1 stoichiometry of both A and B reactive groups.

There are many examples of AB monomers in the polymer literature [104]. In particular there are examples of the Diels–Alder reaction being used in successful AB type approaches to polymer syntheses. There are however several practical problems associated with this type of approach. If the diene or dienophile is highly reactive, the synthesis and purification of the monomer can be very difficult. Furthermore, these reactive monomers can be difficult to store without some partial advancement of the molecular weight. If on the other hand the diene or dienophile is not very reactive under moderate conditions, then low molecular weight polymers are obtained. If more stringent conditions for polymerization are employed, the "retro" Diels–Alder reaction and other side reactions can become more important and therefore lead to the formation of a low molecular weight polymer.

The use of benzocyclobutene as the source of the diene in a Diels–Alder polymerization offers a unique solution to the problems described above. Benzocyclobutene containing monomers can be stored indefinitely at room temperature without concern for further advancement of the molecular weight. It is only when benzocyclobutene is heated to temperatures of approximately 200 °C that the reactive diene, o-quinodimethane, is formed at a significant rate and enters into reaction with the dienophile. The only requirement of the dienophile is that it must be stable at these temperatures and not undergo reaction with itself. The most common dienophiles that have been successfully used in the formation of polymers from AB type benzocyclobutene monomers have been acetylenes, olefins and maleimides.

7.2 Benzocyclobutene-Acetylene Polymers

By far most of the work reported in the literature dealing with the reaction of benzocyclobutene with acetylenes has been done by Tan and Arnold [11, 91, 94, 95]. They prepared several monomers and their respective polymers. Figure 35 outlines the synthetic strategies utilized in the preparation of these monomers.

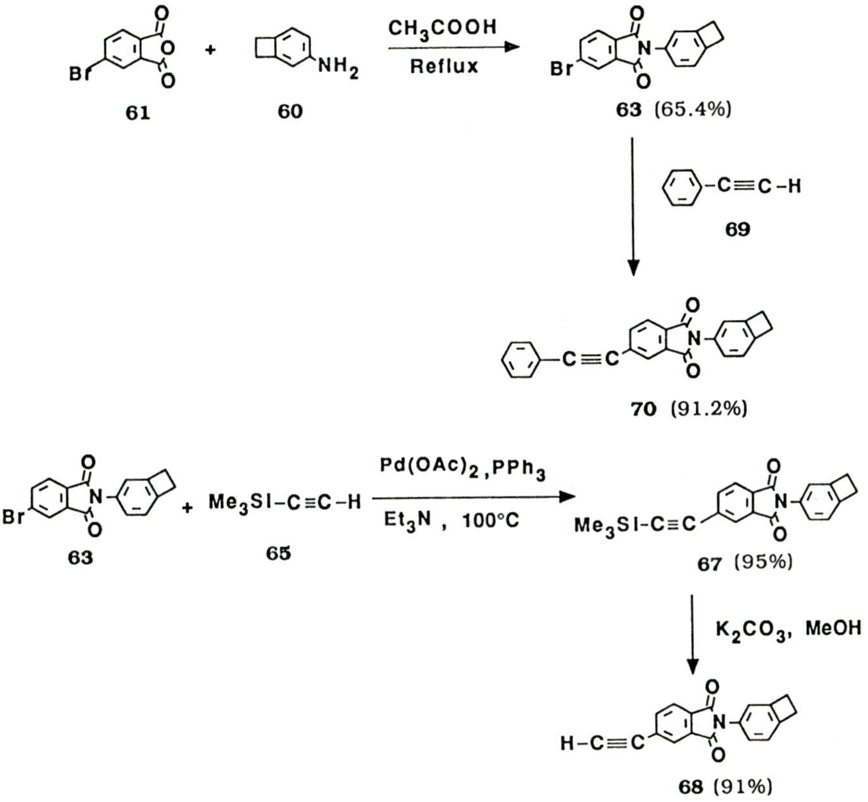

Fig. 35. Syntheses of the acetylenic benzocyclobutene monomers **70, 68, 71** and **72**

4-Aminobenzocyclobutene **60** was first reacted with either 5-bromo-**61** or 5-nitrophthalic anhydride **62** under cyclodehydration conditions to afford the bromo or nitro benzocyclobutene imides **63** and **64** respectively. The second step in the sequence was a displacement of the nitro or bromo substituent by either the acetylenic carbon nucleophile **65** or by the phenate derived from 3-hydroxydiphenylacetylene **66**. In the case of the direct acetylene displacement, the use of 1 mole % of a palladium zero catalyst was required in order obtain to the adduct **67** in excellent yield (95%) [40]. The use of the protected trimethysil-yacetylene **65** was necessary in order to avoid the possibility of reaction occurring at both ends of the acetylene [105]. The trimethylsilyl group was easily removed under mild basic conditions to provide the final N-(4-benzocyclobutenyl)-4-ethynylphthalimide **68** in a yield of 91% [106]. Similarly, replacing the trimethylsilylacetylene with phenylacetylene **69** and carrying out the reaction under similar conditions afforded the N-(4-benzocyclobutenyl)-4-(2-phenylethynyl)-phthalimide **70** in a 91% yield [107].

The formation of the ether linked systems **71** and **72** started with the nitro substituted anhydrides **62** and **73**. Anhydride **73** was first reacted with 4-

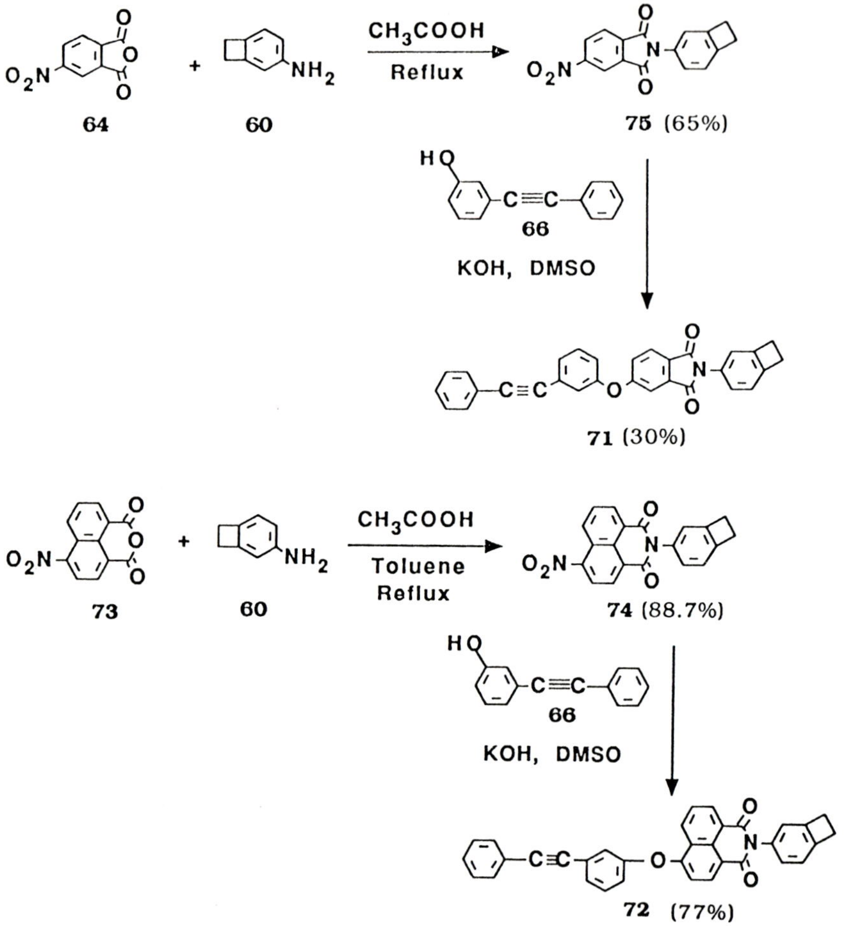

Fig. 35. (Continued) Syntheses of the acetylenic benzocyclobutene monomers 70, 68, 71 and 72

aminobenzocyclobutene **60** to provide the *N*-(4-benzocyclobutenyl)imide **74**. The electron withdrawing nature of the imide functionality activates the nitro group to nucleophilic displacement [104, 108]. In both monomer cases the nucleophile was the potassium salt of 3-hydroxydiphenylacetylene **66** and the reaction was run in DMSO (dimethyl sulfoxide). Reaction of the phenate with *N*-(4-benzocyclobutenyl)-4-nitrophthalimide **75** afforded a poor yield (30%) of the ether adduct **71**, whereas reaction of the same nucleophile with the *N*-benzocyclobutenyl napthalimide **74** provided a good yield (77%) of the desired monomer, **72**. Presumably the improved yield in the naphthalimide example was due to the fact that both of the carbonyl groups of the imide ring activated the aromatic site of substitution to nucleophilic attack whereas, in the phthalimide example, only the carbonyl, which is in a para relationship with the leaving group, could activate this site to substitution. The melting points of these latter

Table 17. Thermal properties of acetylenic benzocyclobutene Polymers

Example	Tg^a (°C)	Isothermal TGA^b (% weight retention)
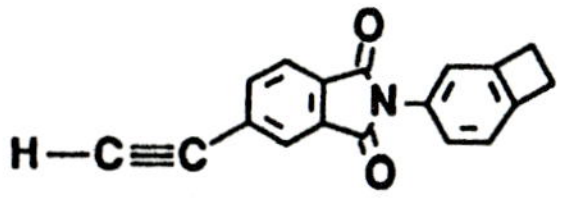 **68**	380	41
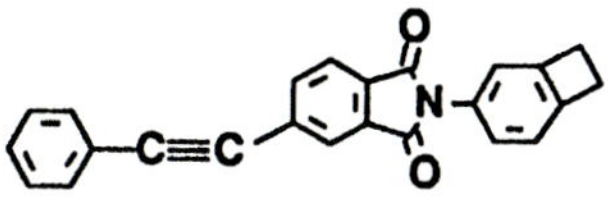 **70**	278	84^c
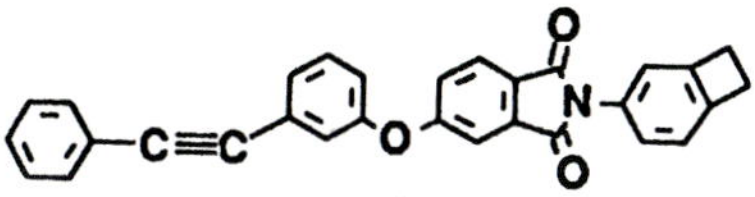 **71**	215	62
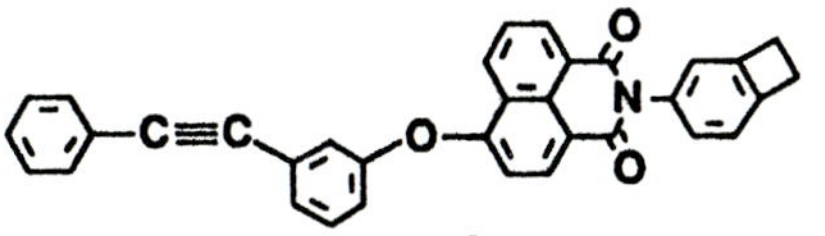 **72**	380	31

^{a} Sample cured at 250 °C for 8 h followed by 350 °C for an additional 8 h: Tg measured by differential scanning calorimetry (DSC).
^{b} Thermal gravimetric analysis (TGA) 200 h at 343 °C in air.
^{c} Tg by TMA was 294 °C.

two monomers were rather high, 200 °C and 238 °C respectively and would likely make it very difficult to process these monomers from the melt without some concomitant, thermally initiated opening of the four membered ring of the benzocyclobutene.

While there was a substantial amount of data reported on the syntheses of these acetylenic benzocyclobutene imide monomers, there was surprisingly little characterization done on the resulting homopolymers [91, 94, 95]. Table 17 details what information is available.

The DSC of primary acetylenic monomer **68** showed the presence of two distinct exotherm peaks. The first peak of 202 °C was attributed to the acetylene reacting separately with a second acetylene while the exothermic peak of 270 °C was proposed to be a benzocyclobutene reacting with a second benzocyclobutene. The resulting polymer was believed to be a highly crosslinked network and to some extent the Tg of 380 °C would support this contention. Interestingly, this homopolymer had poor thermal stability at 343 °C in air, with a retention of 41% of its weight after 200 h.

Replacing the hydrogen in **68** with a phenyl group leads to the secondary acetylenic monomer **70**. It was believed that this disubstituted acetylene would suppress the reaction of acetylene with itself and insure that there was an acetylene functionality available for reaction with the *o*-quinodimethane at 200 °C. The DSC of **68** showed the presence of a single exothermic peak at 263 °C which the authors felt was adequate evidence for the occurrence of a Diels–Alder reaction between the acetylene and benzocyclobutene. Unfortunately they did not report on any control experiments such as that between diphenylacetylene and simple benzocyclobutene hydrocarbon or a monofunctional benzocyclobutene in order to isolate the low molecular weight cycloaddition product for subsequent characterization. The resulting homopolymer of **68** had a Tg of 274 °C and also had the best thermooxidative stability of all of the acetylenic benzocyclobutenes studied (84% weight retention after 200 h at 343 °C in air).

Two other monomers were prepared that incorporated an ether linkage between the acetylene and benzocyclobutene group (**71** and **72**). Monomer **71** is closely related in structure to monomer **70**, with the only difference being the phenylether linkage. Interestingly, while monomer **71** had a melting point close to that of **70**, it displayed significantly different behaviour in the DSC. Monomer **71** had two distinct and separate exotherm peaks at 263 °C and 372 °C. The first was proposed to arise from the reaction of benzocyclobutene with itself while the second exotherm was presumably due to the reaction of an acetylene with a second acetylene. This result was surprising in light of the contention made earlier that monomer **70** reacted selectively in a Diels–Alder fashion. The authors postulated that the ether oxygen made the acetylene an electron rich dienophile and reduced the reactivity of the acetylene toward the *o*-quinodimethane, but again offered no further supporting evidence. The homopolymer resulting from **71** had Tg of 215 °C which was the lowest value obtained for this series of monomers, and retained 62% of its weight of 200 h at 343 °C in air. Out of this series of four acetylenic monomers/polymers, **71** had the second best thermooxidative stability observed.

Monomer **72** was the only molecule in the series to incorporate the naphthalimide functionality. The synthesis of **72** followed from the same chemistry used previously with the only exception being the structure of the starting anhydride. The melting point of **72** was 238 °C which meant that as the monomer was melting the benzocyclobutene was simultaneously undergoing ring opening to form the *o*-quinodimethane which could react with either itself or the acetylene functionality. The Tg of the polymer from **72** was 380 °C and when evaluated as to its thermooxidative stability turned out to be the least thermally stable of all with a retention of only 31% of its weight after 200 h at 343 °C, in air. The authors felt that this thermal instability was due to the "peri" hydrogens of the naphthalene moiety [109].

The order of thermal stability based upon the results of the isothermal TGAs was concluded to be **70** > **71** > **68** > **72**. This same order was proposed to reflect the relative reactivity of the monomers in a Diels–Alder reaction.

Overall these polymers displayed relatively high Tgs and good to fair thermal stability. However, very little other data were provided on the homopolymers which makes a more detailed evaluation of their potential utility difficult. One point which can be addressed is their processability from the melt. The relatively high melting points of these monomers would not provide much of a processing window for their use in the formation of composites. It might be possible to partially advance or B-stage these monomers to amorphous solids which would reduce their crystallinity and possibly lower the Tg enough to afford a better processing window.

7.3 Benzocyclobutene-Alkene Polymers

A second class of AB benzocyclobutene monomers and polymers which have been reported in the literature are those in which a benzocyclobutene and a stilbene double bond are contained within the same molecule (Fig. 36) [4, 5].

As shown in Fig. 36, R^1 and R^2 can be represented by a great variety of functionalities such as hydrogen, alkyl, cycloalkyl, aryl, substituted aryl and heterocyclic groups. Furthermore both electron withdrawing groups such as COOH, COOR, CN and electron donating groups such as OR and NR_2 may be present. X can be either a direct bond from an olefin or bifunctional linking group. Typical X groups can be taken from the partial listing of R^1 and R^2 [4].

In general, the monomers that have been prepared and subsequently homopolymerized do not have a linking group between the double bond and the benzocyclobutene ring and R typically has been an aryl group. Table 18 lists the structures of those monomers which have been reported and Fig. 37 details the general process for their preparation.

The chemistry used to prepare these olefin containing monomers was the same as that described for the preparation of many of the aforementioned AB acetylene benzocyclobutenes [106]. The only exception was that terminal alkenes were used as the starting materials with $Pd(OAc)_2/P(o\text{-Tolyl})_3$ as the catalyst. In general the reactions proceeded well, affording the monomers in yields ranging from 39 to 78%. The monomers had melting points below the onset of their polymerization temperature (200 °C) and could be readily polymerized from the melt to yield clear transparent castings. Preliminary characterization of the polymers by DSC showed the presence of a well defined Tg, and further evaluation of the polymers by TGA indicated that these materials had very good short term thermal stability at temperatures greater than 350 °C in air. Interestingly, the Tgs of the polymers were all within a 3 °C range of one another (Table 19). The presence of the different aromatic groups in the

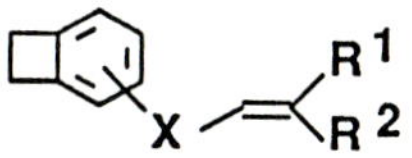

Fig. 36. Generic structure of AB type alkenylbenzocyclobutene monomers

Table 18. Properties of AB-benzocyclobutene/alkene monomers and homopolymers

Monomer	M.P. (°C)	Yield (%)	Tg[1,2] (°C)
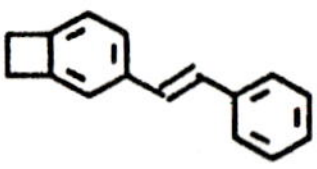 78a		77.8	220
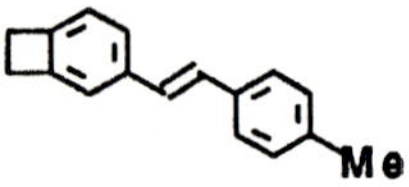 78c	90–92	84	221
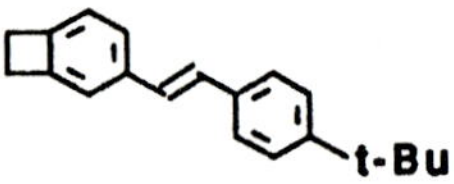 78b		42.6	218
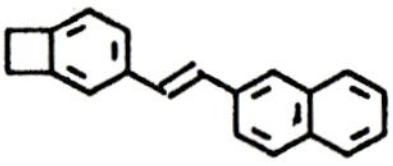 78e		44.0	220
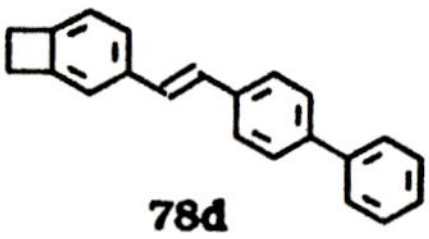 78d		38.5	

[1] Tg is of the homopolymer.
[2] Polymer samples were prepared by heating the monomer at 170 °C until it melted whereupon the temperature was raised to 200 °C and held for 1 h. The temperature was riased to 220 °C and held for 1 h before finally heating the sample at 250 °C for 3 h.

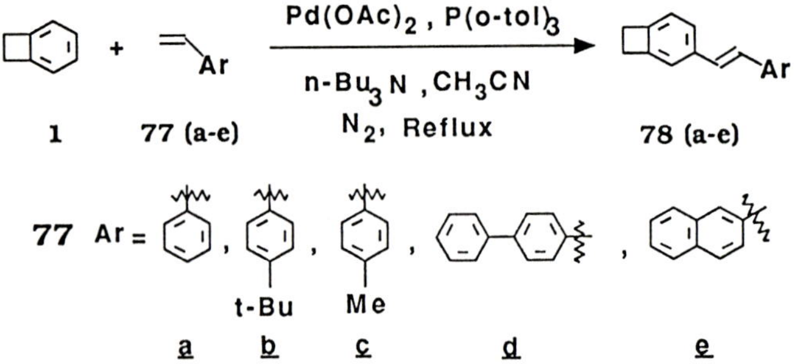

Fig. 37. Synthesis of 1-aryl-2-(4-benzocyclobutenyl)ethylene monomers **78 (a-e)**

monomers did not appear to cause a difference in the Tgs of the resulting polymers. No further characterization data was presented on these polymers, although a recent publication by the authors claimed that these homopolymers had relatively low molecular weight values when measured by GPC (polystyrene reference) [110].

In this same study it was proposed that the predominant reaction pathway in these types of molecules was a Diels–Alder reaction between the benzocyclobutene of one monomer with an olefin contained in a second monomer leading to the overall formation of a linear polymer (Fig. 38). There did not appear to be a significant amount of crosslinking occurring early on in the polymerization reaction. In fact the authors speculated that as the reaction proceeded, the benzocyclobutene or olefin might react via a different mechanism that did not lead to the production of higher molecular weight materials or crosslinking. However, they did not offer any direct evidence to support this speculation.

In a separate study, small concentrations of bisbenzocyclobutene 11 were copolymerized in with monomer 78a in an effort improve the mechanical properties of the original homopolymer from 78a (Fig. 39) [5].

As can be seen in Table 19, incorporation of 1 mol% of the bisbenzocyclobutene 11 provided a copolymer with an increase in Tg from 220 °C to 245 °C, flexural strength and modulus of 142.8 MPa and 3.29 GPa respectively and a fracture toughness (K_{1c}) of 1.76 MPa × $m^{0.5}$. Copolymers containing increased levels of the bisbenzocyclobutene crosslinking agent had higher Tgs (287 °C), consistant flexural strengths (159–166 MPa) and moduli (3.31 GPa), but the fracture toughness fell off precipitously (0.2 MPa × $m^{0.5}$). Several other bisbenzocyclobutenes have been used as crosslinking agents for the AB alkene benzocyclobutenes with similar results.

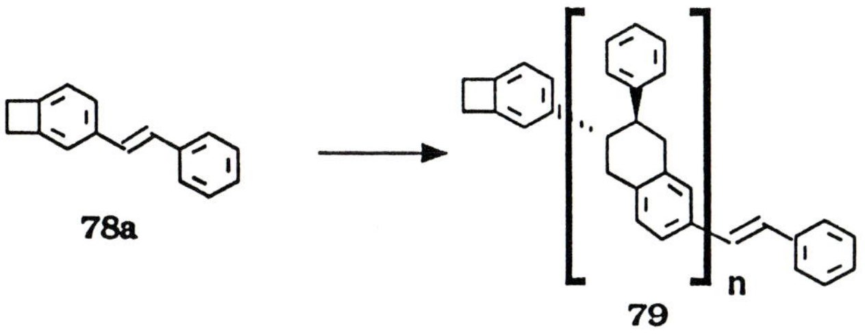

78a   **79** n

Fig. 38. Proposed structure of the Diels–Alder polymer derived from 1-(4-benzocyclobutenyl)-2-phenylethylene 78a

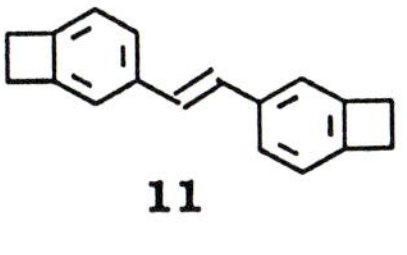

11   Fig. 39. Structure of 1,2-bis(4-benzocyclobutenyl)ethylene 11

Table 19. Properties of copolymers of 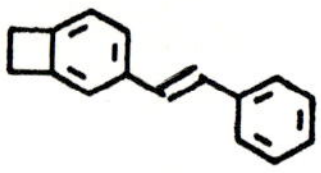 and

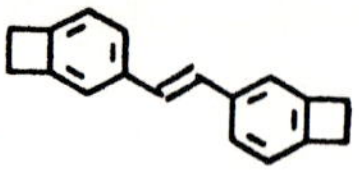

78a   **11**

Entry	Moles 78a Moles 11	Tg (°C)	Flex. strength (MPa)	Flex. modulus (GPa)	K_{1c} (MPa × in$^{0.5}$)
1.	100 78a	220	–	–	–
2.[a]	99/1	245	143	3.29	1.76
3.	95/5	265	169	3.12	1.27
4.	90/10	287	161	3.33	0.20

[a] Percent elongation = 6.4.

Unfortunately, it was not clear why these AB types of systems did not give high molecular weight polymers, but from the results presented above, it would appear that the incorporation of crosslinking agents, such bisbenzocyclobutenes can provide copolymers with good thermomechanical and mechanical properties including good fracture toughness values.

7.4 Benzocyclobutene-Maleimide Polymers

The third major area of AB benzocyclobutene technology has dealt with the reaction of benzocyclobutene with maleimide to form a substantially linear homopolymer (Fig. 40).

As before, the homopolymer forming reaction involved the generation of an o-quinoidimethane from benzocyclobutene which in turn reacted with the double bond of a maleimide functionality on a second monomer to give chain extended AB material via a Diels–Alder pathway. The polymers prepared to date using this type of approach have been shown to have excellent thermal and mechanical properties and have been considered as potential candidates for use as matrix resins for advanced composites. This area has been investigated by the researchers at both the Air Force Wright Laboratories and the Dow Chemical Co.

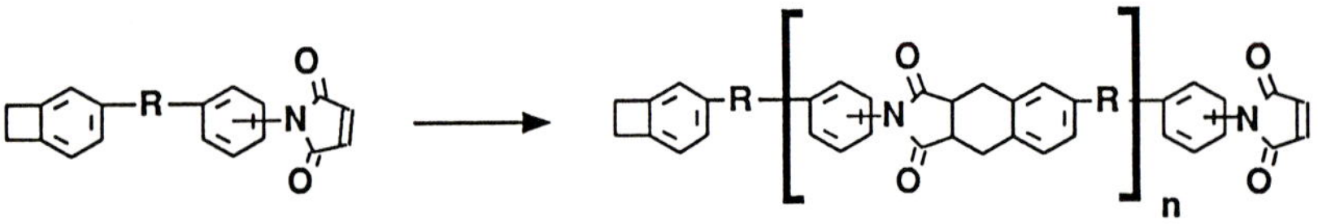

80   **81**

Fig. 40. Proposed generic structure for the polymer derived from the Diels–Alder polymerization of a maleimidobenzocyclobutene AB monomer **80**

Tan and Arnold have reported on the synthesis of several novel maleimido benzocyclobutene monomers and their conversion into homopolymers (Fig. 41) [14, 103].

The simplest AB benzocylobutene maleimide monomer, **84** was prepared from 4-aminobenzocyclobutene **60** and maleic anhydride **82**. The monomer was obtained as a crystalline solid (m.p 77 °C) in 64% yield. Monomer **84** had a relatively high vapor pressure such that when heated at 148 °C and held at this temperature for 1 h, it lost approximately 40% of its mass. Further, when it was heated between 200 and 238 °C, a 71% weight loss was observed [14]. The

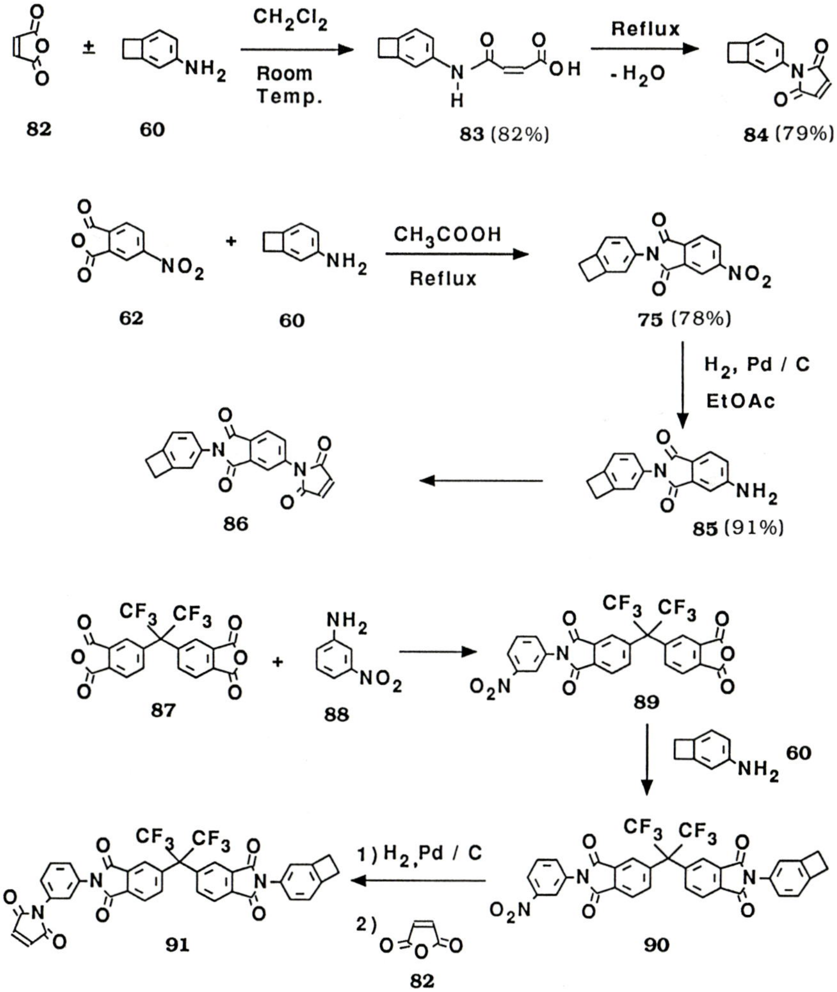

Fig. 41. Syntheses of maleimidobenzocyclobutene AB monomers **84, 86** and **91**

polymer from monomer **84** was successfully prepared in a DSC experiment
(10 °C/min in nitrogen) when the cell was held under positive nitrogen pressure
(3.4 × 10^6 Pa). The resulting homopolymer had a Tg of 287 °C when the DSC
was rerun. The monomer volatility problem apparently caused the researchers
to stop any further work on this material.

Another AB monomer which was prepared, **86**, combined a maleimide and
benzocyclobutene with a phthalimide linking group. The phthalimide was
expected be a very thermally stable structural unit and therefore it was not
surprising to find that the resultant homopolymer had a Tg of 328 °C and
showed a 10% weight loss in a TGA at 498 °C. Unfortunately, **86** also had a
relatively high melting point (230 °C) which would not offer a good processing
window of this material from the melt.

The monomer **91** was prepared in a multistep process and the authors did
not quote the yield obtained for the final product (Fig. 41). In the first step the
dianhydride **87**, was reacted with *m*-nitroaniline **88** to form the mono imide
anhydride **89** without any of the bis imide product being reported. Once this
material was isolated the remaining anhydride functionality was reacted with 4-
aminobenzocyclobutene **60** to form the *N*-benzocyclobutenyl imide, **90**. The
nitro group was reduced to the amine (H$_2$, 10% Pd/C) which in turn was reacted
with maleic anhydride to afford the final AB monomer, **91**. Polymerization of **91**
was carried out in a DSC (10 °C/min to 450 °C) [14]. Monomer **91** had a melting
point of 99 °C and the final homopolymer had a Tg of 257 °C [14]. A TGA of the
homopolymer indicated that at 508 °C the polymer suffered a 10% weight loss.

In another study, all of the above monomers were separately polymerized as
a solution in 1-methyl-2-pyrrolidinone (NMP) at a concentration of 5–6%
solids, and at 202 °C, overnight [14]. If the solutions were heated longer than
this, they were reported to either deposit solids or else form gels. The gels from **84**
and **86** were not soluble in organic solvents but did swell. On the other hand the
polymer from **91** was partially soluble in NMP (approximately 50% of the
polymer mass) and this soluble fraction had an inherent viscosity of 0.34 dL/g at
a concentration of 0.39 g/dL. The Tg of this soluble fraction was determined to
be 291 °C. No further data was provided about this latter material or any of the
other monomer/polymers discussed above.

These same authors also reported on several other types of AB maleimide
benzocyclobutene monomers. The syntheses of these materials are shown in
Fig. 42 [103].

Additionally, monomer **17** was prepared by the same synthetic sequence
used for **95** with the exception being that 4-nitrobenzoyl chloride was used as the
starting material (Fig. 43).

Monomer **95** was shown to have a melting point of 93 °C by DSC
(10 °C/min) and an exothermic peak at 261 °C. A rescan of the DSC sample
showed that the Tg was 249 °C. The TGA indicated that the major weight loss
occured at 430 °C with 10% weight loss occuring at 470 °C. Interestingly, the
only reference to **17** was shown in both a table and TGA trace with no
indication as to how the monomer/polymer behaved in a DSC polymerization

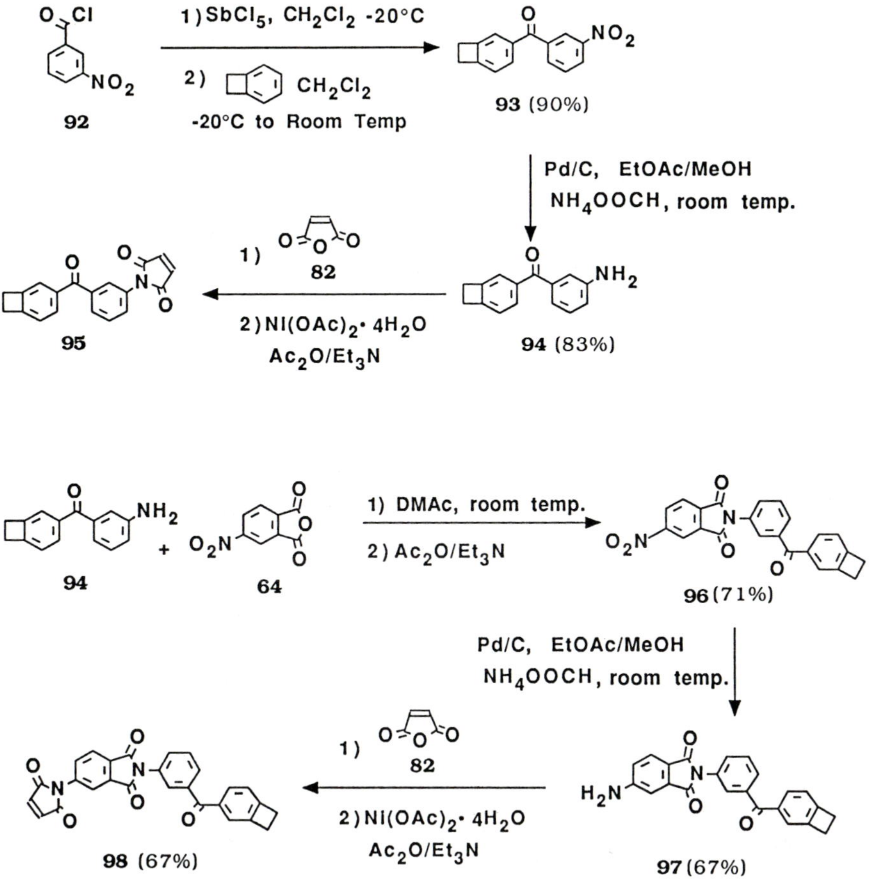

Fig. 42. Syntheses of ketone linked maleimide benzocyclobutenes **95** and **98**

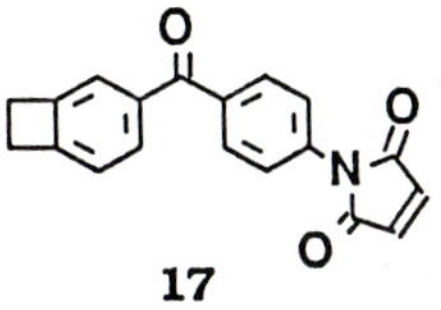

17

Fig. 43. Structure of 4-benzocyclobutenyl-4-maleimidophenyl ketone **17**

experiment. No information was given describing the simple physical properties of this homopolymer.

The phthalimide keto linked **98** monomer was reported to have a Tg of 44 °C and a polymerization exotherm at 257 °C. The homopolymer of **98** upon rescanning in the DSC had a Tg of 258 °C. The TGA (in air) of this material indicated a major weight loss at 403 °C followed by a 10% loss in weight at

452 °C. Again, there was no further characterization data presented for the polymer.

AB maleimide benzocyclobutene monomers and polymers have also been actively studied and developed by researchers at the Dow Chemical Co. The first AB maleimide type benzocyclobutene monomer prepared and polymerized by Kirchhoff was the benzocyclobutenyl maleamic acid **106** [6]. The detailed synthesis of **106** is shown in Fig 44.

In this synthetic sequence, displacement of the benzylic chloride from *o*-chlorobenzylchloride **100**, with ethyl cyanoacetate anion afforded 3-(2-chlorophenyl)-2-carboethoxypropionitrile **101** in 68% yield after distillation. Saponification of the ethyl ester followed by thermal decarboxylation and fractional vacuum distillation, provided 3-(2-chlorophenyl)propionitrile **102** in 92% yield. Intramolecular cyclization was accomplished with sodium amide in liquid ammonia to afford 1-cyanobenzocyclobutene **103** as a colorless liquid in a

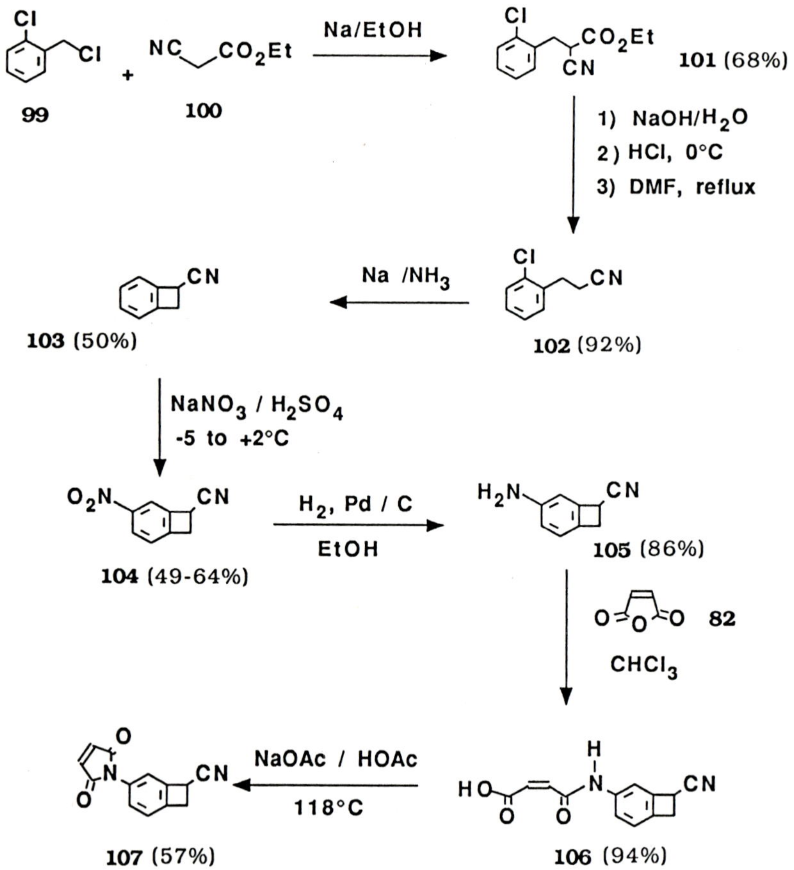

Fig. 44. Synthesis of l-cyano-4-maleimidobenzocyclobutene **107**

50% yield [111]. Nitration of 1-cyanobenzocyclobutene, **103** was carried out using sodium nitrate in concentrated sulfuric acid and provided 5-nitro-1-cyanobenzocyclobutene **104** as a crystalline solid in 64% yield. It is interesting to note that the nitration of 1-cyanobenzocyclobutene **103** proceeded in higher yield than that observed for the corresponding nitration of the parent hydrocarbon. The parent hydrocarbon under these reaction conditions yield as a major byproduct the ring opened alkyl *o*-nitrite with the yield of 4-nitrobenzocyclobutene being approximately 20–30%. The cyano group apparently reduces the susceptibility of the four member ring to open under protic conditions. The reduction of the nitro group was accomplished with hydrogen at ambient pressure, 10% palladium on charcoal as the catalyst and provided the amine, **105** in 86% yield. Formation of the maleamic acid **106** was achieved by reacting the amine **105** with maleic anhydride in chloroform at room temperature. The yield of maleamic acid **106** was 94%. The final AB maleimide benzocyclobutene **107** was obtained by cyclodehydration of the maleamic acid with sodium acetate-glacial acetic acid and was isolated in 57% yield.

The first polymer prepared was from the ammonium salt of the maleamic acid **106**. Acid **106** was reacted with ammonium hydroxide to form the water soluble ammonium maleamate salt which was in turn cast as a film upon glass [6]. Evaporation of the water followed by thermal curing of the salt at 175 °C provided a pale yellow coating which adhered quite well to the glass and was not affected by a variety of organic solvents. It also showed thermal stability up to 320 °C.

Another approach to the preparation of AB maleimide benzocyclobutenes was to use the parent hydrocarbon as the primary building block. One of the first monomers prepared directly from the hydrocarbon was **17**, which was a material that was later reported on by researchers at Air Force Wright Laboratories [103].

The UDRI synthesis began by reacting benzocyclobutene with 4-nitrobenzoyl chloride **108** in the presence of a stoichiometric amount of the Lewis Acid, antimony pentachloride, at low temperature. The route pursued by Dow researchers utilized higher temperatures and 1 mole % of ferric oxide as the catalyst (Fig. 45) [46, 47, 102].

In this reaction the benzocyclobutene was used both as the reactant and solvent. The heterogeneous catalyst could be filtered away from the reaction

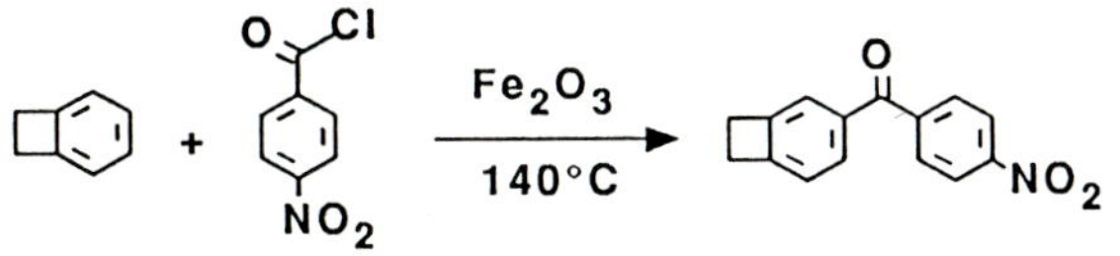

Fig. 45. Catalytic benzoylation of benzocyclobutene **1** with 4-nitrobenzoyl chloride **108** in the presence of ferric oxide

mixture and the excess benzocyclobutene recovered by distillation leaving the crude nitro product **15** which was purified by recrystallization from *n*-hexane. Subsequent reduction of the nitro group to the amine was carried out using hydrogen and a palladium catalyst on charcoal to provide the amine in 95–100% yield. Formation of the amic acid with maleic anhydride followed by cyclodehydration afforded the final imide **17** in 70% yield. The melting point of this monomer was 148 °C and it had a single exothermic peak at 260 °C, by DSC. Cooling of the DSC sample to room temperature and rescanning with the same temperature program indicated the presence of a Tg at 310 °C.

Neat resin castings of the homopolymer from **17** were prepared under thermal conditions with a final temperature of 250 °C (Table 20). The polymer from **17** had a room temperature flexural strength of 207 MPa, room temperature modules of 3.24 GPa, elongation of 6% and a surprisingly high value for its compact tension fracture toughness (K_{1c}) of 1.59 MPa × m$^{0.5}$ or G_{1c} of 780 J/m^2 [30, 112]. By comparison, Compimide® 796, a polyimide resin candidate for application as a matrix resin for composites has a Tg of > 300 °C, flexural strength and modulus (at room temperature) of 76 MPa and 4.64 GPa, elongation of 1.7% and a G_{1c} of 63 J/m^2.

Several different AB maleimide benzocyclobutenes were prepared and Table 21 lists these monomers and homopolymers along with their thermal and

Table 20. Properties of poly

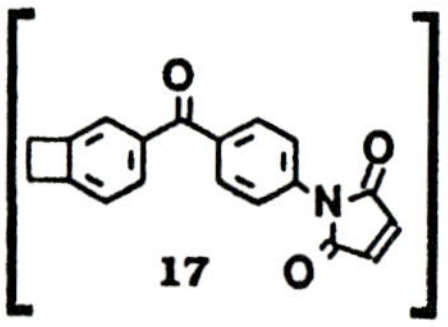

Density (g/cc)	1.30
Tg (°C)	317
Flex. strength (MPa)	207
Flex. modulus (GPa)	3.24
Flex. modulus 400 °F/Wet (GPa)	1.92
Coeff. of linear thermal expansion (μm/m °C)	
Below Tg	43
Above Tg	193
Thermogravimetric analysis (TGA), 5% wt. loss temp. (°C)	
[Air]	460
[N$_2$]	470
Elongation	6
K_{1c} (MPa × m$^{0.5}$)	1.59
G_{1c} (J/m^2)	780
Water uptake % (at equilibrium)	4.2
Dielectric constant	3.15
Dissipation factor	0.0026

Table 21. Properties of AB benzocyclobutene maleimide homopolymers derived from:

Entry	Tg (°C)	Flexural modulus (GPa)	K_{1c} (MPa × m$^{0.5}$)	G_{1c} (J/m^2)
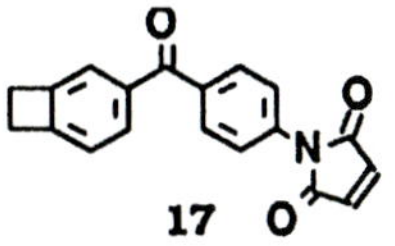 **17** MP: 148 °C	317	3.25	1.59	780
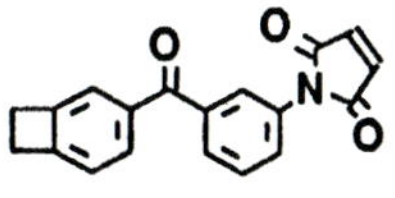 **95** MP: 119 °C	251		0.86	215
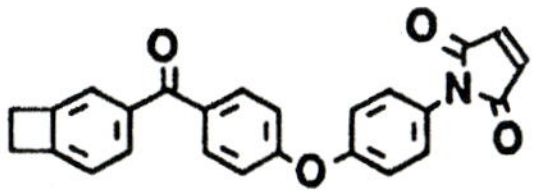 **109a** MP: 157 °C	270	3.16	1.81	1330
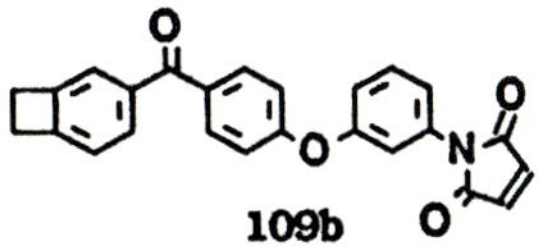 **109b** MP: 95 °C	230		> 2.75	> 3000
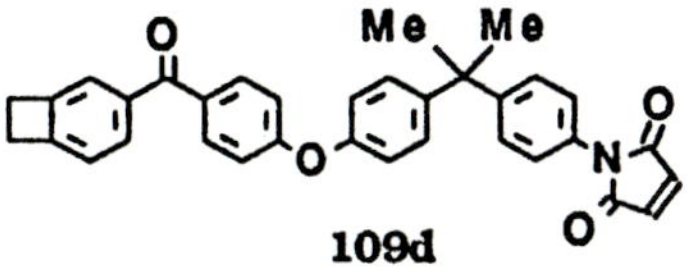 **109d** Viscous liquid	230	3.08	1.85	980
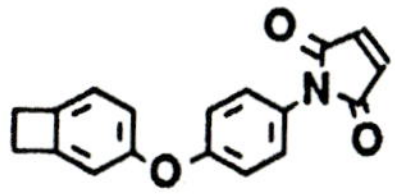 **114a** MP: 113 °C	260	3.50	2.31	1530

Table 21. *Contd.*

Entry	Tg (°C)	Flexural modulus (GPa)	K_{1c} (MPa × m$^{0.5}$)	G_{1c} (J/m^2)
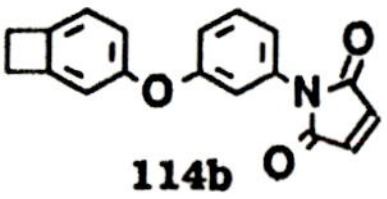 **114b** MP: 103 °C	226		1.26	455
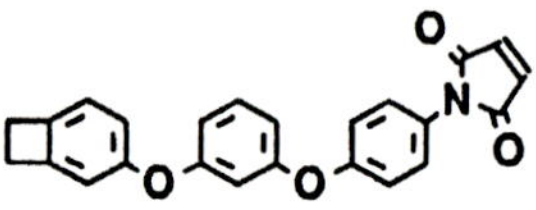 **119** MP: 126 °C	202	3.71	> 2.75	> 3000

mechanical properties [112]. Figure 46 outlines the general synthetic sequence of reactions used in the preparation of these materials.

Monomers **111** (a–d), were prepared from the common starting material **15** by a potassium phenate displacement of the aromatic nitro group. The yields of the keto-ether amine products ranged from 90 to 100% and were of sufficient purity after extractive work up to be utilized directly in the synthesis of the various maleimide monomers. Imidization of the aminobenzocyclobutenes was accomplished using standard reaction conditions (maleic anhydride to form the amic acid followed by cyclodehydration with acetic anhydride and triethylamine) and provided the maleimide products in yields ranging from 60 to 90%.

Another series of monomers that was prepared began with the displacement of an unactivated aryl halide with phenate anion in the presence of a copper catalyst [113–115]. Figure 46 outlines the basic route followed for the preparation of **114a**. The sequence of reactions started with 4-bromobenzocyclobutene, **2** which was reacted with the phenate of *p*-acetamidophenol, **112a** in the presence of copper (I) chloride as a catalyst, to afford the ether linked product **113a** in a yield of 50–75%.

The acetyl group of **113a** was removed by treating with aqueous acid, followed by base to provide the free amine. Final conversion to the maleimide **114a** was accomplished following the same reaction conditions described earlier (maleic anhydride followed by acetic anhydride and triethylamine at room temperature).

The preparation of monomer **119** required a different approach. 4-Benzocyclobutenyl-3-hydroxyphenyl ether **115** was reacted with 4-nitro-chlorobenzene **116** under basic conditions to afford the nitro adduct **117** [114].

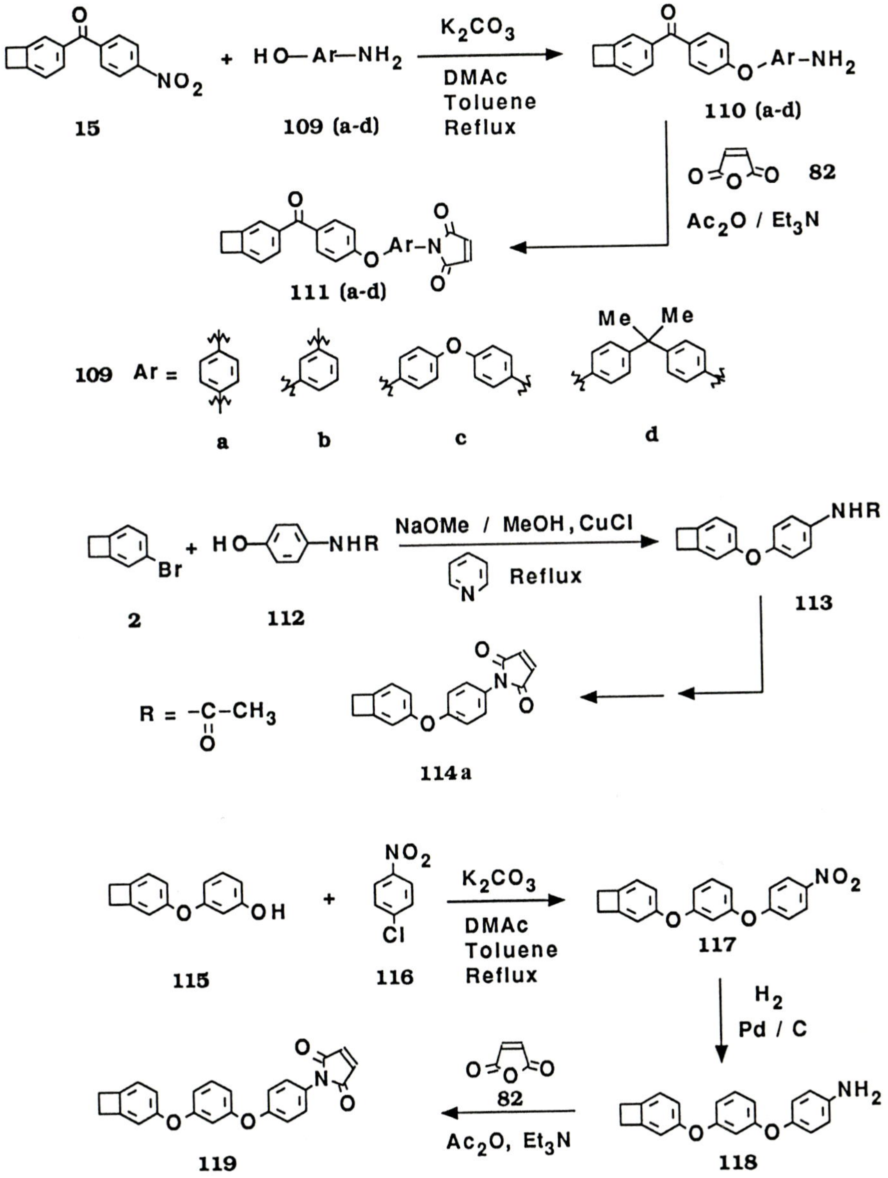

Fig. 46. Syntheses of maleimidobenzocyclobutene AB monomers containing ether linkages

Reduction of the nitro group to the amine **118** was achieved by hydrogenation over palladium on charcoal under moderate pressure. Again conversion to the final maleimide **119** was achieved by reacting the amine with maleic anhydride followed by cyclodehydration with acetic anhydride and triethylamine at room temperature.

With the exception of monomer **111d,** all of the monomers were crystalline solids with melting points ranging from 95 to 157 °C (Table 21). When molten, they formed low viscosity liquids which allowed these monomers to be readily processed into their homopolymers at higher temperature (200 °C).

All of the monomers in Fig.46 were polymerized into neat resin castings using a thermal cure cycle with a maximum temperature of 250 °C. The polymers were then subjected to preliminary thermal and mechanical property screening and a summary of their physical properties is shown in Table 21. All of the polymers had Tgs above 200 °C (DSC) and 5% weight losses (TGA) in the range between 430 and 470 °C (air and nitrogen). Room temperature flexural moduli and strengths were 3.10–3.52 GPa and 152–207 MPa respectively.

One of the most interesting properties of these materials were the outstanding fracture toughness values that most of the homopolymers exhibited (Table 21). The polymer prepared from **17** had a K_{1c} of 1.59 MPa $\times$ m$^{0.5}$, (G_{1c} = 780 J/m^2). Interestingly, the regioisomer, **95**, had a lower K_{1c} (0.84 MPa $\times$ m$^{0.5}$) along with a lower Tg (251 °C) than structure **17**. Similar behavior was observed for the polymers derived from the ether linked monomers, **114a** and **114b**. Again the only difference between the monomers was the *para-meta* relationship of the ether with the maleimide group. The *meta*-linked material had a lower Tg and lower fracture toughness values than the *para*-linked monomer/polymer.

The effect of meta- and para-regioisomer relationships in a polymer backbone, upon the physical properties of polyarylates and polyamides, has been extensively investigated [116]. In general, varying amounts of a *meta* isomer are incorporated into the backbone of the polymer in order to reduce the rigidity of the all *para*-linked material. This has been shown to result in lowering the Tg, melting point and modulus of these polymers with the overall effect of improving processing parameters.

Therefore it was somewhat surprising when this trend was reversed when both a ketone and an ether linkage were incorporated between the maleimide and the benzocyclobutene (**109a** vs **109b**). The Tg dropped from 270 °C to 230°C as would have been expected, but the fracture toughness for the meta isomer was larger than the para-isomer (**109a**,1330 J/m^2 vs **109b** > 3000 J/m^2). There is no explanation at this time for this result.

The combination of relatively low Tgs (< 230 °C) and high fracture toughnesses of the homopolymers from **111b** and **119** led to the successful formation of films by heating under pressure. The ability to form films would seem to suggest that these two polymers may be more similar to thermoplastic, or lightly crosslinked thermoplastic materials rather than thermoset resins. The difference between these materials and the AB olefin benzocyclobutene (monomer **78a**) described in an earlier section, is the apparent formation of higher molecular weight polymers with the AB maleimide benzocyclobutenes.

Of all of the AB benzocyclobutene maleimide homopolymers prepared, **114a** appeared to have the best combination of thermal and mechanical properties as well as ease of fabrication (Table 21). This monomer/polymer was used as a

Table 22. Properties of graphite fibre[a] composite using poly
                                         as a matrix resin

114a

Tensile strength (MPa)	1080
Tensile Modulus (GPa)	71.8
Flexural strength (MPa)	1311
Flexural modules (GPa)	68.3
Short beam shear strength (MPa)	92
Compressive Strength (MPa)	856
Compressive strength after impact (MPa)	332
Open-hole compressive strength (MPa)	
25 °C, Dry	293
177 °C, Wet[b]	202
203 °C, Wet	182

[a] Celion G30-500 8HS.
[b] Wet = 2 weeks at 170 °F and 85% relative humidity.

matrix resin for the preparation of composite panels by resin transfer molding techniques. The composite panels' mechanical and thermal properties were determined and some of these results are given in Table 22.

As can be seen from the results, the composite formed from monomer/-polymer **114a** with Celion® G30-500 8HS fabric exhibited excellent mechanical properties [28]. To a first approximation it would appear that the inherent fracture toughness of the matrix resin has been carried over to the composite panels. The CAI (compressive strength after impact) and OHC (open hole compression) tests are a direct measurements of the toughness of the composite part. the value of 332 MPa for the CAI compares very favorably to the value of 300 MPa typical for the thermoplastic composites. The OHC values under hot–wet test conditions would seem to indicate that the composite has very good retention of its mechanical properties at both 177 °C and 203 °C.

In conclusion, the AB benzocyclobutene monomers can be polymerized to form polymers with a broad range of mechanical properties. The properties of the polymers depend not only upon the type of reactive functionalities but also the nature of the linking group between functionalities. Based upon the properties presented for these homopolymers, it would seem that a broad spectrum and combination of unique thermal and mechanical properties can be obtained from these relatively simple molecules.

Acknowledgement. The authors would like to thank the Dow Chemical Company for its continuing support and encouragement in the development of the benzocyclobutene technology. Many of the people cited in the references have contributed significantly to the results reported in this review. K.J.B. would like to thank NASA-Langley for their support in the development of much of the AB maleimide benzocyclobutene technology (NAS-1-18841). R.A.K. would like to thank the Materials Laboratory (AFWL/MLBP) AF Wright Laboratories for partial support of the investigation into benzocyclobutene thermosets (F33615-85-C-5092).

8 References

1. Kirchhoff RA, Bruza KJ, Carriere CJ, Rondan NG, Sammler RL (1991) J Macromol Sci Chem 11 & 12: 1079
2. Kirchhoff RA, Baker CE, Gilpin JA, Hahn SF, Schrock AK (1986) Proceedings of 18th International SAMPE Conference, p 478
3. Kirchhoff RA, (1985)U S Pat 4,540,763
4. Kirchhoff RA, Schrock AK, Hahn SF (1988) US Pat 4,724,260
5. Kirchhoff RA, Schrock AK, Hahn SF, US (1988) 4,783,514
6. Kirchhoff RA, US (1987) 4,638,078
7. Kirchhoff RA, US (1989) 4,826,997
8. Kirchhoff RA, US (1991) 4,999,449
9. Kirchhoff RA, Schrock AK, Gilpin JA, US (1987) 4,642,329
10. Hahn SF, Kirchhoff RA, US (1988) 4,730,030
11. Arnold FE, Tan L-S (1986) 31st International SAMPE Symposium, p 968
12. Tan L-S, Arnold FE (1985) Polym Prepr (Amer Chem Soc Div Polym Chem) 26: 176
13. Tan L-S, Arnold FE (1986) Polym Prepr (Amer Chem Soc Div Polym Chem) 27: 453
14. Tan L-S, Soloski EJ, Arnold FE (1986) Polym Prepr (Amer Chem Soc Div Polym Chem) 27: 240
15. Finkelstein H (1910) Chem Ber 43: 1528
16. Cava MP, Napier DR (1956) J Am Chem Soc 78: 500
17. Jensen FR, Coleman WE (1958) J Am Chem Soc 80: 6149
18. Klundt IL (1970) An excellent review of the basic chemistry of benzocyclobutene Chem Rev 70: 471
19. Thummel RP (1980) A review on the synthesis and reactions of benocyclobutenes and related compounds Acc Chem Res 13: 70
20. Oppolzer W (1978) A review on the use of intramolecular ring closures of benzocyclobutenes to synthesize natural products, Synthesis, p 794
21. Kametani T (1979) A review on the use of benzocyclobutenes in natural product synthesis, Pure & Appl Chem 51: 747
22. Iwatsuki S (1984) A review on the polymerization of o-quinodimethane compounds Berlin, Heidelberg, New York, Advances in Polymer Science 58: 93
23. Charlton JL, Alauddin MM (1987) A review on the synthesis and reactions of o-quinodimeth-anes, Tetrahedron 43: 2873
24. McCullough JJ (1980) A review on o-xylylenes and isoindenes as reaction intermediates, Acc Chem Res 13: 270
25. Funk RL, Vollhardt KPC (1980) A review on the intramolecular ring closure reactions of o-xylylenes insteroid synthesis, Chem Soc Rev 9: 41
26. Martin N, Seoane C, Hanack M (1991) A review on the synthesis and reactions of o-quinodemethanes, Org Prep Proc Int 23: 239
27. Boekelheide V (1980) A review of cyclophanes with numerous examples involving be-nzocyclobutenes, Acc Chem Res 13: 65
28. Tan L-S, Arnold FE (1985) Polym Prepr (J Am Chem Soc Div of Poly Chem) 26: 178
29. Tan L-S, Arnold FE (1986) Poly Prepr (J Amer Chem Soc Div of Poly Chem) 27: 240
30. Bishop MT, Bruza KJ, Laman SA, Lee WM, Woo EP (1992) Poly Prepr (J AM Chem Soc Div of Poly Chem) 33: 362
31. Tan L-S, Arnold FE, Soloski EJ (1988) J Poly Sci, Poly Chem 26: 3103
32. Tan L-S, Soloski EJ, Arnold FE (1987) Poly mater Sci Eng 56: 650
33. Corley LS (1990) US 4,927,907
34. Corley LS (1990) US 4,973,636

35. Quarderer JG, Stone FC, Beitz M, O'Donnell PM, US 4,851,606
36. March J (1985) Advanced Organic Chemistry, 3rd edn, New York
37. Liu M-B (1990) US 4,891,455
38. Patai S (1969) The Chemistry of Carboxylic Acids and Esters, Interscience, New York
39. Scho nb erg A, Bartoletti I, Heck RF (1979) J Org Chem 39: 3318
40. Heck RF (1990) Palladium Reagents in Organic, Synthesis Academic Press, New York
41. Patai S (1972) The Chemistry of Acyl Halides, interscience, New York, p 35–68
42. Plevyak JE, Heck RF (1978) J Org Chem 43: 2454
43. Heck RF (1982) Organic Reactions 27: 345
44. Rylander L (1967) Catalytic Hydrogenation over Platinum Metals, Academic Press, New York, p 168–202
45. Searle NE (1948) US 2,444,536
46. Bruza KJ, Kirchhoff RA (1989) US 4,795,827
47. Bruza KJ, Kirchhoff RA (1989) US 4,825,001
48. Bruza KJ, Bonk PJ, Harris RF, Kirchhoff RA, Stokich TM, McGee RL Jr, DeVries RA (1991) Procedings of 36th International SAMPE Symp, p 457
49. Bogen G W, Lyssy ME, Monnerat GA, Woo EP, (1988) SAMPE Jorunal 24(6): 19
50. Kubens R, Schultheis H, Wolf R, Grigat E, Kunstolle (1968) 58(12): 827
51. Grigat E, Putter R (1964) Chem Ber 97: 3012
52. Delano CB, Harrison ES (1978) Soc Adv Matl Proc Eng Series 23: 506
53. Hergenrother PM, Harrison ES, Gosnell PB (1978) Soc Adv Matl Proc Eng Series 23: 506
54. Roth WR, Bierman M, Dekker H, Jochems R, Mosselman C, Hermann H (1978) Chem Ber 111: 3892
55. Michl J, Flyn CR (1973) J Am Chem Soc 95: 5802
56. Pollack SK, Raine BC, Hehre WH (1981) J Am Chem Soc 103: 6308
57. Michl J, Tseng KL (1977) J Am Chem Soc 99: 4840
58. Flyn CR, Michl MJ (1974) J Am Chem Soc 96: 3280
59. Miller RD, Kolc J, Michl J (1976) J Amer Chem Soc 98: 8510
60. Dolbier WJ, Matsui K Jr, Michl J, Horak DV (1977) J Am Chem Soc 99: 3876
61. Warrener RM, Russell RA, Lee TS (1977) Tetrahedron Lett 49
62. Kolghorn H, Meier K, Naturforch (1977) A322a: 780
63. Gleitcher GJ, Newkirk DD, Arnold JC (1978) J Am Chem Soc 95: 2526
64. Baudet J (1971) J Chim Phys-Chim Biol 68: 191
65. Ito Y, Nakatsuka M, Saegusa T (1982) J Am Chem Soc 104: 7609
66. Baird NC (1972) J Am Chem Soc 94: 4941
67. Trahanovsky WS, Chou C-H, Fisher D, Gerstein BC (1988) J Am Chem Soc 110: 6579
68. Huisgen R, Seidl H (1964) Tetrahedron Lett 3381
69. Wong PK (1986) US 4,622,375
70. Jensen FR, Coleman WE, Berlin AJ (1962) Tetrahedran Lett, p 15
71. Cava MP, Deona AA (1959) J Am Chem Soc 81: 4266
72. Errede LA (1961) J Am Chem Soc 83: 949
73. Errede LA (1961) J Poly Sci, XLIX, p 253
74. Ito Y, Nakatsuka M, Saegusa T (1982) J Am Chem Soc 104: 7609
75. Ito Y, Nakatsuka M, Saegusa T (1980) J Am Chem Soc 102: 863
76. Ito Y, Nakatsuka M, Saegusa T (1981) J Am Chem Soc 103: 476
77. Tan L-S, Arnold FE (1988) J Poly Sci Part A: Polym Chem 26: 1819
78. Bartmann M (1988) US 4,719,283
79. Tan L-S, Arnold FE (1987) US 4,711,964
80. Kirchhoff RA, Hahn SF, US 4,687,823
81. Tan L-S, Soloski EJ, Arnold FE (1987) Poly Mater Sci Eng 56: 650
82. Solomon DH (1972) Step Growth Polymerizations, Marcel Dekker, chap 7
83. Alder K, Stein G (1931) Ann Chem 485: 223
84. Alder K, Stein G (1932) Ann Chem 496: 204
85. Alder K, Stein G (1934) Ann Chem 67: 613
86. Staudinger H (1928) Ann Chem 467: 73
87. Staudinger H, Bruson H (1926) Ann Chem 447: 97
88. Upson RW (1955) US 2,726,232
89. Stille JK, Plummer L (1959) Abst 136th Meeting, Am Chem Soc pp 3–7
90. Tan L-S, Arnold FE (1988) J Poly Sci Chem 26: 3103

91. Tan L-S, Arnold FE (1987) J Poly Sci Chem 25: 3159
92. Tan L-S, Arnold FE (1987) US 4,711,964
93. Denny LR, Soloski EJ (1988) Polym Prepr, Am Chem Soc Div of Poly Chem 29: 194
94. Tan L-S, Arnold FE (1985) Polym Prepr, Am Chem Soc Div of Poly Chem 26: 178
95. Tan L-S, Arnold FE (1987) US 4,675,370
96. References 90 and 91, the precise meaning of wet was not specified.
97. Denny LR, Goldfarb IJ, Farr MP (1988) ACS Symp Ser, 367 (crosslinked Polymers) p 366
98. Denny LR, Goldfarb IJ, Farr MP (1987) Poly Mat Sci Eng 56: 656
99. Lee CY-C, and Goldfarb IJ (1981) Polymer Eng and Sci 21: 787
100. Kirchhoff RA (1989) US 4,826,997
101. Kirchhoff RA (1990) US 4,965,329
102. Bruza KJ (1990) US 4,996,288
103. Tan L-S, Arnold FE (1990) US 4,916,235
104. Williams FJ, Relles HM, Manello JS, and Donahue PE (1977) J Org Chem 42: 3419
105. Sonogashira K, Tohda Y, and Hagihara N (1975) Tetrahedron Lett p 4467
106. Takahashi S, Kuroyama Y, Sonogashira K, and Hagihara N (1980) Synthesis, p 627
107. Weir JR, Patel BA, Heck RF (1980) J Org Chem 45: 4926
108. Williams FJ, and Donahue PE (1977) J Org Chem 42: 3414
109. Reinhardt BA, Tsai TT, and Arnold FE (1984) New Monomers and Polymers, Culbertson BM, Pittman CR ed., Plenum Press, New York, pp 41–43
110. Hahn SF, Martin SJ, and McKelvy ML (1992) Macromolecules 25: 1539
111. Skorcz JA, Kaminski FE, Baumgarten HL, ed., (1973) Org Syn Collective Vol V, John Wiley & Sons, New York, p 263
112. Bruza KJ and Kirchhoff RA (1992) The Adhesion Society Proceddings Fifteenth Annual Meeting Boerio FJ, ed. Hilton Head South Carolina, February 16–19, pp 154–155
113. Bacon RGR and Stewart OJ (1965) J Chem Soc 4953
114. Williams AL, Kinney RE, and Bridger RF (1967) J Org Chem 30: 2501
115. Lindley J (1984) Tetrahedron 40: 1433
116. Stevenson DR, and Mulvaney JE (1972) J Polym Sci A-1 10: 2713
117. Lee W, Yalvac S, Laman S, McGee RM (1992) Proceesings of the 37th International SAMPE Symposium and Exhibition 679

Received May 1993

Poly(Arylene Ether)s Containing Heterocyclic Units

Paul M. Hergenrother[1], John W. Connell[1], Jeff W. Labadie[2]
and James L. Hedrick[2]
[1] NASA Langley Research Center, Hampton, VA 23681-0001, USA
[2] IBM Almaden Research Center, 650 Harry Road, San Jose, CA 95120-6099, USA

Since 1985, a major effort has been devoted to incorporating heterocyclic units within the backbone of poly(arylene ether)s (PAE). Heterocyclic units within PAE generally improve certain properties such as strength, modulus and the glass transition temperature. Nucleophilic and electrophilic aromatic substitution have been successfully used to prepare a variety of PAE containing heteorcyclic units. Many different heterocyclic families have been incorporated within PAE. The synthetic approaches and the chemistry, mechanical and physical properties of PAE containing different families of heterocyclic units are discussed. Emphasis is placed on the effect variations in chemical structure (composition) have upon polymer properties.

1 Introduction

Poly(arylene ether)s (PAE) are one of the most popular families of high performance engineering thermoplastics. Their popularity is due to their attractive combination of properties such as relatively low cost, good processability, excellent chemical resistance, high thermal stability and good mechanical performance. Several PAE are commercially available from various companies throughout the world. Some of the more common commercially available PAE are Amoco's Udel P1700 and Radel R polysulfone, ICI's Victrex polyetheretherketone, and BASF's Ultrapek polyetherketone. Poly(arylene ether)s are used in a variety of applications as adhesives, composite matrices, coatings, fibers, films, moldings and membranes. These polymers are generally prepared by aromatic nucleophilic displacement [1] or electrophilic reactions [2, 3]. Other routes have also been used to prepare PAE such as oxidative coupling [4], Ullmann ether synthesis [5], Scholl reaction [6], nickel-catalyzed coupling of aromatic chlorides [7] and ring opening of cyclic oligomers [8].

During the last several years, a significant effort has been devoted to incorporating heterocyclic units into the backbone of PAE. When heterocyclic units are placed within the arylene ether polymer chain, certain properties such as strength, modulus and glass transition temperature (T_g) generally increase. Aromatic nucleophilic displacement and electrophilic reactions have been used to synthesize poly(arylene ether)s containing heterocyclic units (PAEH).

This article will discuss the chemistry and properties of PAEH with emphasis on developments during the last 7 years. No attempt was made to list all the PAEH made over this period since that would be beyond the scope of this article. Attention will focus on the synthetic approaches to various heterocyclic families, chemical structure/property relationships and the mechanical and physical properties.

2 Historical Perspective

Work on all-aromatic heterocyclic polymers began earnestly in the early 1960's to satisfy the demands primarily of the aerospace and electronic industries. These polymers were found to exhibit excellent thermal stability, superb chemical resistance and high mechanical properties but difficult processability. As work progressed on these polymers, various means were investigated to improve their processability (primarily solubility to allow solution casting to form films and compression moldability to obtain adhesively bonded panels and composites) without severely compromising other properties. One of the approaches involved the incorporation of flexible groups such as an ether group between benzene rings in the backbone of the polymer. As the ether or arylene ether content increases in an amorphous polymer, the T_g decreases and the solubility and compression moldability generally increase (melt viscosity decreases).

Numerous examples exist in the literature of aromatic heterocyclic polymers containing relatively high arylene ether content. The amorphous polyquinoxaline [9] of structure 1 is an excellent example. This polymer was prepared from the reaction of the appropriate bisglyoxal and 3,3',4,4'-tetraaminobiphenyl. Another example is the semi-crystalline polyimide [10] of structure 2 which was synthesized from the reaction of the appropriate diamine and dianhydride. The last representative example in structure 3 is a more popular material, a polyetherimide commonly referred to as General Electric's Ultem 1000. One of the synthetic routes to the polymer of structure 3 was via aromatic nucleophilic displacement [11]. Under appropriate reaction conditions, the nitro groups of a 4,4'-bis(nitrophthalimide) were displaced by the proper bisphenolate to yield high molecular weight polyetherimide. Instead of referring to the polymer of structure 3 as a poly(arylene ether) even though arylene ether groups were formed during the propagation step, it is classified as a polyetherimide.

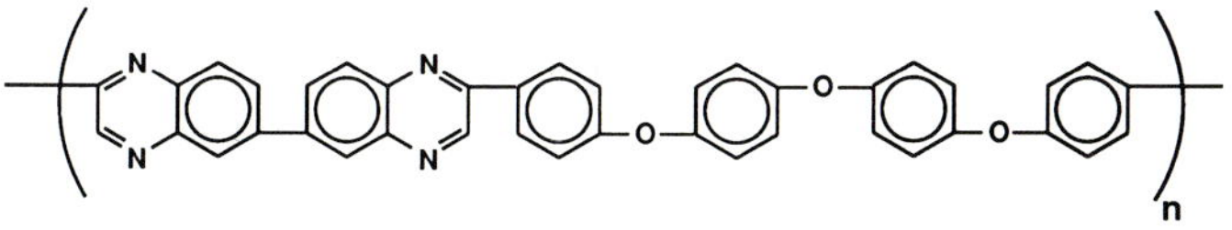

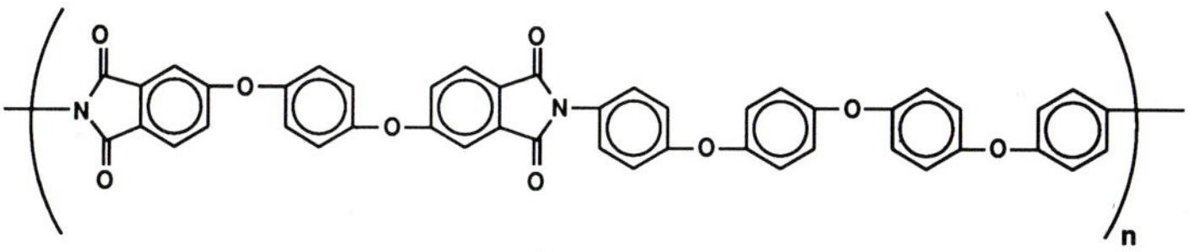

1

2

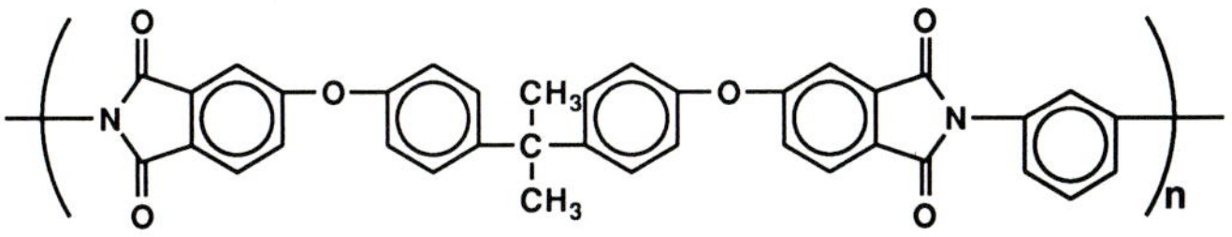

3

The first report of the successful synthesis of high molecular weight PAEH by aromatic nucleophilic displacement reaction appears to be from the reaction of 2,5-di(4-fluorophenyl)-1,3,4-oxadiazole and 2,2-di(4-hydroxyphenyl)propane (bisphenol A) [1]. A polymer with a relative viscosity of 0.50 dL/g and a T_g of 180 °C was obtained. The 1,3,4-oxadiazole unit activates the fluoro group sufficiently to allow displacement by the bisphenolate. The corresponding dichloro compound gave low molecular weight polymer. Another PAEH with a T_g of 180 °C and a crystalline melt temperature (T_m) of 250 °C was prepared from the reaction of 3,6-dichloropyridazine with bisphenol A [1].

The term poly(arylene ether)s as applied to heterocyclic polymers containing ether connecting groups is a matter of personal preference. There are no formal classification rules that establish the general polymer family. These polymers could also be classified according to the heterocyclic unit. For example, PAEH containing quinoxaline or benzimidazole units could be referred to simply as polyquinoxalines or polybenzimidazoles. A broader grouping, namely PAEH, was selected for polymers containing heterocyclic units which were prepared by the more traditional arylene ether polymer synthesis, particularly aromatic nucleophilic displacement or electrophilic reactions.

3 Poly(Arylene Ether)s Containing Heterocyclic Units Prepared by Aromatic Nucleophilic Displacement

Poly(arylene ether) synthesis via nucleophilic aromatic substitution involves the step-growth polymerization of an activited dihalo or dinitro monomer with a bisphenolate, where generation of the ether linkage is the polymer-forming reaction. Effective activating groups are electron withdrawing groups, e.g. sulfone and ketone, and are substituted either ortho or para to the leaving group in the electrophilic monomer. In the case of PAEH synthesis, the heterocyclic unit can be present in either the activated dihalo or dinitro compound or in the bisphenolate. Facile displacement occurs primarily due to the ability of the activating group to stabilize the negative charge developed through the formation of a Meisenheimer complex which is generally the rate-determining step. In addition, polar aprotic solvents play an important role in stabilizing the Meisenheimer complex and thus are required for facile nucleophilic aromatic substitution.

The synthesis of PAEH via aromatic nucleophilic displacement reaction generally involves the reaction of stoichiometric quantities of the monomers (an aromatic dihydroxy compound and an activated aromatic difluoro, dichloro or dinitro compound) plus an excess of powdered potassium carbonate (10–20 mol %) in aprotic polar solvents such as N,N-dimethylacetamide (DMAc) or N-methylpyrrolidinone (NMP) at a concentration of 20% solids (based upon the weight of the monomers and volume of the solvent). Toluene is normally used to remove the water formed during the early stage of the reaction by azeotropic distillation. Alternatively, N-cyclohexyl pyrrolidinone (CHP) can be used as a solvent or cosolvent since at elevated temperatures it is immiscible with water. The reaction is stirred and heated to ~ 135–$160\,°C$ over 2–4 hours and then heated to and maintained at ~ 155–$190\,°C$, depending upon the solvent, under nitrogen for several hours (time depends upon the reactivity of the monomers). During this period, a noticeable increase is observed in the solution viscosity. In some cases, especially with semi-crystalline polymers, higher boiling solvents such as diphenyl sulfone or tetramethylenesulfone(sulfolane) are required to keep the polymer in solution and drive it to high molecular weight. The reaction temperature in diphenyl sulfone may be 200 to 350 °C. In DMAc or NMP, even some amorphous polymers precipitate from solution prior to high molecular weight formation. Polymer isolation from DMAc or NMP reactions generally involves dilution, filtration and quenching in acidic water or dilution and quenching in acidic water followed by thoroughly washing the precipitate. The diphenyl sulfone reactions are generally diluted and poured into a mixture of acetone and acetic acid followed by thoroughly washing the precipitate successively with water and methanol.

A preferred synthetic procedure to PAEH concerns the formation of the bisphenolate salt followed by the addition of the activated difluoro, dichloro or dinitro monomer. As an example, the heterocyclic bisphenol is stirred in a mixture of toluene and an aprotic polar solvent such as DMAc, NMP or diphenyl sulfone at ~135–140 °C for several hours in the presence of ~10 mol % excess of powdered anhydrous potassium carbonate (stoichiometric amount of sodium or potassium hydroxide can be used) under a Dean-Stark trap in a nitrogen atmosphere. Water is removed by azeotropic distillation. A stoichiometric quantity of the difluoro monomer is then added to the slightly cooled reaction mixture. The toluene is removed and the reaction is stirred at ~155 °C in DMAc for one to several hours. Polymer isolation is performed as previously described. This procedure minimizes hydrolysis of the difluoro monomer, gel formation and molecular weight equilibration of the polymer.

If desired, controlled molecular weight PAEH can be prepared by upsetting the stoichiometry in favor of one of the monomers. An end-capper such as *m*-cresol for fluoro terminated arylene ethers or 4-fluorobenzophenone for hydroxy terminated arylene ethers can be used to terminate the polymer chains.

3.1. PAE Containing Quinoxaline Units

As with several PAEH, PAE containing quinoxaline units were prepared by two different approaches. One route involved the reaction of an aromatic bis(fluoroquinoxaline) compound with various bisphenolates as depicted in Eq. (1)[12]. The quinoxaline unit as with many other heterocyclic units was

$$X = C(CF_3)_2, C(CH_3)_2 \text{ and } C(C_6H_5)_2$$

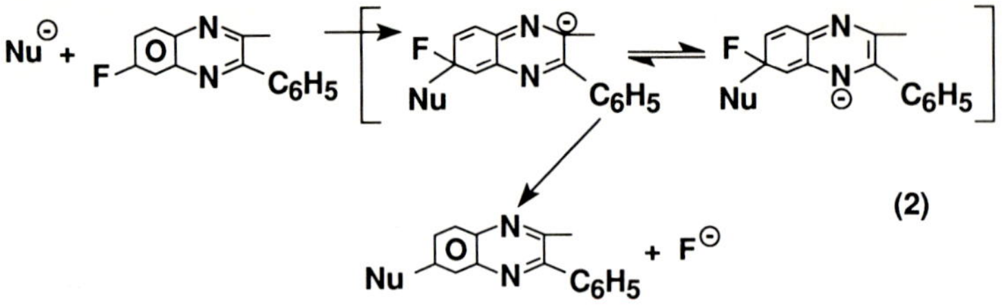

$$(2)$$

Nu = nucleophile

sufficient to activate the fluoro groups for displacement. In addition, the quinoxaline ring can stabilize the negative charge developed in the transition state through a Meisenheimer-like complex (Eq. (2)), similar to more conventional activating groups like sulfone or carbonyl.

The T_gs of representative polymers as measured by differential scanning calorimetry (DSC) at a heating rate of 10 °C/min are presented in Table 1. The polymers exhibited excellent thermal stability as shown by weight loss rate as low as 0.02 wt. %/hour as determined by isothermal aging at 400 °C in nitrogen. After the initial reports, additional work was disclosed on polymers exhibiting T_gs as high as 315 °C and thin films with tensile strength, modulus and elongation at room temperature of 108 MPa, 2.8 GPa and 13%, respectively [13]. The mechanical properties of the poly(arylene ether phenylquinoxaline)

Table 1. Glass transition temperature of poly(arylene ether phenylquinoxaline)s

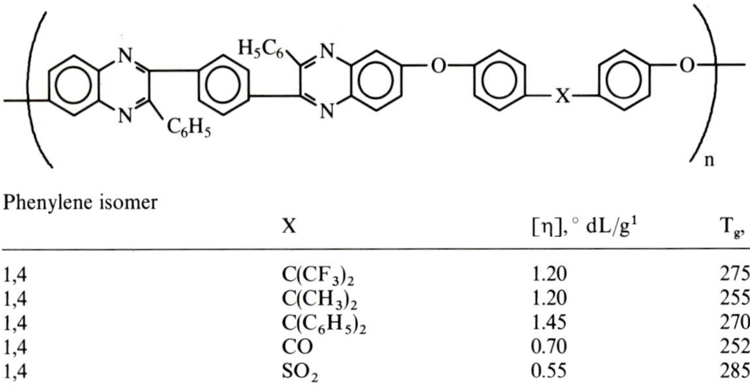

Phenylene isomer	X	$[\eta]$,° dL/g[1]	T_g, °C[2]
1,4	$C(CF_3)_2$	1.20	275
1,4	$C(CH_3)_2$	1.20	255
1,4	$C(C_6H_5)_2$	1.45	270
1,4	CO	0.70	252
1,4	SO_2	0.55	285
1,4	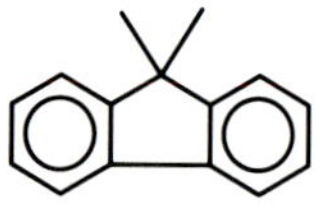	1.10	315
1,3	$C(CF_3)_2$	0.44	240
1,3	$C(CH_3)_2$	1.23	230
1,3	$C(C_6H_5)_2$	0.65	250

[1] Determined in NMP at 25 °C
[2] Determined in DSC at a heating rate of 10 °C/min

films were dependent upon the final drying temperature. Films cast from NMP and dried below 300 °C were brittle whereas NMP cast films dried to 350 °C were flexible [13]. The higher drying temperature presumably was required to remove the tenaciously bound NMP.

Other PAE containing quinoxaline units in Table 2 were prepared from the reaction of bisphenol A with various substituted 2,3-di(4-fluorophenyl) quinoxalines [14]. The fluoro atom in the 4-position of the phenyl group was found to be sufficiently activated towards nucleophilic aromatic substitution to yield polymers with moderate molecular weights. Thus, the pyrazine ring activates fluoro on the annulated benzo ring and also on the pendent phenyl groups, albeit to a lesser extent in the latter case. The extent of activation can be assessed by both ^{1}H and ^{19}F NMR spectroscopy. Due to its high natural abundance and sensitivity to NMR detection, ^{19}F served as an excellent probe for determining reactivity [14]. This work was done as part of an effort to develop fundamental information on high performance materials for potential use in the electronics industry.

The second synthetic route to PAE containing quinoxaline units involved the reaction of an aromatic dihydroxy quinoxaline or aromatic bis(hydroxy-quinoxaline) with activated aromatic difluoro compounds (Eq. (3)) [15]. The dihydroxy quinoxaline and bis(hydroxyquinoxaline) monomers were readily prepared from the condensation of 1,2-diaminobenzene with 4,4'-dihydroxyben-zil and aromatic bis(o-diamines) with 4-hydroxybenzil, respectively. The T_gs of a series of PAE containing quinoxaline units are presented in Tables 3 and 4. For these polymers, the trend for the T_g is sulfone > carbonyl > terephthaloyl- > isophthaloyl. This trend holds for most polymer families when polymers of similar molecular weights are compared. Several polyphenylquinoxalines of the same chemical structure as those in Table 3 were also prepared by the poly-

Table 2. Glass transition temperatures of poly(arylene ether quinoxaline)s

X	[η], dL/g[1]	T_g, °C
H	0.55	195
CF$_3$	Insoluble	216
COC$_6$H$_5$	0.40	209
SO$_2$C$_6$H$_5$	0.45	255

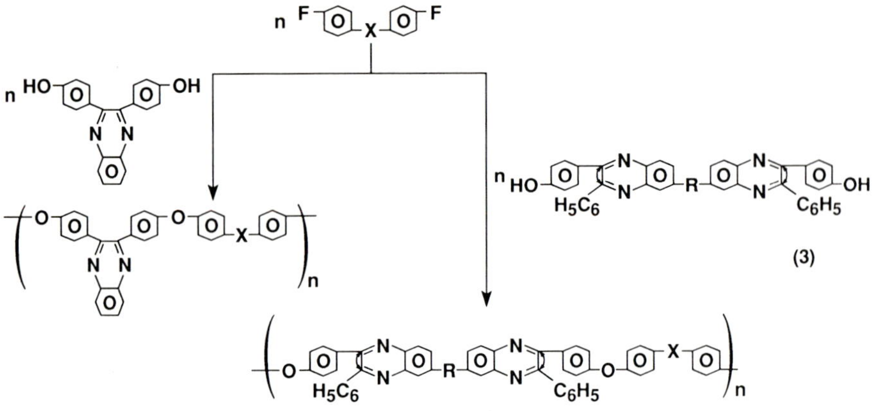

where X = activating groups in Table 3
R = nil, O and CO

(3)

condensation of aromatic bis(o-diamines) and aromatic bis(phenyl-1,2-dicarbonyl) compounds [16]. Polymers of the same chemical structure exhibited similar T_gs. In Table 4, the configurationally ordered terephthaloyl polymer exhibited a crystalline melt temperature (T_m) of 365 °C. The terephthaloyl containing polymers in Table 3 were not crystalline because of configurational disorder. Thin films of the third polymer in Table 3 gave tensile strength, tensile modulus, and break elongation at 25 °C of 78.6 Mpa, 2.43 GPa and 7.7% and at 177 °C of 45.2 MPa, 1.72 GPa and 65%, respectively. This work was performed to develop basic information on chemical structure/ property relationships to help design better performing materials for high performance structural applications in the aerospace industry.

An AB monomer, 2-(4-hydroxyphenyl)-3-phenyl-6-fluoroquinoxaline [17], was prepared from the reaction of 4-hydroxybenzil and 1,2-diamino-4-fluorobenzene and subsequently polymerized under aromatic nucleophilic displacement conditions in NMP. The resultant polymer exhibited an intrinsic viscosity of 1.23 dL/g, a T_g of 247 °C and thin film tensile strength and modulus at room temperature of 107 Mpa and 3.2 GPa, respectively [17]. The same AB monomer was also copolymerized with hydroquinone and 4,4′-difluorodiphenyl sulfone to yield a series of copolymers with interesting properties [17]. The same AB monomer was prepared and polymerized by other researchers to yield a polymer with an intrinsic viscosity of 0.65 dL/g and a T_g of 255 °C [18].

Poly(arylenethioether phenylquinoxaline)s have been prepared from the reaction of a bis(chloroquinoxaline) compound with 4,4′-dimercaptodiphenyl ether [19]. The polymer of structure 4 had an unusually high intrinsic viscosity of 4.0 dL/g, a T_g of 243 °C and good solubility in NMP and chloroform. The greater nucleophilicity of the thiophenolate anion versus the phenolate anion allowed the formation of high molecular weight polymer from the bis(chloroquinoxaline) monomer. The dichloro monomer is much less reactive than the corresponding difluoro monomer.

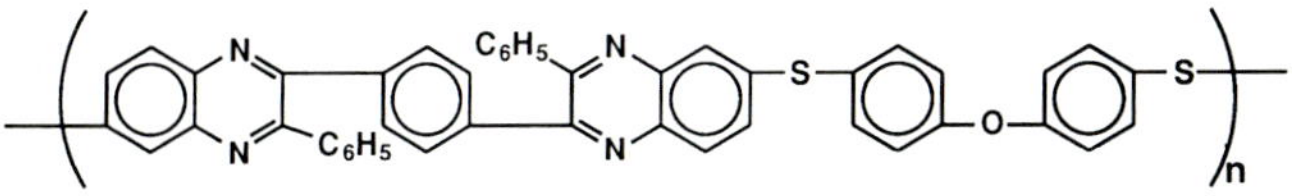

4

Table 3. Glass transition temperatures of poly(arylene ether phenylquinoxaline)s

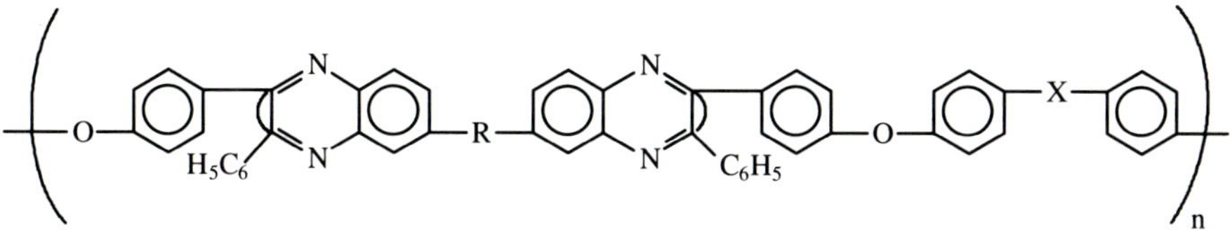

R	X	$[\eta_{inh}]$, dL/g	T_g, °C
nil	SO$_2$		283
nil	CO	0.80	252
nil	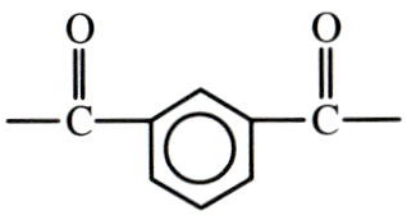	1.09	240
O	SO$_2$	0.34	240
O	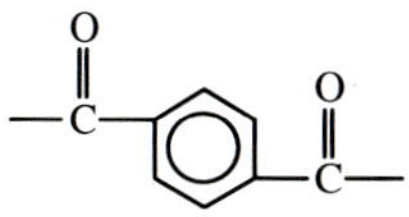	0.45	226
O	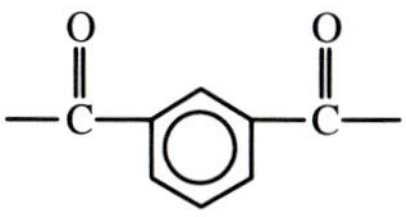	0.46	213
CO	SO$_2$	0.69	268
CO	CO	1.30	253
CO	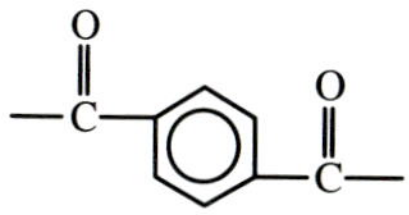	–	255
CO	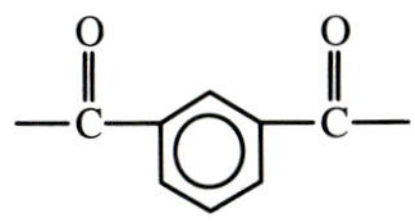	0.61	235

Table 4. Glass transition temperatures of poly(arylene ether quinoxaline)s

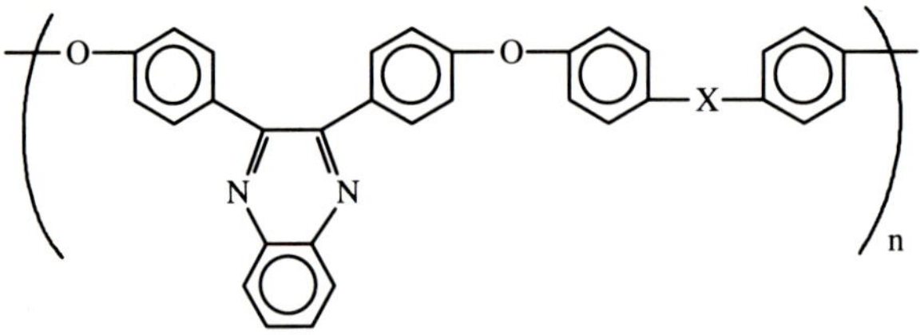

X	$[\eta_{inh}]$, dL/g	T_g, °C
SO$_2$	0.54	240
CO	0.58	209
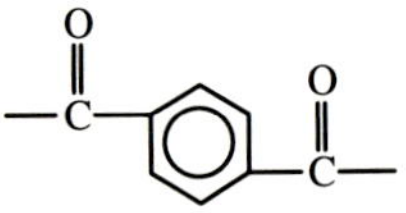	0.83	208(365)*
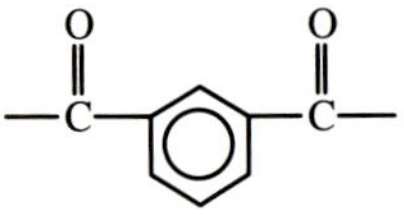	0.50	179

* Crystalline melt temperature (T$_m$)

3.2 PAE Containing Phenylimidazole Units

Poly(arylene ether)s containing phenylimidazole units were initially prepared from the reaction of 2-phenyl-4,5-di(4-hydroxyphenyl)imidazole and various activated aromatic difluoro compounds according to Eq. (4) [20]. The dihydroxyimidazole was first prepared from the reaction of 4,4′-dimethoxybenzil, benzaldehyde and ammonium acetate in refluxing acetic acid followed by

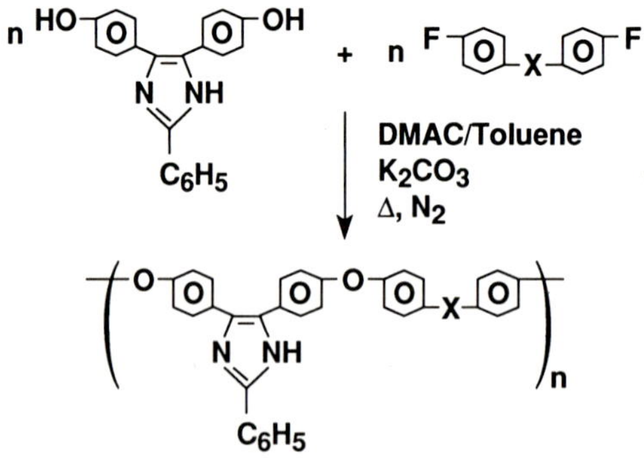

where X = activating groups in Table 5

demethylation of the dimethoxy compound with hydrogen bromide in acetic acid. Later work used 4,4′-dihydroxybenzil in a one step synthesis to prepare 2-phenyl-4,5-di(4-hydroxyphenyl)imidazole. The T_gs and thin film tensile properties for a series of PAE containing phenylimidazoles are given in Table 5. The T_g trend as seen earlier for the quinoxaline series is evident again, namely sulfone > carbonyl > terephthaloyl > isophthaloyl. In addition, as seen in Table 5, the phenylphosphine oxide group in a polymer provides a higher T_g than the sulfone group. The film elongation is low presumably due to the presence of foreign particles (e.g. gel, dust particles, etc.). The solutions used to cast these films as well as many other films reported herein were not filtered nor were the films dried in a dust-free environment.

A controlled molecular weight version of the terephthaloyl based polymer in Table 5 with an inherent viscosity of 0.59 dL/g exhibited number average ($\bar{M}_n$) and weight average ($\bar{M}_w$) molecular weights of 82,000 and 250,000 g/mole (Daltons), respectively [21]. These molecular weights may be higher than expected. The molecular weights were determined by gel permeation (size exclusion) chromatography/differential viscometry using universal calibration. The DMAc polymer solutions used for the molecular weight measurement contained lithium bromide. Lithium bromide is commonly used to suppress polyelectrolyte effect through hydrogen bonding. Melt viscosity of this polymer (ηinh, 0.51 dL/g) was between 10^5 and 10^6 pascal seconds at 300 °C under an angular frequency of 0.1 radians/second. This polymer exhibited good performance in adhesive and composite specimens. For example, titanium-to-titanium lap shear specimens gave tensile shear strengths (ASTM D1002) of 33.3 MPa at 25 °C, 25.5 MPa at 177 °C and 21.0 MPa at 200 °C [21]. The adhesive specimens were fabricated by heating to 300 °C under 1.4 MPa and maintaining under these conditions for 1 h.

The isophthaloyl based polymer in Table 5 was evaluated as a high-precision surface film that can be coated with highly reflective aluminium [22]. The compliant film bonded well to a parabolic, graphite-phenolic honeycomb/ epoxy composite panel typical of the type considered for use in the fabrication of large precision reflectors for astrophysical and optical communication systems. The poly(arylene ether phenylimidazole) film showed no change in the T_g or tensile properties after radiation doses up to 1000 Mrads. A key to the excellent performance of this polymer in this application was the good adhesion to the epoxy composite after thermal cycling. Imidazoles can react with epoxies and hydrogen bond to cured epoxies. They are commonly used as catalysts to cure epoxies. In this regard, work is underway to develop poly(arylene ether phenylimidazole)s as toughening agents for epoxies [23].

Further work on poly(arylene ether phenylimidazole)s led to a series of polymers containing two phenylimidazole units per mer unit (Eq. (5)) [24]. The dihydroxy monomers, 2,2′-(1,3- and 1,4-phenylene)bis[(4-phenylhydroxy)-5-phenylimidazole] were prepared from the reaction of iso or terephthaldehyde, 4-hydroxybenzil and ammonium acetate in refluxing acetic acid. High molecular weight polymers as represented in Table 6 were readily prepared in DMAc.

Table 5. Properties of poly(arylene ether)s containing phenylimidazole

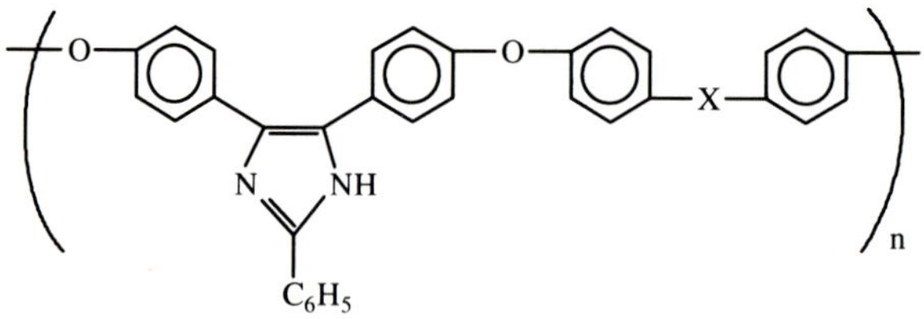

X	η_{inh}, dL/g[1]	T_g, °C[2]	23 °C Thin film tensile properties[3]		
			Strength, MPa	Modulus, GPa	Elong., %
O‖P(C$_6$H$_5$)	0.24	318	–	–	–
SO$_2$	0.41	277	–	–	–
CO	0.61	259	91.7	2.79	5.0
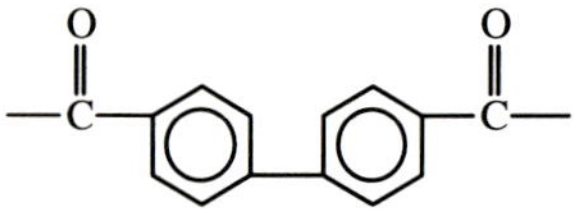	0.53	258	–	–	–
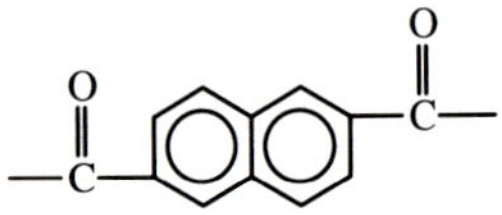	0.40	248	–	–	–
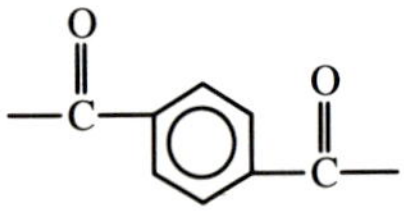	0.89	248	97.9	2.81	6.0
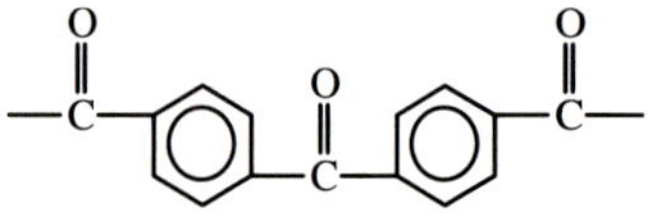	0.49	239	95.1	2.69	6.3
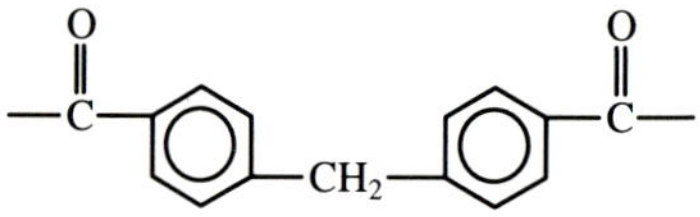	0.58	231	–	–	–
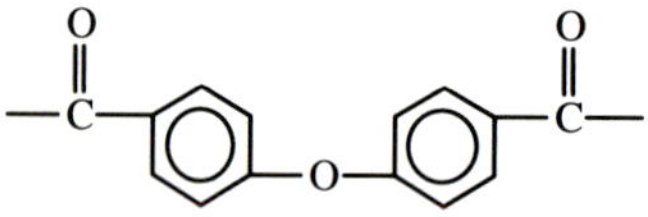	0.64	230	–	–	–
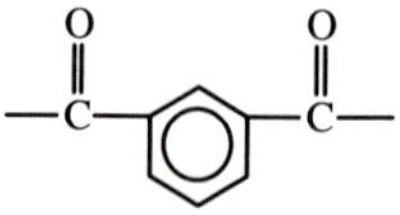	0.55	230	82.7	2.50	4.0

[1] Determined on 0.5 % solutions in DMAC at 25 °C
[2] Determined by DSC at a heating rate of 20 °C/min.
[3] Solution cast unoriented films, properties determined according to ASTM D882

Table 6. Properties of poly(arylene ether phenylimidazole)s

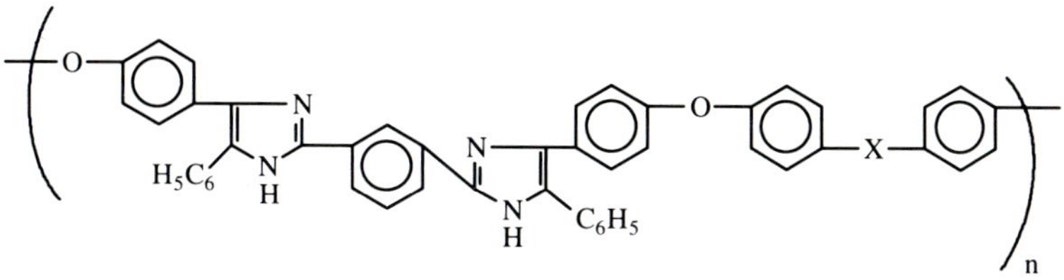

X	η_{inh}, dL/g[1]	T_g, °C[2]	23 °C Thin film tensile properties[3]		
			Strength, MPa	Modulus, GPa	Elong., %
SO$_2$	0.50	292	–	–	–
CO	0.69	282	117.9	2.94	6.7
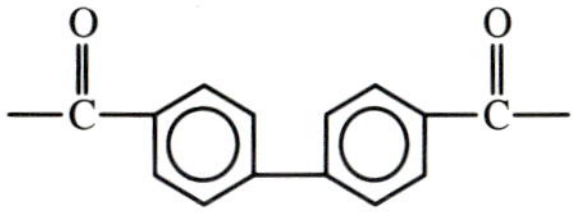	0.61	276	110.3	3.13	7.5
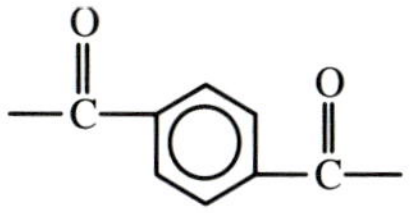	0.46	262	–	–	–
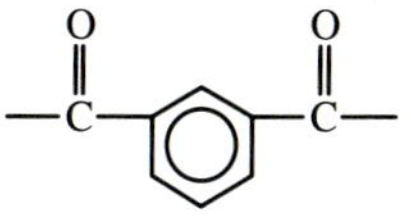	0.89	259	–	–	–
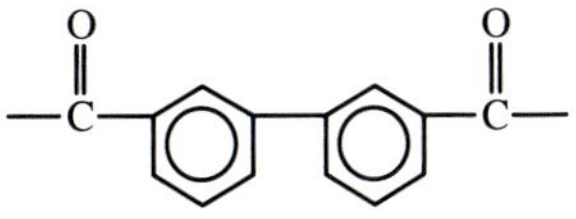	0.56	250	107.5	2.73	6.0

[1] Determined on 0.5% DMAC solutions at 25 °C
[2] Determined by DSC at a heating rate of 20 °C/min
[3] Solution cast unoriented films

Number average and weight average molecular weights of 115,000 and 394,000 g/mol, respectively, were measured for the isophthaloyl based polymer (η_{inh} = 0.65 dL/g) in Table 6. Again the molecular weights are higher than anticipated presumably due to molecular aggregation from strong hydrogen bonding. The DMAc polymer solutions used for the molecular weight measurements contained lithium bromide. Preliminary adhesive properties of the iso-phthaloyl based polymer in Table 6 were attractive as evidenced by

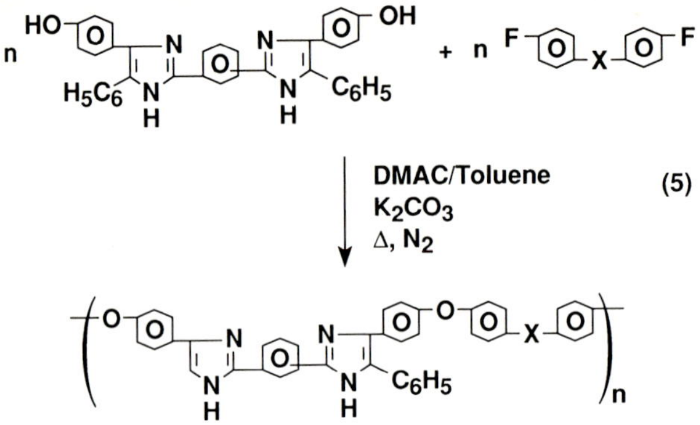

where X = activating groups in Table 6

(5)

titanium-to-titanium tensile shear strengths of 27.6 MPa at 23 °C and 24.8 MPa at 177 °C [24].

The T_gs of poly(arylene ether phenylimidazole)s containing pendent groups are given in Table 7 [25]. Within each of the 4 series of polymers, the T_g trend is the same as previously pointed out, namely sulfone > carbonyl > terephthaloyl > isophthaloyl. In addition, within the last three series of polymers in Table 7, the T_g trend is CF_3 > CH_3 > C_2H_5. The series with the highest T_gs was the first one in Table 7 where both R and R' substituents on the pendent phenyl ring were CF_3 groups.

3.3 PAE Containing Benzoxazole Units

Arylene ether polymers containing benzoxazole units were initially prepared from the reaction of bis(fluorophenylbenzoxazoles) and various aromatic dihydroxy compounds as depicted in Eq. (6) [26, 27]. One of the benzoxazole

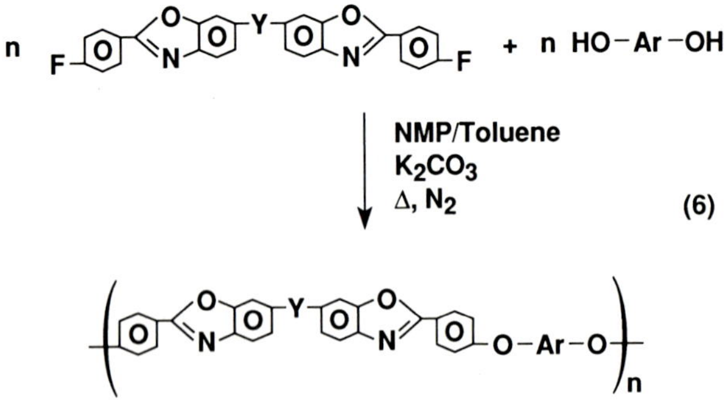

(6)

where Y and Ar are defined in Table 8

Table 7. Glass transition temperatures of poly(arylene ether phenylimidazole)s containing pendent groups

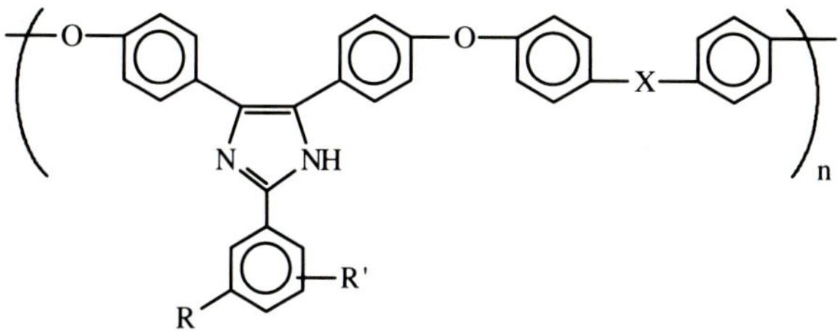

X	R	R'*	η_{inh}, dL/g[1]	T_g, °C[2]
SO$_2$	CF$_3$	3-CF$_3$	1.69	296
CO	CF$_3$	3-CF$_3$	0.59	270
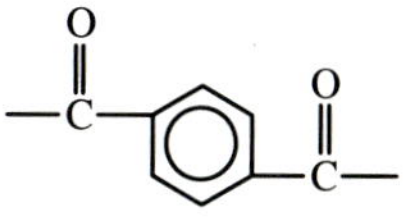 —C(O)—C$_6$H$_4$—C(O)— (para)	CF$_3$	3-CF$_3$	0.39	259
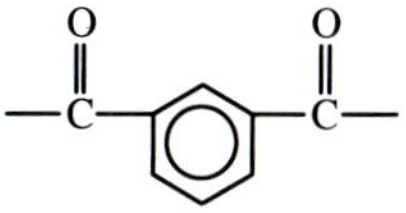 —C(O)—C$_6$H$_4$—C(O)— (meta)	CF$_3$	3-CF$_3$	0.46	234
SO$_2$	H	4-CF$_3$	0.41	284
CO	H	4-CF$_3$	0.52	263
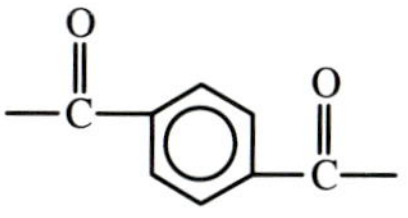 —C(O)—C$_6$H$_4$—C(O)— (para)	H	4-CF$_3$	0.71	249
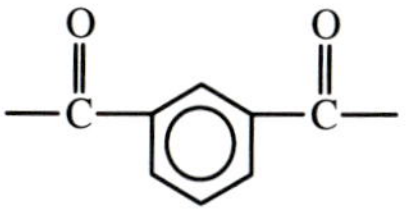 —C(O)—C$_6$H$_4$—C(O)— (meta)	H	4-CF$_3$	0.52	233
SO$_2$	H	4-CH$_3$	0.34	275
CO	H	4-CH$_3$	0.44	254
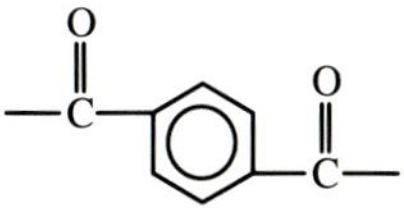 —C(O)—C$_6$H$_4$—C(O)— (para)	H	4-CH$_3$	0.59	245
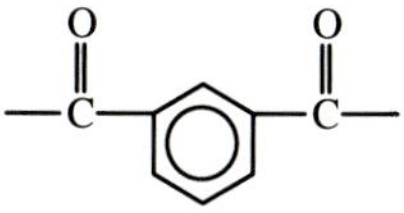 —C(O)—C$_6$H$_4$—C(O)— (meta)	H	4-CF$_3$	0.43	227
SO$_2$	H	4-C$_2$H$_5$	0.44	271
CO	H	4-C$_2$H$_5$	0.41	243
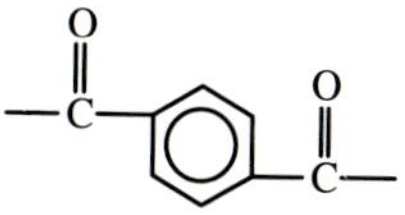 —C(O)—C$_6$H$_4$—C(O)— (para)	H	4-C$_2$H$_5$	0.42	236
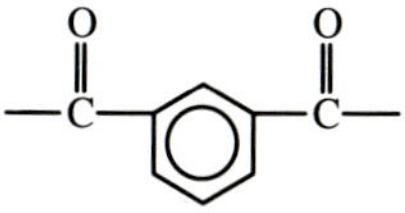 —C(O)—C$_6$H$_4$—C(O)— (meta)	H	4-C$_2$H$_5$	0.67	218

* Number indicates substituent position on the ring

monomers, 2,2'-di(4-fluorophenyl)-6,6-bibenzoxazole, was prepared in essentially quantitative yield from the reaction 3,3'-dihydroxy-4,4'-diaminobiphenyl and 4-fluorobenzoic acid in CHP at 260 °C under nitrogen [27]. Proton NMR was used to demonstrate that the benzoxazole unit was a sufficient electron-withdrawing group to activate the fluoro group for displacement. As a result of chemical shifts in the 2 and 6 protons of the 4-fluorophenyl group attached in the 2 position of the benzoxazole ring, the electron-withdrawing effect of the 2-benzoxazolyl group was determined to be similar to that of a carbonyl group [27]. When 2,2'-di(4-fluorophenyl)-6,6'-benzoxazole was polymerized with certain aromatic hydroxy compounds in the presence of potassium carbonate in NMP, precipitation occurred prior to the formation of high molecular weight polymer. This was circumvented by using CHP in place of NMP at 250 °C.

The T_gs of various poly(arylene ether benzoxazole)s are presented in Table 8. The effect of a bulky group to increase T_g is evident where the polymers containing the 9,9-diphenylfluorene moiety have the highest T_gs. Thin films of the second polymer in Table 8 gave tensile strength, modulus and elongation at ambient temperature of 91 MPa, 2.4 GPa and 5%, respectively [26]. Typical of most amorphous thermoplastics, the polymers in Table 8 when placed under stress exhibited poor resistance to acetone or methylene chloride, undergoing severe stress crazing and cracking [27].

Poly(arylene ether benzoxazole)s were also synthesized from the reaction of bis[(4-hydroxyphenyl)benzoxazole]s and activated aromatic difluoro monomers as shown in Eq. (7) [28, 29]. The bis[(4-hydroxyphenyl)benzoxazole]s were readily prepared by condensation of the appropriate bis(o-aminophenol) (e.g. 3,3'-dihydroxy-4,4'-diaminobiphenyl) with phenyl-4-hydroxybenzoate in diphenyl sulfone at ~260 °C. Under proper conditions, the less expensive 4-hydroxybenzoic acid can be used in place of the phenyl ester to provide high yields of the desired bis[(4-hydroxyphenyl)benzoxazole]s. As presented in Table 9, the hexafluoroisopropylidene (6F) containing polymers were amorphous. These polymers were prepared in DMAc. However, the polymers derived from 6,6'-bis[2-(4-hydroxyphenyl)benzoxazole] were prepared in diphenyl sul-

DMAC or $(C_6H_5)_2$ SO_2/Toluene
K_2CO_3
Δ, N_2 (7)

where X and Y are defined in Table 9

Table 8. Glass transition temperatures of poly(arylene ether benzoxazole)s

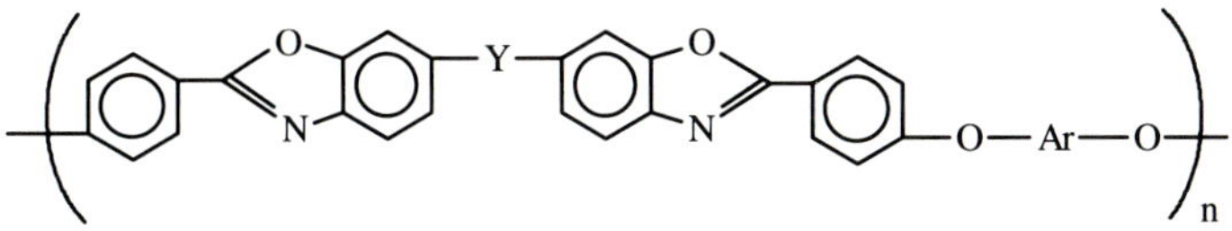

Y	Ar	$[\eta]$, dL/g*	T_g, °C
nil	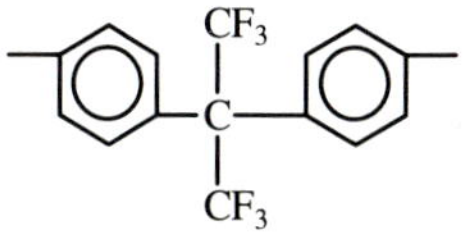 —C₆H₄—C(CF₃)₂—C₆H₄—	0.49	230
—C(CF₃)₂—	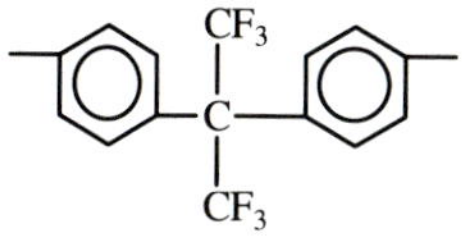 —C₆H₄—C(CF₃)₂—C₆H₄—	0.51	241
nil	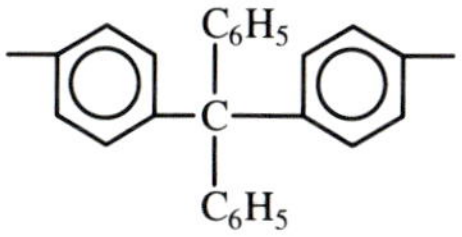 —C₆H₄—C(C₆H₅)₂—C₆H₄—	0.53	245
—C(CF₃)₂—	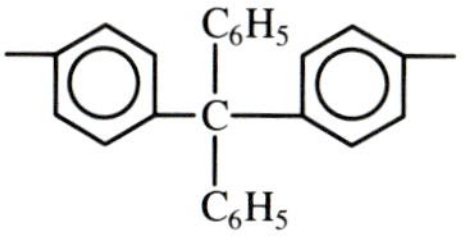 —C₆H₄—C(C₆H₅)₂—C₆H₄—	0.87	259
nil	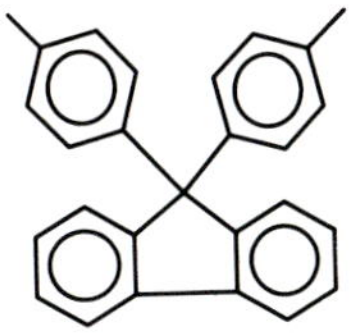 (dimethyl-substituted fluorenylidene bisphenyl)	1.04	303
—C(CF₃)₂—	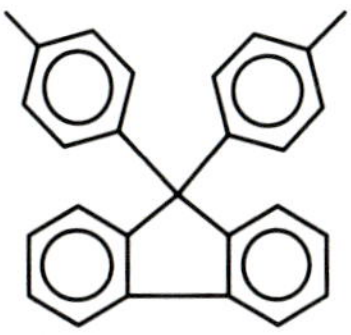 (dimethyl-substituted fluorenylidene bisphenyl)	0.65	291

* Intrinsic viscosity in CHP at 25 °C

fone at 260 °C. When polymerizations were attempted in DMAc at ∼155 °C, the polymers precipitated from solution prior to high molecular weight formation. As seen in Table 9, three of these polymers were semi-crystalline. The DSC curves and the wide angle X-ray (WAXS) diffractograms of the polymers isolated directly from the polymerization reaction are presented in Figs. 1 and 2, respectively. When the semi-crystalline polymers were heated above their T_ms

Table 9. Properties of poly(arylene ether benzoxazole)s

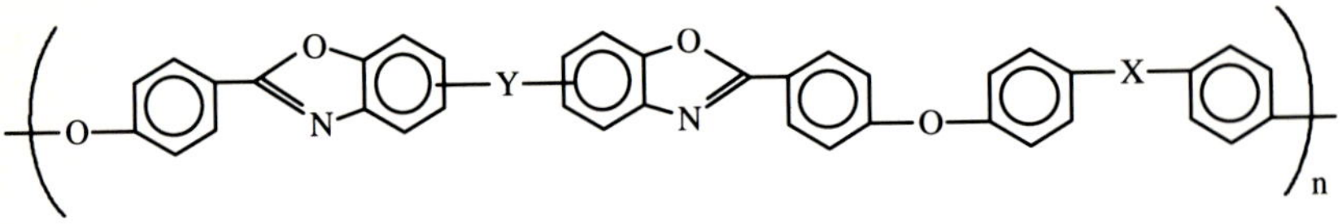

Y[1]	X	η_{inh}, dL/g	$T_g(T_m)$, °C[2]	Temperature (°C) 5% wt.loss in[2]	
				Air	N_2
—C(CF$_3$)$_2$—	SO$_2$	0.86	275	500	507
—C(CF$_3$)$_2$—	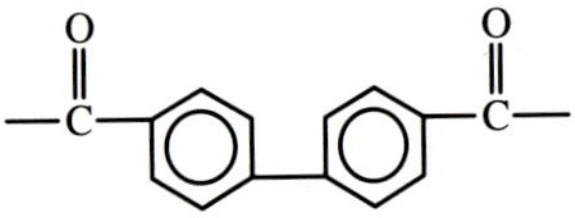 —C(=O)—C$_6$H$_4$—C$_6$H$_4$—C(=O)—	0.72	247	494	530
—C(CF$_3$)$_2$—	CO	0.67	244	499	529
—C(CF$_3$)$_2$—	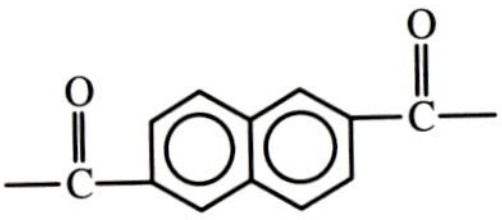 [naphthalene diketone]	0.49	243	488	525
—C(CF$_3$)$_2$—	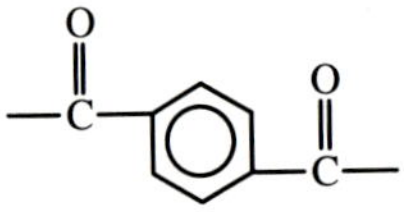 [phenylene diketone]	0.92	239	489	524
—C(CF$_3$)$_2$—	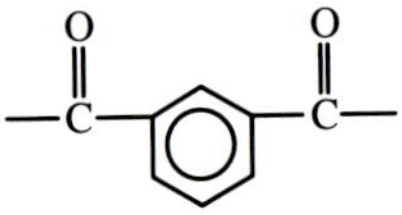 [meta-phenylene diketone]	0.40	217	492	517
—C(CF$_3$)$_2$—	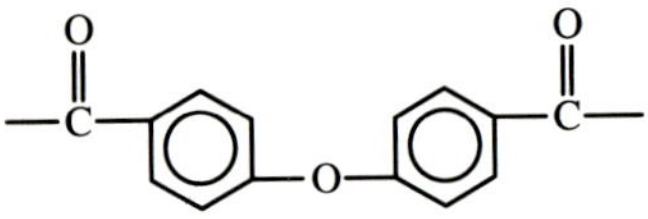 [diphenyl ether diketone]	0.62	211	478	528
—C(CF$_3$)$_2$—	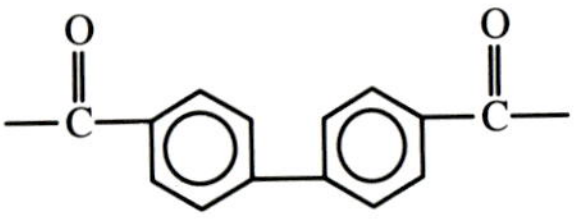 [biphenyl diketone]	0.41	204	488	526
nil	CO	0.50[3]	239(415)	449	505
nil	SO$_2$	0.40[3]	237	470	477
nil	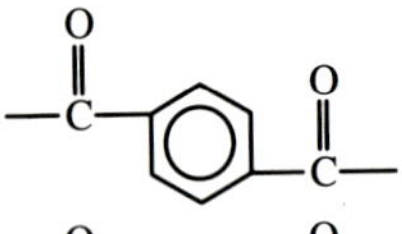 [phenylene diketone]	0.40[3]	227(395, 436)	446	501
nil	 [phenylene diketone]	1.04[3]	205(363)	475	530

[1] C(CF$_3$)$_2$ located in 5,5′ position, nil located in 6, 6′ position
[2] By thermogravimetric analysis at a heating rate of 2.5 °C/min.
[3] Determined on 0.5% H$_2$SO$_4$ solutions at 25 °C

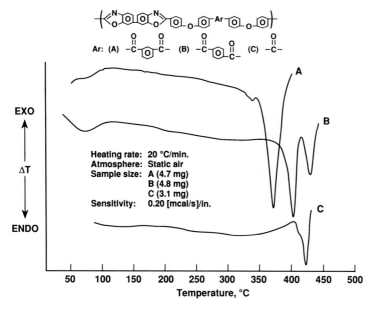

Fig. 1. Differential scanning calorimetric curves of PAEBs

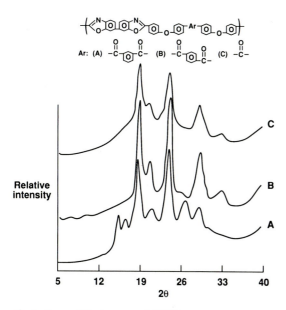

Fig. 2. X-ray diffractograms of PAEBs

and slowly cooled, they became amorphous and the crystallinity could not be regained by annealing at various temperatures between 300 and 350 °C for several hours. No work was reported on attempts to induce crystallinity using solvents.

The 6F containing poly(arylene ether benzoxazole)s in Table 9 generally exhibited higher temperatures of 5% weight loss by thermogravimetric analysis than the polymers without the 6F group. Solution cast unoriented thin films of the 6F polymers gave room temperature tensile strength, modulus and elongation that ranged from 84.8 to 102 MPa, 1.94 to 2.40 GPa and 9.0 to >15%, respectively [29].

3.4 PAE Containing Benzothiazole Units

Several poly(arylene ether benzothiazole)s were prepared from the reaction of 2,6-di(fluorophenyl)benzo[1, 2-d:4,5-d'] bisthiazole and various aromatic dihydroxy compounds in a mixture of NMP and CHP [30]. Again the heterocyclic unit activated the fluoro group adequately to permit displacement and the formation of high molecular weight polymer. The bis(fluorophenyl)heterocycle was formed from the reaction of 1,4-dimercapto-2,5-diaminobenzene dihydrochloride and excess 4-fluorobenzoyl chloride in the presence of triethylamine in chloroform. The resultant dithioesterdiamide was subsequently ring closed by heating to 260 °C. The T_gs of three polymers are given in Table 10. Random copolymers were prepared by copolymerizing the bis(fluorophenyl)bisthiazole monomer with various amounts of 4,4'-difluorodiphenyl sulfone and 2,2-bis(4-hydrophenyl)hexafluoropropane [30]. The copolymers showed better solubility and higher toughness than the polymers in Table 10.

3.5 PAE Containing Pyrazole Units

The reaction of bis(4-hydroxyphenyl)pyrazoles with activated aromatic difluoro and dichloro compounds under aromatic nucleophilic displacement conditions

Table 10. Glass transition temperatures of poly(arylene ether benzothiazole)s

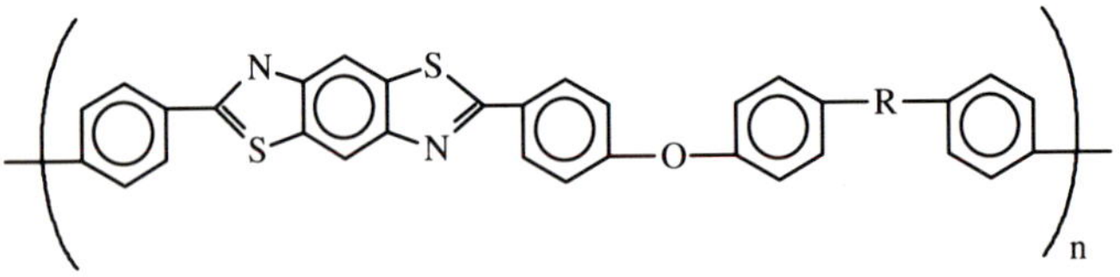

R	$[\eta]$, dL/g*	T_g, °C
$C(CF_3)_2$	0.83	240
$C(CF_3)_2$	Insol.	248
$C(C_6H_5)_2$	0.48	250

* Determined in NMP at 25 °C

in DMAc or NMP provided high molecular weight poly(arylene ether pyrazole)s according to Eq. (8) [31, 32]. The pyrazole bisphenols were prepared by multistep routes. As an example, 1,3-di(4-methoxyphenyl)-1,3-propanedione was prepared from the reaction of methyl-4-methoxybenzoate and 4-methoxyphenyl benzyl ketone in the presence of sodium hydride. Hydrazine was then condensed with the diketone to yield 3,5-di(methoxyphenyl)pyrazole which was demethylated to yield 3,5-di(4-hydroxyphenyl)pyrazole [31]. The relationship of chemical structure to T_g is shown in Table 11. The first series of polymers where R and R′ are hydrogen exhibit the highest T_gs. The NH in the polymers can participate in strong intermolecular association through hydrogen bonding. Polymers with groups capable of hydrogen bonding and also containing bulky groups would normally be expected to display even higher T_gs. The third series in Table 11 represents such polymers but their T_gs are lower than those in the first series. Apparently the bulky phenyl groups sterically hinders and thereby partially disrupts the hydrogen bonding.

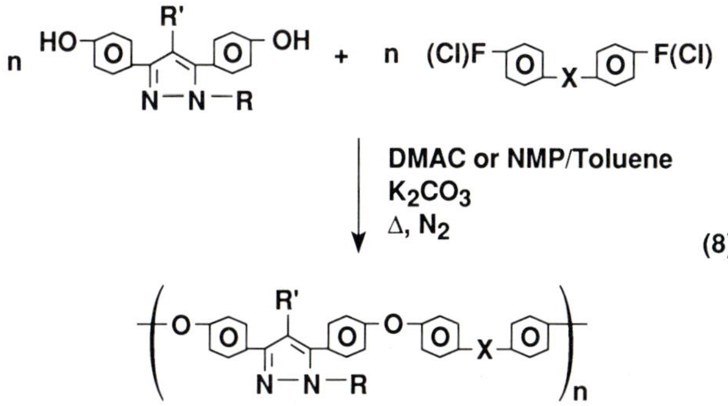

where R, R′ and X are defined in Table 11

The second polymer in the second series in Table 11 had the following mechanical properties at room temperature: tensile modulus of 2.76 GPa, titanium-to-titanium tensile shear strength of 24.8 MPa, and fracture toughness (G_{Ic}, critical strain energy release rate) as determined on compact tension specimens (ASTM E399) of ~ 5000 J/m^2 [32]. For comparison, Udel P1700 polysulfone from Amoco under identical test conditions gave a G_{Ic} of 2300 J/m^2.

3.6 PAE Containing 1,3,4-Oxadiazole Units

The 1,3,4-oxadiazole unit was initially incorporated into PAE in 1967 by reacting 2,5-di(4-fluorophenyl)-1,3,4-oxadiazole with bisphenol A [1]. The same route and same difluoro monomer were used more recently to prepare a series of poly(arylene ether 1,3,4-oxadiazole)s [33]. The corresponding dinitro or dichloro compounds, 2,5-di(4-nitro or chlorophenyl)-1,3,4-oxadiazole [33], were

Table 11. Glass transition temperatures of poly(arylene ether pyrazole)s

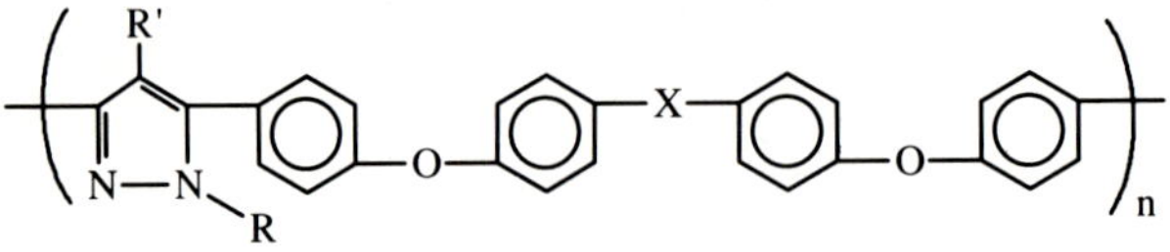

R	R'	X	η_{inh}, dL/g	T_g, °C
H	H	SO_2	0.77	260
	H	CO	0.68	229
	H	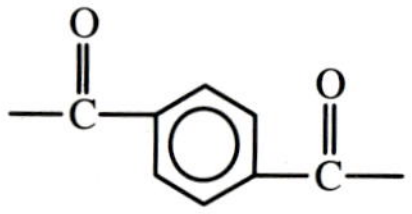 —C(=O)—C6H4(1,4)—C(=O)—	0.47	235
	H	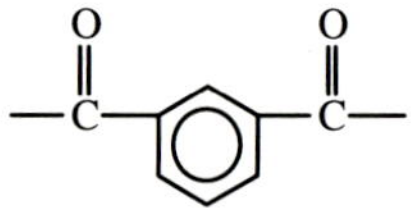 —C(=O)—C6H4(1,3)—C(=O)—	0.67	205
C_6H_5	H	SO_2	0.44	235
	H	CO	1.97	205
	H	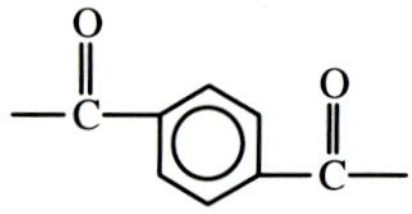 —C(=O)—C6H4(1,4)—C(=O)—	1.24	205
	H	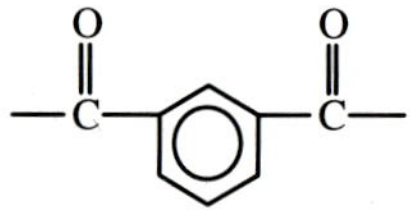 —C(=O)—C6H4(1,3)—C(=O)—	0.91	190
H	C_6H_5	SO_2	0.80	237
	C_6H_5	CO	0.70	208
	C_6H_5	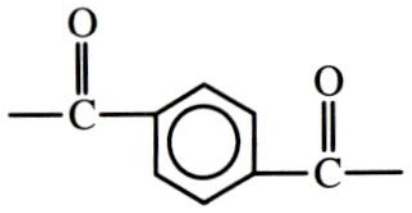 —C(=O)—C6H4(1,4)—C(=O)—	0.79	200
	C_6H_5	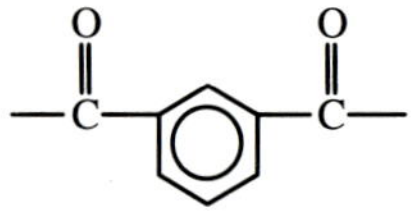 —C(=O)—C6H4(1,3)—C(=O)—	0.70	188
C_6H_5	C_6H_5	SO_2	1.00	243
	C_6H_5	CO	0.46	205
	C_6H_5	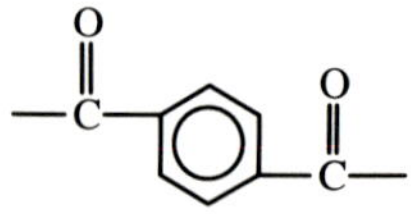 —C(=O)—C6H4(1,4)—C(=O)—	1.13	210
	C_6H_5	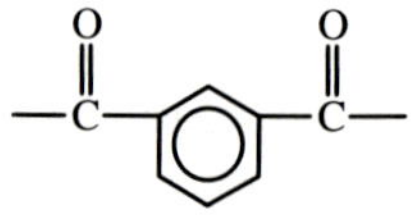 —C(=O)—C6H4(1,3)—C(=O)—	1.40	196

also successfully used to prepare high molecular weight poly(arylene ether-1,3,4-oxadiazole)s. Poly(arylene ether oxadiazole)s have also been prepared from the reaction of 2,5-di(4-hydroxyphenyl)-1,3,4-oxadiazole and activated aromatic difluoro and dichloro compounds in diphenyl sulfone as depicted in Eq. (9) [34, 35]. Polymerization in DMAc or NMP resulted in low molecular weight polymer because the polymer precipitated before high molecular weight was attained. 2,5-Di(4-hydroxyphenyl)-1,3,4-oxadiazole was prepared by melt condensation of 4-hydroxybenzoic hydrazide and phenyl-4-hydroxybenzoate. If the reaction is stopped after heating for ~1 h at ~210 °C, the major product is 1,2-di(4-hydroxybenzoyl)hydrazine. This is a common intermediate for the synthesis of both di(hydroxyphenyl)oxadiazole and triazole monomers. When the 1,2-di(4-hydroxybenzoyl)hydrazine is heated to ~300 °C, cyclodehydration occurs to yield 2,5-di(4-hydroxyphenyl)-1,3,4-oxadiazole [35]. Aniline hydrochloride can be reacted with 1,2-di(4-hydroxybenzoyl)hydrazine at ~250 °C to yield 3,5-bis(4-hydroxyphenyl)-4-phenyl-1,2,4-triazole [35].

where X is defined in Table 12

The T_gs and in some cases, the T_ms of several poly(arylene ether-1,3,4-oxadiazole)s are reported in Table 12. The last five polymers in Table 12 display the same T_g trend as seen for other polymers, namely phenylphosphine oxide > sulfone > carbonyl > terephthaloyl > isophthaloyl. The terephthaloyl polymer could be heated above the T_m, and subsequently quenched to the amorphous form, and then annealed at ~330 °C to induce crystallinity. Once the T_ms of the carbonyl and isophthaloyl polymers were exceeded, crystallinity could not be reintroduced by annealing at 300 to 330 °C for several hours. The T_g and T_m of the isophthaloyl polymer are abnormally close.

Transient T_ms have been observed in other systems (i.e. polyimides) and appear to be due to synthetic conditions coupled with chemical structure. Under the same synthetic conditions, polymers of certain chemical structures form transient crystalline phases whereas others do not. For example, when the polyamide acid from the reaction of 3,3′,4,4′-benzophenonetetracarboxylic di-

Table 12. Glass transition temperature of poly(arylene ether 1,3,4-oxadiazole)s

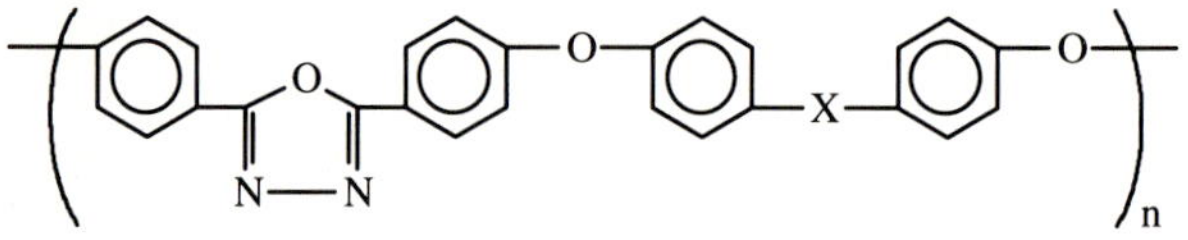

X	Viscosity, dL/g	$T_g(T_m)$, °C	Ref.
—C(C$_6$H$_5$)$_2$—	0.44[1]	190	32
—C(CH$_3$)$_2$—	0.45[1]	201	32
—C(CF$_3$)$_2$—	0.50[1]	210	32
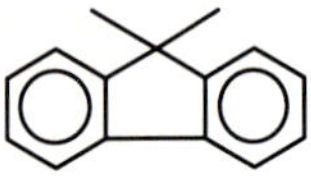	0.55[1]	220	32
	1.38[2]	242	34
—SO$_2$—	1.02[2]	226	34
—CO—	1.57[3]	205 (325)	34
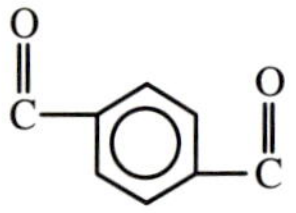	1.71[3]	201 (390)	34
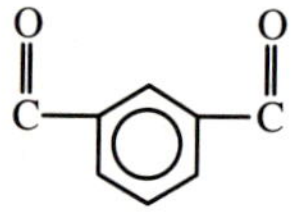	1.53[2]	182 (265)	34

[1] Intrinsic viscosity in NMP at 25 °C
[2] Inherent viscosity on 0.5% solution in _m_-cresol at 25 °C
[3] Inherent viscosity on 0.5% solution in H$_2$SO$_4$ at 25 °C

anhydride and 3,3'-diaminobenzophenone is converted to polyimide with acetic anhydride and triethylamine, a fine powder of polyimide precipitates. The powder displays a strong DSC endotherm at ~300 °C which upon proper heating, moves to the thermodynamic T_m of ~340 °C. After exceeding the T_m at 340 °C, the polyimide is amorphous and the crystallinity cannot be regained by thermal annealing.

The isophthaloyl based polymer in Table 12 gave unoriented thin film tensile strength, modulus and elongation at room temperature of 114.4 MPa, 3.55 GPa and 4.4%, respectively [35]. The relatively high modulus for the amorphous film was unexpected. Titanium-to-titanium tensile shear strengths of the isophthaloyl based polymer after 1000 hours of exposure at 200 °C in air were 34.1 MPa at 23 °C, 22.2 MPa at 150 °C and 7.6 MPa at 177 °C [35]. The strength at 177 °C was surprising since the T_g was only 182 °C. DSC and WAXS analyses revealed that the adhesive was amorphous.

3.7 PAE Containing 1,2,4-Triazole Units

Poly(arylene ether 1,2,4-triazole)s were synthesized from the reaction of 3,5-di(4-hydroxyphenyl)-4-phenyl-1,2,4-triazole with three different activated aromatic difluoro monomers [34, 35]. The polymer from the triazole bisphenol and 1,4-di(4-fluorobenzoyl)benzene exhibited an inherent viscosity of 3.40 dL/g (0.5% solution in *m*-cresol at 25 °C), a T_g of 216 °C and a T_m of 377 °C [35]. Solution cast amorphous unoriented thin films of this polymer gave 23 °C tensile strength, modulus and elongation of 87.6 MPa, 2.7 GPa and 7.8% respectively. No work was performed to induce crystallinity in the film.

   Poly(arylene ether triazole)s have also been prepared by heterocyclic-activated displacement polymerization [36]. The 1,2,4-triazole unit sufficiently activated, albeit weakly, aryl fluorides for nucleophilic displacement. Several 3,5-bis(4-fluorophenyl)-4-aryl-1,2,4-triazoles were polymerized with various bis-phenols to yield polymers with T_gs from 185 to 230 °C [36]. The 1,2,4-triazole unit appears to be one of the more weakly activating heterocycles towards nucleophilic substitution polymerization.

3.8 PAE Containing Benzimidazole Units

Poly(arylene ether benzimidazole)s have received more attention than any other PAE containing heterocyclic units. This is due primarily to their unique combination of properties even at relatively low molecular weights and their potential for use in several high performance applications. The initial report in 1991 involved polymers from the reaction of 3 different bis[(4-hydroxyphenyl)benzimidazole]s with various activated aromatic difluoro monomers as shown in Eq. (10) [37]. The bis[(4-hydroxyphenyl)benzimidazole]s were prepared from the reaction of aromatic bis(*o*-diamines) and phenyl-4-hydroxybenzoate in diphenyl sulfone. The use of 4-hydroxybenzoic acid would obviously reduce the

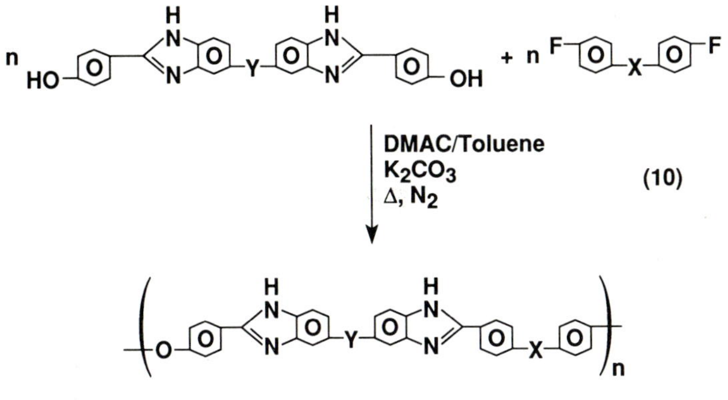

$$(10)$$

where X and Y are defined in Table 13

cost of the bisphenol monomer. The bis[(4-hydroxyphenyl)benzimidazole]s failed to yield sharp melting points in spite of several recrystallizations. These materials are hygroscopic, tend to be amorphous and complex with the solvent. As an example, the material of structure 5 after two recrystallizations from DMAc gave a broad melting point (by differential thermal analysis at a heating rate of 10 °C/min, endotherm began at ~380 °C and peaked at 398 °C) and an elemental analysis whose carbon content was 1.3% less than the theoretical value. This batch of monomer however gave high molecular weight polymers upon reaction with various activated aromatic difluoro monomers.

5

The T_gs of several poly(arylene ether benzimidazole)s are presented in Table 13. The high T_gs were attributed to strong intermolecular association through hydrogen bonding. This strong intermolecular association obviously influences other properties such as solubility, moisture absorption and mechanical performance. Unoriented solution-cast thin film gave tensile strength and modulus, as reported in Table 14, significantly higher than films from other PAEH.

To assess the effect of molecular weight on polymer properties, various endcapped controlled molecular weight polymers were prepared and characterized [38, 39]. The polymers were prepared by upsetting the stoichiometry in favor of the difluoro monomer and endcapping with 2-(4-hydroxyphenyl)-benzimidazole. Other endcapping agents such as phenol can also be used. In Table 15, the properties of polymers prepared by upsetting the stoichiometry by 1.5, 3, 5 and 7 mole % are compared with those of a polymer made at exact stoichiometry. In general, the higher molecular weight polymers exhibited better properties although not dramatically higher than the lower molecular weight materials except for some scattering due to sample preparation. For example, film elongation varies primarily because care was not exercised to exclude foreign particles (i.e. gel, dust particles, etc.). High molecular weight polymers are expected to exhibit higher properties such as elongation and toughness than low molecular weight materials.

The effect of drying atmosphere on the properties of a poly(arylene ether benzimidazole) film is shown in Table 16. After drying at 330 °C, the air dried film exhibited a darker color, higher T_g, higher strength and modulus at 232 °C and lower elongation at 232 °C than the nitrogen dried film. This difference is attributed to a higher degree of crosslinking in the air dried film. The nitrogen dried film in Table 16 is DMAc insoluble and has a T_g 5 °C higher than that of the same polymer in Table 15. The major difference is thermal history. The 3% stoichiometric upset polymer in Table 15 was dried at ~150 °C in air and

Table 13. Glass transition temperatures of poly(arylene ether benzimidazole)s

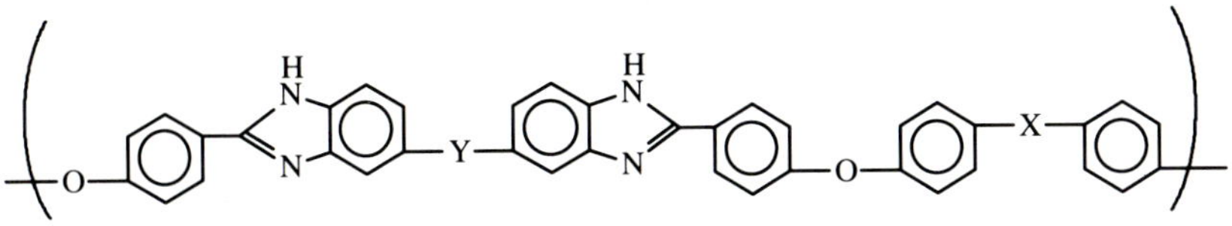

Y	X	$[\eta_{inh}]$, dL/g[1]	T_g, °C
nil	SO$_2$	1.87	352
O	SO$_2$	1.42	322
CO	SO$_2$	0.93	ND[2]
nil	CO	1.11	307
O	CO	1.34	294
CO	CO	0.93	ND
nil	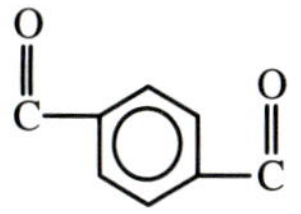 (para diketophenylene)	1.19	295
O	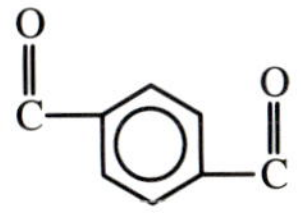 (para diketophenylene)	1.23	282
CO	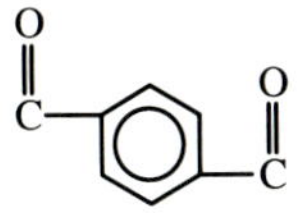 (para diketophenylene)	0.79	276
nil	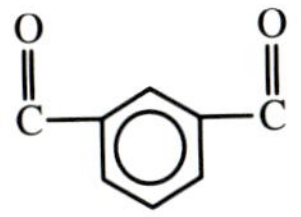 (meta diketophenylene)	1.99	276
O	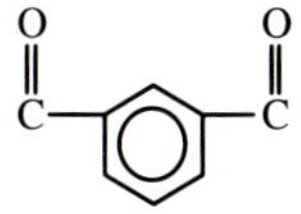 (meta diketophenylene)	1.79	269
CO	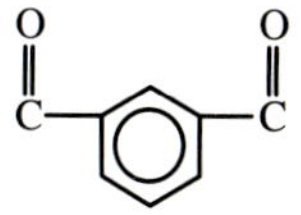 (meta diketophenylene)	1.43	264

[1] Inherent viscosity on 0.5% solutions in DMAC at 25 °C
[2] ND = not detected

remained soluble in DMAc. Whereas the nitrogen dried film in Table 16 was exposed to 330 °C for 1 h and became insoluble in DMAc. During the drying at 330 °C, some crosslinking and/or ordering evidently occurred to increase the T_g and render the film insoluble in DMAc.

The air dried poly(arylene ether benzimidazole) films displayed excellent resistance to methylene chloride, hydraulic fluid (Skydrol) and jet fuel (HyJet IV)

Table 14. Unoriented thin film properties of poly(arylene ether benzimidazole)s

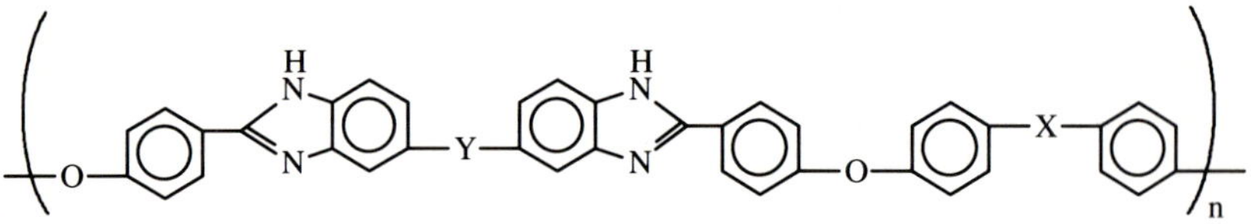

Y	X	$[\eta_{inh}]$, dL/g	Test temp, °C	Tensile properties* Strength, MPa	Modulus, GPa	Elong., %
nil	SO_2	1.87	23	155.2	4.5	9
			232	102.8	2.6	7
O	SO_2	1.42	23	129.7	3.9	8
			232	84.8	2.2	9
nil	CO	1.11	23	157.2	4.5	12
			232	97.2	2.3	20
O	CO	1.34	23	135.9	4.0	7
			232	77.9	2.0	13
nil	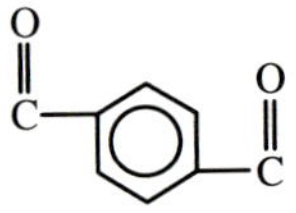 (para-dicarbonyl phenyl ether structure)	1.19	23	139.3	4.2	14
			232	80.0	2.2	18
O	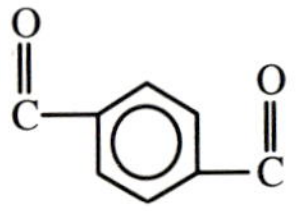 (para-dicarbonyl phenyl ether structure)	1.23	23	121.4	3.7	18
			232	71.0	1.9	33
nil	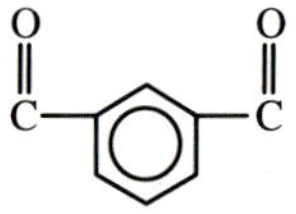 (meta-dicarbonyl phenylene structure)	1.99	23	125.5	4.1	14
			232	85.5	2.8	7
O	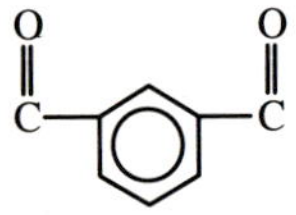 (meta-dicarbonyl phenylene structure)	1.79	23	126.9	4.1	6
			232	59.2	2.2	6
CO	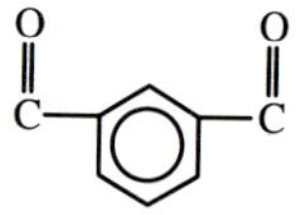 (meta-dicarbonyl phenylene structure)	1.43	23	135.2	4.2	6
			232	70.3	2.4	4

* Film dried at 100, 200 and ~ 50 °C above their T_g in the flowing air

[38]. In addition, the films showed no change in properties after 1000 h at 200 °C in air. Adhesive and composite properties were promising. As an example, unidirectional AS-4 composites exhibited flexural strength and modulus (ASTM-D790) as high as 1910 MPa and 135.8 GPa at 23 °C and 1124 MPa and 100 GPa at 232 °C, respectively [38]. Compressive strength and modulus (ASTM-D3410 Procedure B) at 23 °C were 1150 MPa and 115.8 GPa, respectively.

Table 15. Properties of poly(arylene ether benzimidazole)s

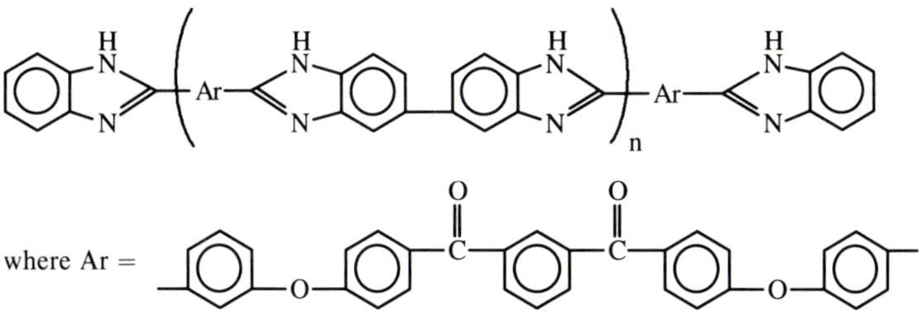

where Ar =

Property	Stoichiometric imbalance, mol %				
	0	1.5	3	5	7
$\bar{M}_n$ (calculated)	–	46367	23008	13665	9661
η_{inh}, dL/g	1.99	1.65	1.42	0.84	0.55
T_g, °C (powder)	276	274	274	269	266
Temp. of 5%wt. loss (air/N$_2$), °C	476/515	467/511	467/507	477/510	470/472
Tensile strength, MPa, 23/232 °C	125/85	125/81	129/83	118/65.5	114/71
Tensile modulus, GPa, 23/232 °C	4.1/2.8	3.6/2.6	4.0/2.8	3.5/2.3	3.6/2.6
Elong. at break, %, 23/232 °C	14/7	12/5	30/12	13/8	22/10
G_{Ic}, J/m^2	–	860	1000	–	475

Table 16. Properties of films[1] dried in air and nitrogen

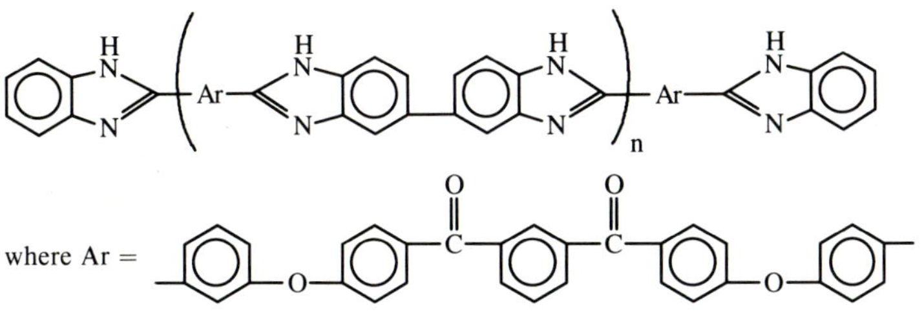

where Ar =

Property	Air	Nitrogen
Color	Amber	Yellow
T_g, °C (film)[2]	284	279
DMAC solubility	Insoluble	Insoluble
Tensile strengh, MPa, 23/232 °C	125.5/82.9	129.6/59.2
Tensile modulus, GPa, 23/232 °C	4.1/2.8	4.2/2.1
Elong. at break, %, 23/232 °C	47/12	40/96

[1] Polymer prepared at 3% stoichiometric imbalance, $\bar{M}_n$ (calculated), $\sim$23,000 g/mol, η_{inh} (0.5% DMAC solution, 25 °C) = 1.42 dL/g
[2] Film cure cycle: 50, 100, 150, 200, 250 for 0.5 h each and 330 °C for $\sim$1 h

The molecular weight of an endcaped poly(arylene ether benzimidazole) prepared at a stoichiometric imbalance of 7 mol% (η_{inh} = 0.55 dL/g) was obtained by low angle laser light scattering (LALLs) and gel permeation (size exclusion) chromatography/differential viscometry (GPC/DV) on DMAc solutions containing 0.0075 M lithium bromide to suppress aggregate formation

[38]. Weight average molecular weights of ∼68,000 and 87,000 g/mole were obtained by LALLS and GPC/DV, respectively. Number average molecular weight of ∼19,000 g/mol was obtained by GPC. The calculated $\bar{M}_n$ for this polymer was ∼9700 g/mol. Polymer molecules that partake in strong inter-molecular association tend to aggregate in solution and consequently appear to be of higher molecular weight. This explains the discrepancy between the calculated and experimentally determined $\bar{M}_n$s. Difficulty in filtration of the higher molecular weight polymer solutions presumably due to larger aggregates, prevented their analysis.

Two AB monomers [40], 2-(4-fluorophenyl)-5-hydroxybenzimidazole (structure 6) and 1-phenyl-2-(4-fluorophenyl)-5-hydroxybenzimidazole (structure 7), were polymerized under aromatic nucleophilic displacement conditions in CHP to yield polymers with intrinsic viscosity (NMP at 30 °C) as high as 2.6 dL/g [40]. The T_gs of the N-hydrogen benzimidazole and the N-phenylben-zimidazole polymers were 368 and 268 °C, respectively. The large difference in the T_gs was obviously due to hydrogen bonding. The two AB benzimidazole monomers were copolymerized and 2-(4-fluorophenyl)-5-hydroxybenzimidazole was also copolymerized with an AB quinoxaline monomer, 2-phenyl-3-(4-hydroxyphenyl)-6-fluoroquinoxaline [40].

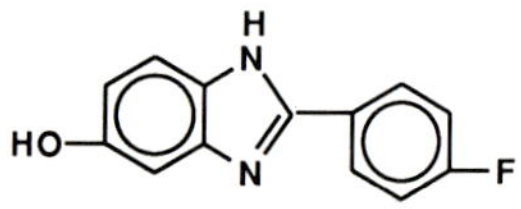

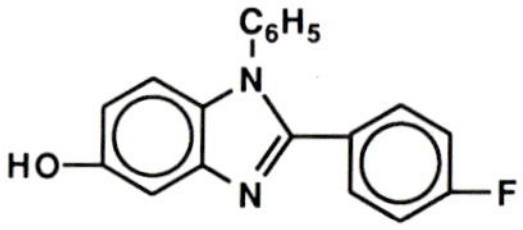

3.9 PAE Containing N-Arylenebenzimidazole Units

A series of poly(arylene ether N-arylenebenzimidazole)s were prepared from two arylenebenzimidazole bisphenols and various activated aromatic difluoro monomers according to Eq. (11) [41]. The di(hydroxyphenyl-N-arylene-

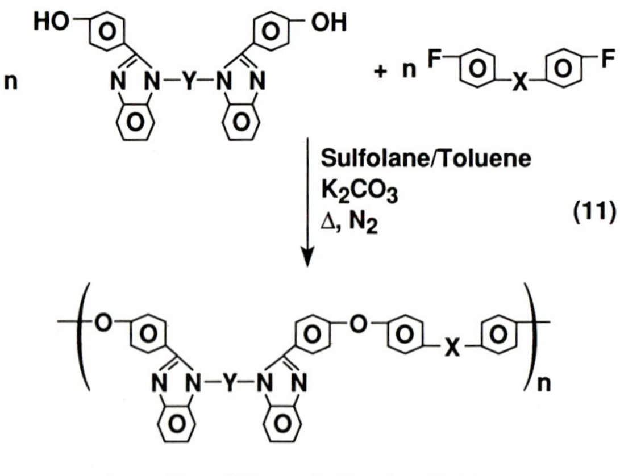

where X and Y are defined in Table 17

benzimidazole) monomers were readily prepared from the reaction of 1,4,-bis(2-aminoanilino)benzene or 4,4'-bis(2-aminoanilino)biphenyl with phenyl-4-hydroxybenzoate in diphenyl sulfone at ∼ 290 °C under nitrogen. Most of the polymers were soluble in m-cresol and NMP. Polymer properties are reported in Table 17. Solution cast, unoriented thin film tensile strengths, moduli and elongations were generally ∼ 82.7 MPa, ∼ 2.45 GPa and ∼ 5%, respectively. Because of the absence of the NH group on the benzimidazole ring, the T_gs and

Table 17. Glass transition temperatures of poly(arylene ether N-arylene benzimidazole)s

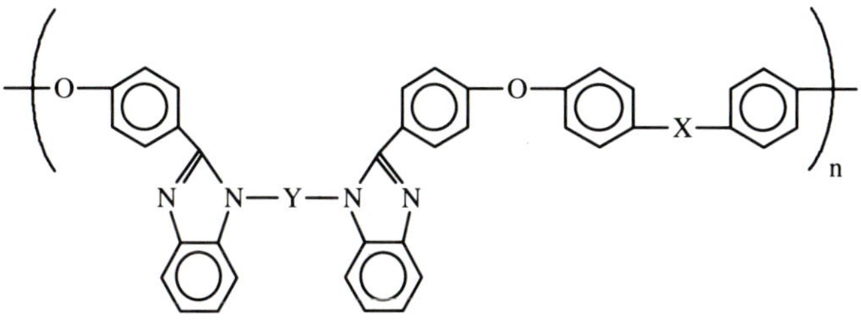

Y	X	η_{inh}, dL/g	T_g, °C
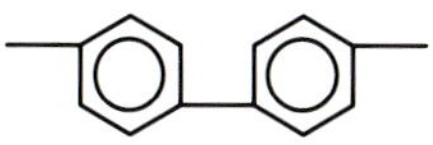	SO_2	0.69	289
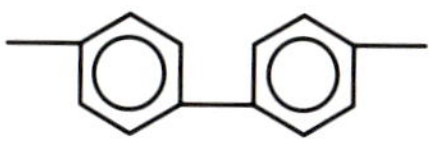	CO	0.37	268
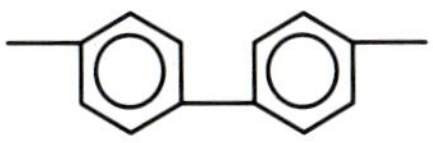	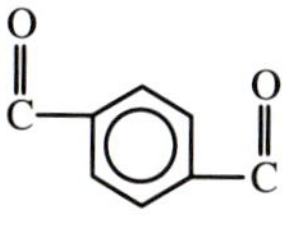	0.62	259
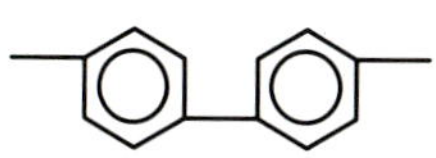	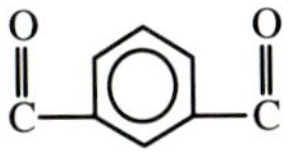	0.59	244
	SO_2	0.77	270
	CO	0.48	242
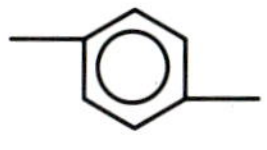	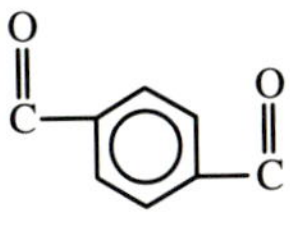	0.86	238
	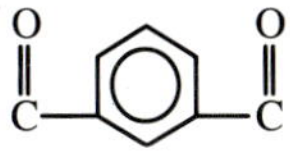	0.75	219

film properties of the poly(arylene ether N-arylenebenzimidazole)s were significantly lower than those of the poly(arylene ether benzimidazole)s (see Tables 13 and 14).

3.10 PAE Containing Furan, Pyrrole and Thiophene Units

A series of poly(arylene ether)s containing furan, pyrrole or thiophene units were prepared from the reaction of the appropriate heterocyclic bisphenol with various activated aromatic difluoro and dichloro monomers as represented in Eq. (12) [42, 43]. The heterocyclic bisphenols were synthesized using a common intermediate, 1,2-bis(4-methoxybenzoyl)-1,2-diphenylethane. The T_gs of the polymers are shown in Table 18. The pyrrole containing polymers exhibited the highest T_gs because of the hydrogen bonding from the NH group. Within each series, the nitrile containing polymers displayed the highest T_gs presumably due to polar interaction and the effect of a bulky group.

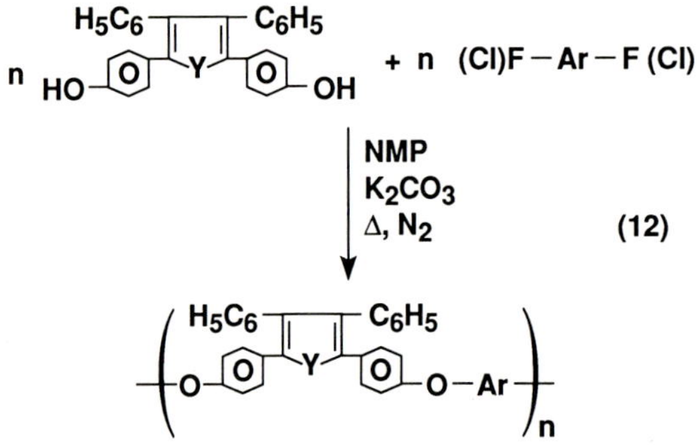

$$n \quad HO-\!\!\left\langle\bigcirc\right\rangle\!\!-\!\!\overset{H_5C_6\quad\quad C_6H_5}{\underset{Y}{\diagup\diagdown}}\!\!-\!\!\left\langle\bigcirc\right\rangle\!\!-OH \;+\; n \;\;(Cl)F-Ar-F\,(Cl)$$

$$\downarrow \quad \begin{matrix} NMP \\ K_2CO_3 \\ \Delta, N_2 \end{matrix} \qquad (12)$$

$$\left(\!\!-O-\!\!\left\langle\bigcirc\right\rangle\!\!-\!\!\overset{H_5C_6\quad\quad C_6H_5}{\underset{Y}{\diagup\diagdown}}\!\!-\!\!\left\langle\bigcirc\right\rangle\!\!-O-Ar-\!\!\right)_{n}$$

where Ar and Y are defined in Table 18

The synthesis of other poly(arylene ether)s containing thiophene units concerned the reaction of two activated halides containing thiophene (structures 8 and 9) with bisphenol A [44, 45]. The polymers from the monomers of structures 8 and 9 and bisphenol A had intrinsic viscosities (NMP, 25 °C) of 1.23 and 0.43 dL/g and T_gs of 158 and 120 °C, respectively.

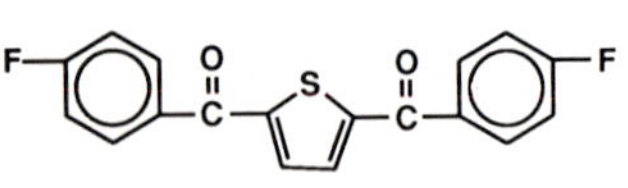

8 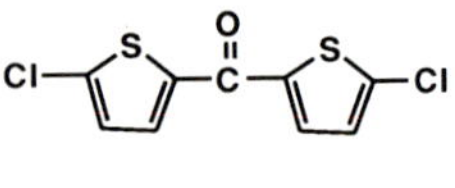   9

Dibenzofuran has also been incorporated into PAE using a unique synthetic approach that requires only the activated dihalo monomer [46]. As an example, 2,8-bis(4-fluorobenzoyl)dibenzofuran (structure 10) was reacted with potassium carbonate in the presence of silica catalyst in diphenyl sulfone at temperatures

Table 18. Glass transition temperatures of poly(arylene ether)s

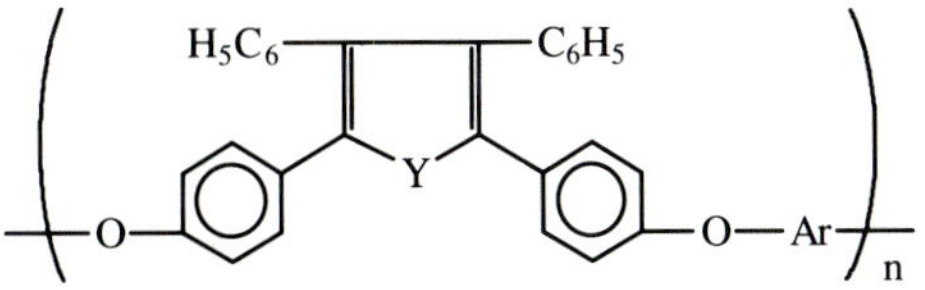

Y	Ar	η_{inh}, dL/g*	T_g, °C
NH	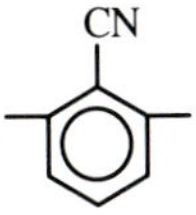 (CN-substituted phenylene)	0.37	250
NH	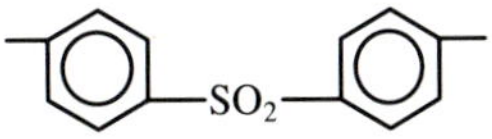 (phenylene–SO₂–phenylene)	0.42	244
NH	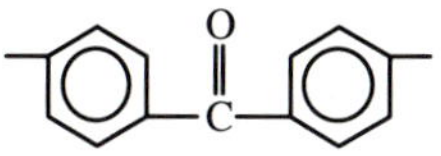 (phenylene–C(=O)–phenylene)	0.73	243
S	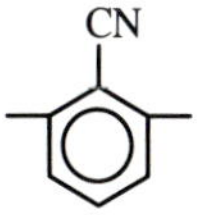 (CN-substituted phenylene)	0.39	217
S	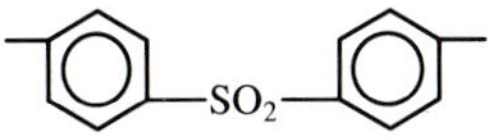 (phenylene–SO₂–phenylene)	0.36	196
S	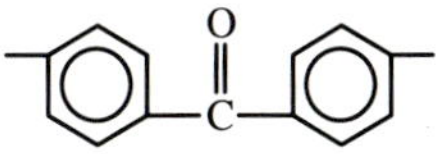 (phenylene–C(=O)–phenylene)	0.46	182
O	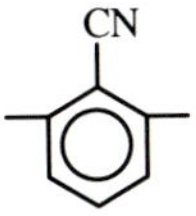 (CN-substituted phenylene)	0.25	238
O	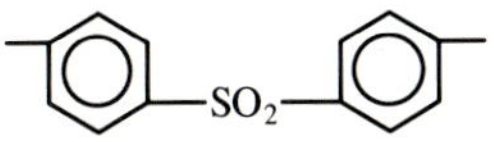 (phenylene–SO₂–phenylene)	0.33	235
O	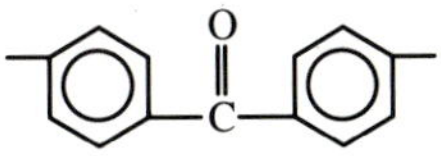 (phenylene–C(=O)–phenylene)	0.23	213

* 0.5% solution in NMP at 30 °C

reaching 300 °C. A polymer was formed having a reduced viscosity of 0.40 dL/g and a T_g of 222 °C [46]. The corresponding dichloro compound can also be used with cuprous oxide as cocatalyst to yield high molecular weight polymer. Structure 10, in essence, functions like an AB monomer. The proposed mechanism [46] involves the reaction of silanol on the silica with the halide to form a

silyl ether. The silyl ether then reacts with the activated fluoro group to form
polymer (silyl group is lost). Amorphous and semicrystalline copolymers containing the dibenzofuran unit were also prepared through aromatic nucleophilic
substitution by reacting the compound of structure 10 with 4,4′-dihydroxy-
benzophenone [46].

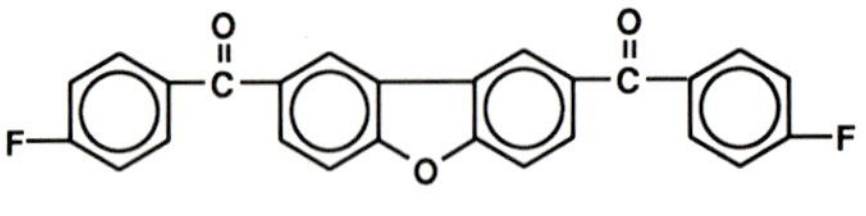

10

3.11 PAE Containing Phthalazine Units

Poly(arylene ether phthalazine)s were prepared by two different routes [47]. As
shown in Eq. (13), one route involved the preparation of precursor poly(arylene
ether diketone)s and subsequent reaction with hydrazine while the other route
concerned the reaction of difluorophthalazine monomers with various bisphenols. The poly(arylene ether phthalazine)s prepared by the two routes exhibited
very similar T_gs. The synthetic route to the difluorophthalazine monomers is also
shown in Eq. (13). A comparison of the properties of a representative series of
poly(arylene ether diketone)s and the corresponding poly(arylene ether

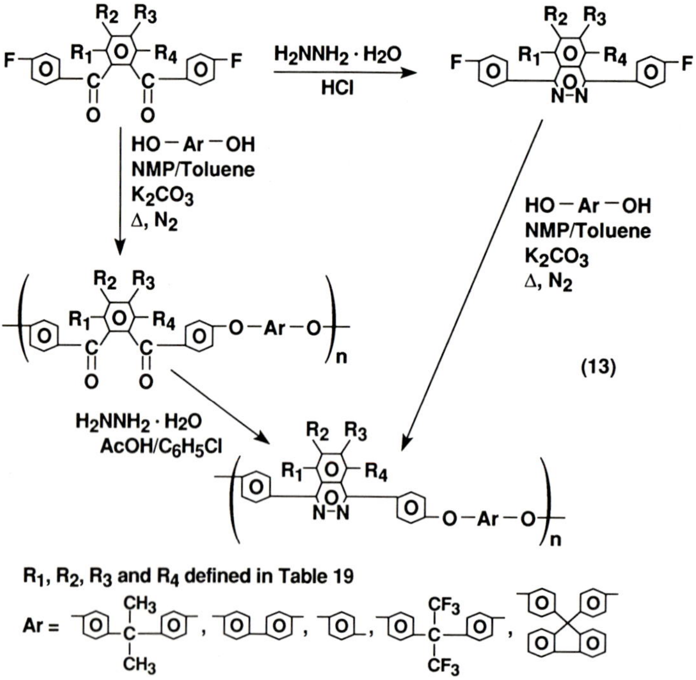

(13)

R_1, R_2, R_3 and R_4 defined in Table 19

phthalazine)s is presented in Table 19. Within the diketone and phthalazine series, the T_g increases with increasing number of pendent phenyl groups. The phthalazine polymers exhibit T_gs higher than the corresponding diketone polymers because of more rigid backbones. Poly(arylene ether phthalazine)s of three different molecular weight levels, end-capped with 3,5-di-tertbutylphenoxy groups were prepared primarily for molecular weight studies using ^{1}H NMR. As an example, a polymer prepared at a degree of polymerization of 70 mer units gave $\bar{M}_n$ by ^{1}H NMR of 65,000 g/mol. End group analysis is usually only accurate to $\sim 25,000$ g/mol using titration techniques. Depending upon the sensitivity of the instrument and the time for the run, NMR can accurately determine molecular weight at substantially high levels. The $\bar{M}_n$ of 65,000 g/mol represents a degree of polymerization of 79. By GPC using polystyrenes as standards, the same polymer showed a $\bar{M}_n$ and $\bar{M}_w$ of 142,000 and 213,000 g/mol respectively [47]. The $\bar{M}_n$ was slightly more than twice the value found by NMR. The 25 °C moduli of the polyphthalazines as reported in Table 19 ranged from 2.32 to 3.98 GPa and did not follow a trend relative to increasing pendent phenyl group content. The polymer containing two pendent phenyl groups per mer unit exhibited the highest modulus.

Table 19. Properties of poly(arylene ether diketone)s and poly(arylene ether phthalazine)s

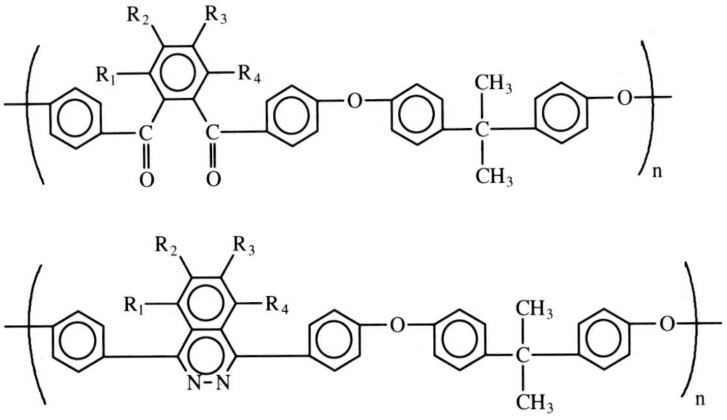

Polymer	R_1	R_2	R_3	R_4	η_{inh}, dL/g	T_g, °C	Modulus, GPa	
							25 °C	200 °C
Diketone	H	H	H	H	0.48	182	–	–
Phthalazine	H	H	H	H	0.72	235	2.74	1.52
Diketone	C_6H_5	H	H	C_6H_5	0.65	221		
Phthalazine	C_6H_5	H	H	C_6H_5	0.89	250	3.98	2.13
Diketone	C_6H_5	C_6H_5	C_6H_5	C_6H_5	0.47	265	–	–
Phthalazine	C_6H_5	C_6H_5	C_6H_5	C_6H_5	0.64	285	2.32	2.10

3.12 PAE Containing Isoquinoline Units

Poly(arylene ether isoquinoline)s in Table 20 were prepared by reacting the precursor poly(arylene ether diketone)s in Eq. (13) with benzylamine in the presence of 1,8-diazabicyclo[5.4.0]undecene (a strong organic base) in refluxing chlorobenzene [48]. Proton NMR studies indicated complete ring closure. The T_gs of the polyisoquinolines in Table 20 followed the same trend as for the polyphthalazines in Table 19, namely the T_g increased with an increase in the pendent phenyl group content. In addition, the T_g increased in going from bisphenol A to hydroquinone to 4,4'-biphenol derived polymers. Several controlled molecular weight fluoroterminated polymers were encapped with 3,5-di-*tert*-butylphenol for molecular weight determination by ^{1}H NMR [48]. A polymer prepared at a theoretical degree of polymerization of 70 gave a $\bar{M}_n$ by ^{1}H NMR of 72,800 g/mol and a degree of polymerization of 81. Using poly-

Table 20. Glass transition temperatures of poly(arylene ether isoquinoline)s

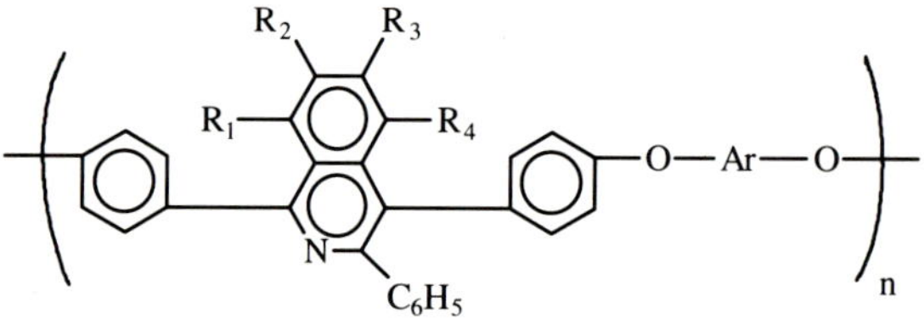

R_1	R_2	R_3	R_4	Ar	η_{inh}, dL/g	T_g, °C
H	H	H	H	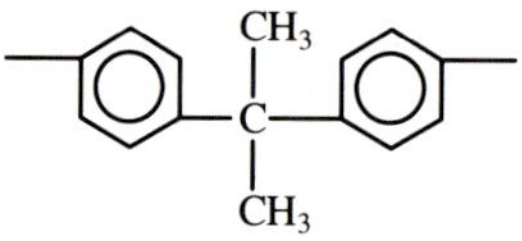	0.59	226
C_6H_5	H	H	C_6H_5	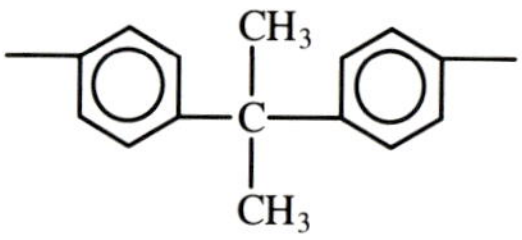	0.68	234
C_6H_5	C_6H_5	C_6H_5	C_6H_5	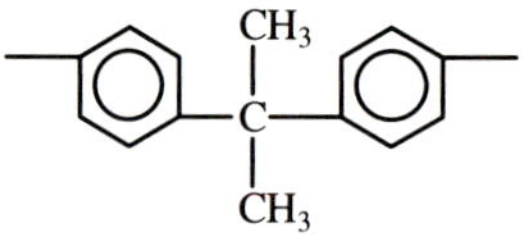	0.65	280
C_6H_5	C_6H_5	C_6H_5	C_6H_5		0.58	297
C_6H_5	C_6H_5	C_6H_5	C_6H_5	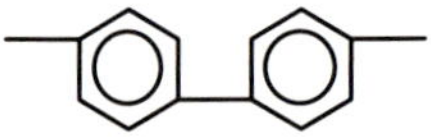	0.90	320

styrene standards, the same polymer by GPC gave $\bar{M}_n$ of 139,000 g/mol and $\bar{M}_w$ of 432,000 g/mol, respectively.

Polyisoquinolines were also prepared by the reaction of difluoroisoquinoline monomers of structure 11 with various bisphenates in N-methylcaprolactam (NMC) at 230 °C in the presence of potassium carbonate. Proton NMR studies have shown the pyridine ring to activate one fluoro group more than the other [48]. Therefore, a high boiling solvent, NMC, was selected to provide a higher reaction temperature for the polymerization. Several polyisoquinolines were prepared by this route and they have essentially the same physical properties as the corresponding polymers prepared from the precursor poly(arylene ether diketone)s.

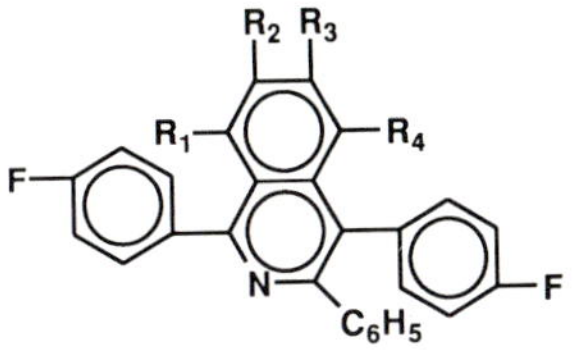

11

3.13 PAE Containing Other Heterocyclic Units

A few other PAEH have been reported. One disclosure concerned PAE containing pendent benzoxazole and benzothiazole units [49]. The activated difluoro benzazole monomers, 2-(2,6-difluorophenyl)benzoxazole or benzothiazole, were prepared from the reaction of 2,6-difluorobenzoyl chloride with 2-aminophenol or 2-aminothiophenol in polyphosphoric acid. By ^{1}H NMR, the benzazole unit was shown to have a greater electron-withdrawing effect on the 2-phenyl group than on the benzo ring [49]. Only a few polymers were reported. As an example, the polymer of structure 12 had an intrinsic viscosity of 0.45 (NMP at 30 °C) and T_g of 170 °C [49].

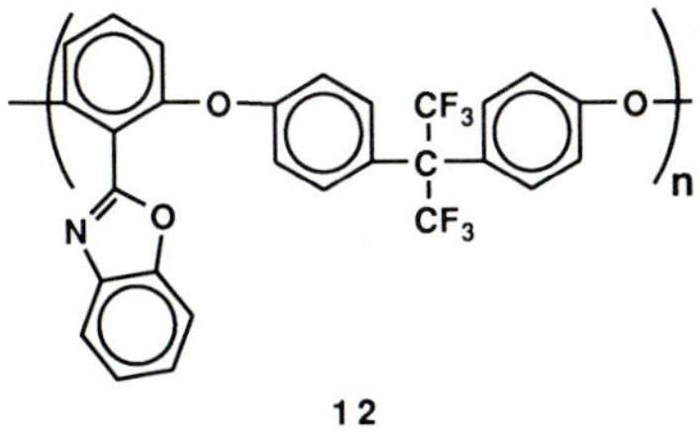

12

The AB monomer in structure 13 was polymerized in CHP in the presence of sodium hydride to yield low molecular polymer [50]. The rigid backbone apparently led to poor solubility and the polymer precipitated before high

13

molecular weight was attained. Another problem common with imide monomers in PAE synthesis is the hydrolytic instability. Care must be exercised in
maintaining a dry system during the polymerization with imide monomers. In
addition, the imide ring more than many other heterocyclic rings is prone to
attack by nucleophiles.

4 Poly(Arylene Ether)s Containing Heterocyclic Units Prepared by Electrophilic Reaction

Relatively few PAEH synthesized via electrophilic reaction have been reported.
The general procedure is described. An excess (0.11 mol) of high purity ($>99.9\%$)
anhydrous aluminum chloride is added with stirring to dry 1,2-dichloroethane
or dichloromethane (25 mL) at -30 to $-15\,°C$. A Lewis base such as DMAc
(0.04 mol) is added to suppress the reactivity of the aluminum chloride. The
dinucleophilic monomer (0.010 mol, generally a diphenoxy compound) and a
stoichiometric quantity of an aromatic diacid chloride are added to the stirred
cold aluminum chloride mixture under an inert atmosphere using dichloromethane (5 mL) as rinse. If desired, the stoichiometry of the monomers can be upset to
control the molecular weight and an end-capper such as 4-phenoxybenzophen-
one for acid chloride terminated arylene ethers can be used. No end-capper is
used for phenoxy terminated arylene ether polymers. The reaction mixture is
stirred and allowed to warm to $20\,°C$ for 4 h forming a gel-like consistency. The
gel is added to a mixture of ice and methanol in a high speed blender. The solid is
isolated, washed well in methanol/water and dried to provide a near-quantitat-
ive yield of polymer.

4.1 PAE Containing Imide Units

As shown in Eq. (14), poly(arylene ether imide)s were synthesized by the Friedel-
Crafts reaction of diphenoxydiimide monomers and iso or terephthaloyl chloride [51]. The diphenoxydiimide monomers were readily prepared by the
condensation of 4-phenoxyaniline with appropriate aromatic tetracarboxylic
dianhydrides. Several poly(arylene ether imide)s of controlled molecular weight
($\bar{M}_n = 40{,}000$ g/mol) prepared by this route are presented in Table 21. As

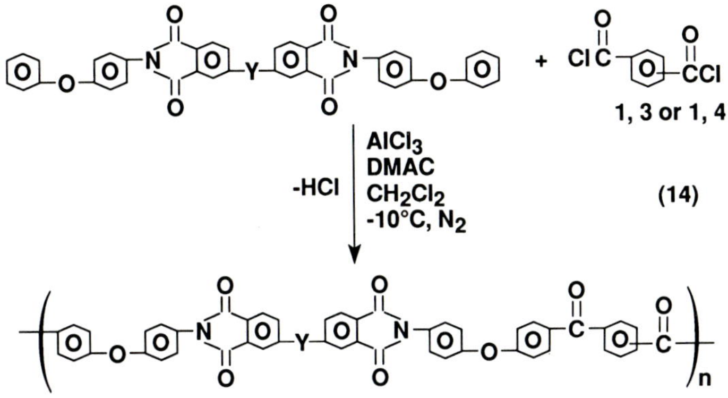

$$(14)$$

where Y is defined in Table 21

normally observed, the polymers containing the more rigid 1,4-isomer exhibited T_gs substantially higher than the corresponding polymers containing the more flexible 1,3-isomer. Some of the polymers in Table 21 have also been prepared by the more traditional route to polyimides, that is the reaction of aromatic diamines with aromatic tetracarboxylic dianhydrides. For example, the reaction of 1,3-bis(4-aminophenoxy-4'-benzoyl)benzene and 3,3',4,4'-benzophenone-tetracarboxylic dianhydride gave a polyimide with a T_g of 222 °C and a T_m of 350 °C [52] which corresponds relatively well with the third polymer in Table 21 having a T_g of 218 °C and a T_m of 358 °C. As another example, the polyimide from the reaction of 1,3-bis(4-aminophenoxy-4'-benzoyl)benzene and 4,4'-hexa-fluoroisopropylidene bis(phthalic anhydride) has a T_g of 236 °C [53] which

Table 21. Properties of poly(arylene ether imide)s

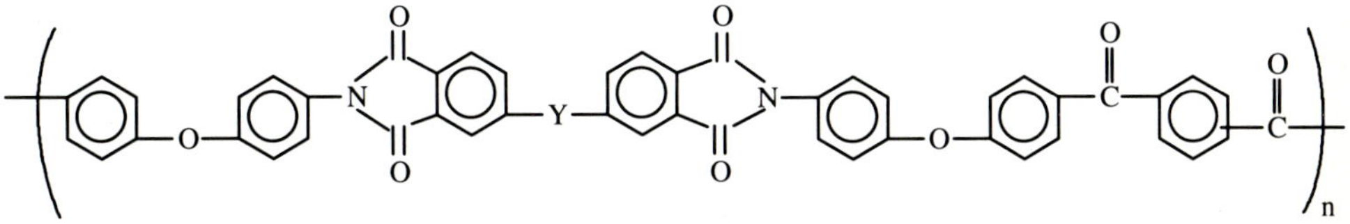

Y	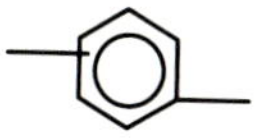 ![phenylene]	η_{inh}, dL/g	$T_g(T_m)$, °C
nil	1,3	0.85	219
	1,4	0.89	242(445)
CO	1,3	1.12	218(358)
	1,4	1.21	246(434)
O	1,4	1.39	219(421)
$-C(CF_3)_2-$	1,3	0.94	233
	1,4	1.55	252

corresponds well to the sixth polymer in Table 21 with a T_g of 233 °C. More extensive work on the properties of poly(arylene ether imide)s prepared via the electrophilic reaction using various amounts of iso and terephthaloyl chloride was reported [54].

4.2 PAE Containing Phenylquinoxaline Units

A poly(arylene ether phenylquinoxaline) of structure 14 was prepared by the aluminum chloride catalyzed reaction of 6,6'-bis[2-(4-phenoxyphenyl)-3-phenylquinoxaline] and isophthaloyl chloride in 1,2-dichloroethane [51]. The polymer had an inherent viscosity of 1.29 dL/g and a T_g of 224 °C. A polymer of the same chemical structure was prepared from the reaction of 3,3',4,4'-tetraaminobiphenyl with 1,3-bis(phenylglyoxalyl-4-phenoxy-4'-benzoyl)-benzene that gave a T_g of 239 °C [16], significantly higher than that prepared by the electrophilic route. In addition, a polymer of the same chemical structure (third polymer in Table 3) prepared via nucleophilic substitution exhibited a T_g of 240 °C.

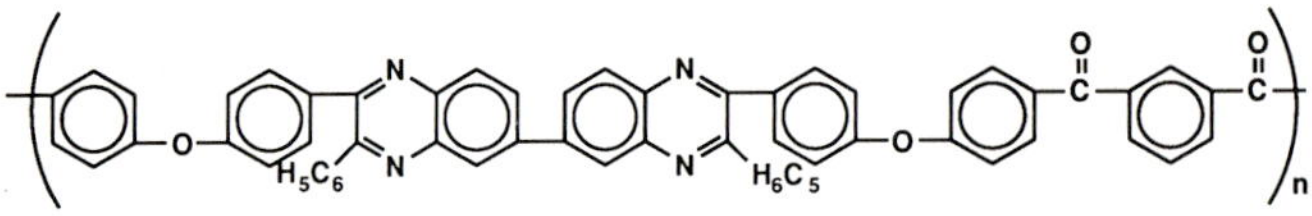

14

5 Arylene Ether/Imide Copolymers

Arylene ether/imide copolymers were prepared by the reaction of various amounts 4,4'-carbonylbis[N-(4'-hydroxyphenyl)phthalimide] and 4,4'-biphenol with a stoichiometric portion of 4,4'-dichlorodiphenyl sulfone in the presence of potassium carbonate in NMP/CHP [55]. To obtain high molecular weight polymer, the temperature of the reaction was kept below 155 °C for several hours before heating to >155 °C in an attempt to avoid undesirable side reactions such as opening of the imide ring. The imide ring is not stable to conditions of normal aromatic nucleophilic polymerizations unless extreme care is exercised to remove water. Special conditions must be used to avoid hydrolysis of the imide as previously mentioned in the section on Other PAE Containing Heterocyclic Units and as practiced in the synthesis of Ultem mentioned in the Historical Perspective section.

Arylene ether/imide block copolymers were prepared as depicted in Eq. (15) from the reaction of amine terminated arylene ether oligomers with theoretical molecular weights of 3110 and 6545 g/mol with anhydride terminated amide

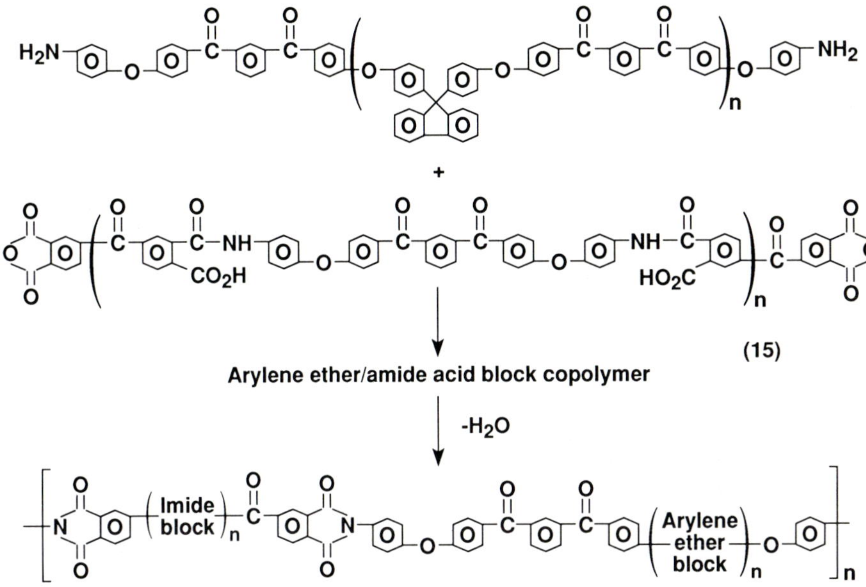

Arylene ether/amide acid block copolymer

(15)

$-H_2O$

acid oligomers with the same theoretical molecular weights [56]. When arylene ether and amide acid oligomers of 3110 g/mol were reacted to form the block copolymers, the resulting cured polymers exhibited single T_gs at a temperature between the T_gs of the individual polymers suggesting compatibility. However, when an arylene ether or an amide acid oligomer of 6545 g/mol was reacted with an oligomer of 3110 or 6545 g/mol, the resulting cured polymer exhibited two T_gs characteristic of incompatibility. Amorphous arylene ether blocks and amorphous and semi-crystalline imide blocks were incorporated within the copolymers [56]. This work was performed in an attempt to incorporate the attributes of each polymer system into one, that is, the excellent processability of the PAE and the high mechanical properties of the polyimides. In general, the mechanical properties (e.g. fracture toughness and thin film tensile properties) followed the rule of mixtures. Other arylene ether/imide copolymer work gave similar results [57, 58].

Arylene either/imide block copolymers were also prepared from the reaction of amino terminated aryl ether ketone and aryl ether ketimine oligomers with molecular weights ranging from 3700 to 12,500 g/mol with the diethyl ester diacid chloride of pyromellitic acid and 4,4′-diaminodiphenyl ether [59]. The resulting poly(amide ether ester)s were found to have distinct advantages over the corresponding poly(amide acid)s such as better stability and solubility and a higher temperature for imidization, important to develop phase separated microstructures. Clear films cast from solutions of the copolymers exhibited tensile modulus of ~ 2.2 GPa and elongation of 33 to 100% [59].

Imide block copolymers derived from PAEH have been investigated as a means of modifying self-adhesion properties of semi-rigid polyimides, e.g.

poly(4,4'-oxydiphenylene pyromellitimide), for thin-film microelectronic applications. The copolymer synthetic scheme was based on the poly(amic ester) precursor route to polyimides [60], where an amine terminated arylene ether heterocyclic oligomer was introduced in the polymerization. The poly(amic ester) approach has the advantage of allowing the use of nonpolar cosolvents to assist block compatibility during polymerization, and, in contrast to polyamic acid approaches, the copolymer could be isolated, purified by selective solvent washes and characterized spectroscopically. Imide copolymers with arylene ether phenylquinoxaline [61, 62] and arylene ether benzoxazoles [63] have been prepared and were shown to have greatly enhanced self-adhesion and reduced solvent swelling characteristics without compromising the dimensional stability or mechanical properties of the polyimide. Although certain properties of the PAEH, e.g. dimensional stability, limit their use as thin-film dielectric materials, incorporation of low levels of these structures as coblocks in semi-rigid polyimides affords materials with a combination of the desirable properties from the imide and PAEH. A similar approach has been carried out with highly fluorinated arylene ether coblocks to afford materials with enhanced dielectric constants and lower water absorption for microelectronic insulator applications [64].

6 Potential Applications

Poly(arylene ether)s containing heterocyclic units exhibit a unique combination of attractive properties that makes them potentially useful in a variety of applications. These properties include chemical resistance, thermal stability and mechanical performance. Potential applications include adhesives, composite matrices, coatings, films, moldings and membranes. Materials must exhibit an attractive combination of processability, performance and price to be widely accepted. Price is obviously volume dependent. However, in certain applications, a high cost material can still be cost-effective if it offers distinct advantage in processability and/or performance. For example, the fabrication of large structural composites such as airplane wing skins is labor intensive and therefore costly. If a higher cost composite matrix resin can offer composite processing advantages (e.g. easier layup, faster cure, less quality control, etc.) over a lower cost material then the cost of the final part may actually be significantly less using the higher cost material. In addition, a more expensive composite material that exhibits higher mechanical properties than a lower cost material can result in cost savings because thinner composites and less material can be used. Similar examples can be cited for use in other applications such as electronic/microelectronic materials. The use of an expensive material with better processability and performance can often be used to provide lower cost components than that of a less expensive material.

Research and development of PAEH is continuing. New polymers with unusual properties will undoubtedly be forthcoming. More emphasis will be directed towards the evaluation of PAEHs for specific applications. Because of the attractive properties offered by these materials, the future looks very promising. Certain members of the PAEH family are expected to become commercially available and be used in a variety of high performance applications.

7 References

1. Johnson RN, Farnham AG, Clendinning FA, Hale WF, Merriam CN (1967) J Polym Sci 5: 2375
2. Vogel HA (1970) 8: 2035
3. Ueda M, Sato MM (1987) Macromolecules 20: 2675
4. Hay AS et. al., (1959) J Am Chem Soc, 81: 6335
5. Vandort HM et al., (1968) Eur Polym J 4: 275
6. Percec V, Nava H (1988) J Polym Sci Pt A Polym Chem 26: 783
7. Colon I, Kwiatkowski GT (1990) 28: 367
8. Colquhoun HM, Dudman CC, Thomas M, O'Mahoney CA, Williams DJ (1990) J Chem Soc Chem Comm 336
9. Hergenrother PM, Kiyohara DE (1970) Macromolecules 3: 387
10. Takekoshi T, Anderson PP, US Pat 4,599,396 (1986) To General Electric Co.; Paper presented at 33rd IUPAC Intl. Sym. Macromolecules, Montreal, Canada, July, 1990 (Book of Abstracts, Session 1.5.2)
11. Takekoshi T, Wirth JG, Heath DE, Kochanowski JE, Manello JS, Webber MJ (1980) J Polym Sci Polym Chem Ed 18: 3069
12. Hedrick JL, Labadie JW (1988) Polym Matl Sci Eng Proc 59: 42: Macromolecules, 21(6): 1883
13. Hedrick JL, Labadie JW (1990) 23(6): 1561
14. Hedrick JL, Twieg R (in press)
15. Connell JW, Hergenrother PM, (1988) Polym Prepr 29(1) 172: Polymer, (1992); 33(7): 3739
16. Bass RG, Waldbauer RO Jr., Hergenrother PM (1988) 29(1): 292
17. Harris FW, Korleski JE (1989) Polym Matl Sci Eng Proc 61: 870
18. Labadie JW, Hedrick JL, Boyer SK (1992) J Polym Sci Pt A Polym Chem 30: 519
19. McKean DR, Labadie JW (1991) Polym Prepr 32(3): 195
20. Connell JW, Hergenrother PM (1989) Polym Matl Sci Eng Proc 60: 527
21. Connell JW, Hergenrother PM (1991) J Polym Sci Pt A Polym Chem Ed 29: 1667
22. Tompkins SS, Funk JG, Bowles DE, Towell TW, Connell JW (1992) Comp Matl For Optical and Electro-Optical Instruments Conf. Proc., 137
23. Roberts-McDaniel PD, Connell JW, Hergenrother PM, Poster paper presented at Symposium on Recent Advances in Polyimides and Other High Performance Polymers, sponsored by Div Polym Chem, Sparks, NV, January (1993)
24. Connell JW, Hergenrother PM (1990) High Perf Polymers 2(4): 211
25. Connell JW, Croall CI (1991) Polym Prepr (1993) 32(2): 162
26. Hilborn JG, Labadie JW, Hedrick JL (1989) Polym Matl Sci Eng Proc 60: 522
27. Hilborn JG, Labadie JW, Hedrick JL (1990) Macromolecules 23: 2854
28. Smith JG Jr, Connell JW, Hergenrother PM (1991) Polym Prepr 32(1): 646
29. Smith JG Jr, Connell JW, Hergenrother (1992) Polymer, 33(8): 1742
30. Hedrick JL (1991) Macromolecules, 24(23): 6361
31. Bass, RG, Srinivasan KR (1991) Polym Prepr 32(1): 619
32. Bass RG, Srinivasan KR, Smith JG Jr (1991) 31(2): 160
33. Hedrick JL, Twieg R (1992) Macromolecules, 25: 2021
34. Connell JW, Hergenrother PM, Wolf P (1990) Polym Matl Sci Eng Proc 63: 366
35. Connell JW, Hergenrother PM, Wolf P (1992) Polymer, 33(16): 3507

36. Carter KR, Miller RD, Hedrick JL, Macromolecules (in press)
37. Smith JG Jr, Connell JW, Hergenrother PM (1991) Polym Prepr 32(3): 193
38. Hergenrother PM, Smith JG Jr, Connell JW (1993) Polymer, 34(4): 856
39. Hergenrother PM, Smith JG Jr Connell JW (1992) Polym Prepr 33(1): 411
40. Ahn BH (1992) Ph.D. thesis, (Harris FW, advisor) University of Akron, Akron, OH
41. Smith JG, Jr., Connell JW, Hergenrother PM (1992) Polym Prepr 33(1): 1098
42. Jeong H, Iwaski K, Kakimoto M, Imai Y (1991) Polym Prepr Japan, 40(1): E360
43. Kakimoto M, Tokyo Institute of Technology, Tokyo, Japan, private communication, July 1992, to be published
44. DeSimone JM, Stompel S, Samulski ET (1991) Polym Prepr 32(2): 172
45. DeSimone JM, Sheares VV, Samulski ET (1992) 33(1): 418
46. Fukawa I, Tanabe R (1992) J Polym Sci Pt A Polym Chem 30: 1977
47. Singh R, Hay AS (1992) Macromolecules 25: 1025
48. Singh R, Hay AS (1992) 25: 1033
49. Carter KR, Jonsson H, Twieg R, Miller RD, Hedrick JL (1992) Polym Prepr 33(1): 388
50. Sundar RA, Mathias LJ (1992) 33(2): 142
51. Horner PJ, Whiteley RH (1991) J Mater Chem 1(2): 271
52. Hergenrother PM, Wakelyn NJ, Havens SJ (1987) J Polym Sci Pt A Polym Chem 25: 1093
53. Hergenrother PM, Havens SJ (1989) 27: 1161
54. Borrill CJ, Whitely RH (1991) J Mater Chem 1(4): 655
55. Johnson BC, McGrath JE (1984) Polym Prepr 25(2): 49
56. Jensen BJ, Hergenrother PM, Bass RG (1991) High Perf Polym 3(1): 3 and 13
57. Jensen BJ, Working DC (1992) Polym Prepr 33(1): 1082
58. Jensen BJ, Havens SJ (1992) 33(1): 1084
59. Hedrick JL, Volksen W, Mohanty DK (1992) J Polym Sci Pt A Polym Chem 30: 2085
60. Volksen W, Yoon DY, Hedrick JL (1992) IEEE Trans-CHMT, 15: 107
61. Hedrick JL, Labadie JW, Russell TP (1991) Macromolecules 24: 4559
62. Hedrick JL, Labadie JW (1990) High Perf Polym 2: 3
63. Hedrick JL, Hilborn J, Palmer TD, Labadie JW, Volksen W (1990) J Polym Sci Pt A Polym Chem 28: 2255
64. Labadie JW, Sanchez MI, Cheng YY Hedrick JL (1991) Symp Matls Res Soc 227: 43

Received May 1993

Condensation Polyimides: Synthesis, Solution Behavior, and Imidization Characteristics

W. Volksen

IBM Almaden Research Center, 650 Harry Road, San Jose,
CA 95120-6099, USA

Polyimides, in particular those derived from fully aromatic monomers, represent the most important class of high temperature polymers. In part, this is a direct result of their excellent combination of physical properties in addition to their outstanding thermo-oxidative stability. As a result, this class of polymers has seen relatively good success in a variety of diverse commercial applications. Although polyimides were initially intended for applications such as wire coatings and free standing films, their potential use as dielectric insulators and passivation layers in microelectronic applications has spurred a renewed interest in these materials. In light of the new data generated by these research efforts, it is the objective of this review to tie together some of the early reports with the more recent literature data concerning the synthetic aspects of condensation polyimides, their solution behavior, and imidization characteristics. Its aim is to provide an up-to-date picture of the important factors involved in polyimide preparation and leave more detailed information to be gathered by the readers from the many key references cited if they desire. This will hopefully provide the readers with a more complete as well as correct picture of the rudiments of polyimide chemistry in addition to hopefully stimulating research interests in those areas which are less completely understood.

Advances in Polymer Science, Vol. 117
© Springer-Verlag Berlin Heidelberg 1994

1 Introduction

Polyimides (Scheme 1) are condensation polymers composed of organic residues linked via cyclic imide functionalities, which are generally derived from the reaction of organic diamines and organic tetracarboxylic acids or derivatives thereof. If these structures are fully aromatic, then one obtains the high performance materials of today, which are capable of providing high thermal stability, excellent mechanical properties, high softening temperatures and good electrical properties. Historically, polyimides date back to 1908 to the probable first mention of polyimides which was in a report by Bogert and Renshaw [1]. They disclosed that 4-aminophthalic anhydride, a relatively stable compound, does not melt upon heating, but evolves water at elevated temperatures with the possible formation of a "polymolecular imide", see Scheme 2. The next report relevant to modern polyimide chemistry was by Brandt in 1958, again relating to 4-aminophthalic anhydride, in which he described the reduction of 4-nitrophthalic anhydride in dioxane solution [2]. The author reported that the freshly formed 4-aminophthalic anhydride reacts with the 4-nitrophthalic anhydride to form an amide-acid and upon completion of the reaction yields an oligomer of about 8–10 units, see Scheme 3. Although the author did not speculate on the low degree of polymerization, the reaction is presumably terminated by anhydride hydrolysis from the water formed in the reduction process. Although these early literature reports may not be the sole examples, they nevertheless represent the basis of modern day polyimide chemistry, as practiced for most aromatic polyimides.

Modern polyimides are most likely an outgrowth of trying to find increased utilization of aromatic polycarboxylic acids as prepared by oxidation of polymethylbenzenes, a major component in petroleum feedstocks. Therefore, it is not too surprising that the first commercial application of these materials appeared

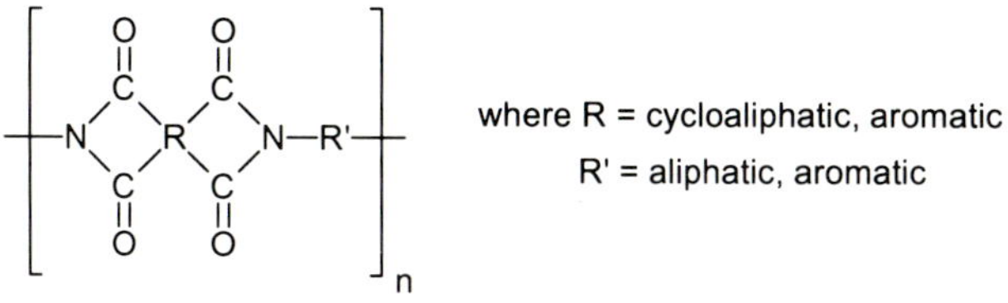

where R = cycloaliphatic, aromatic
R' = aliphatic, aromatic

Scheme 1

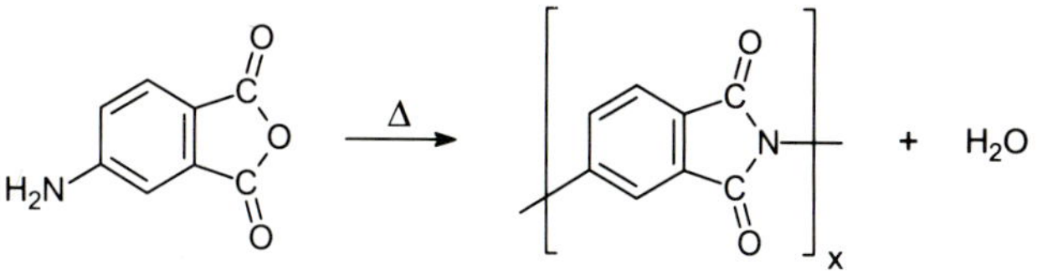

+ H_2O

Scheme 2

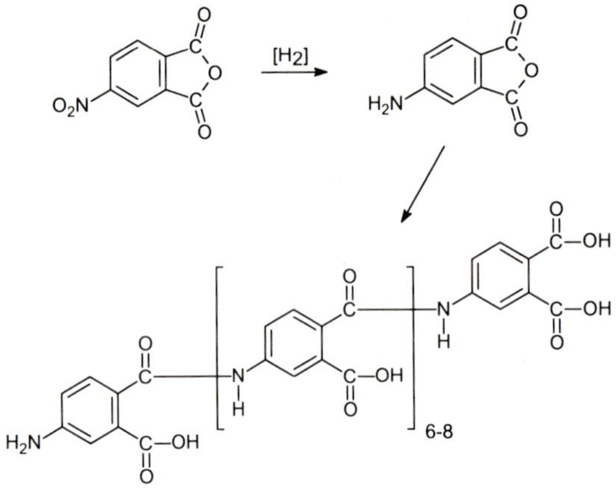

Scheme 3

in the form of aromatic aliphatic polyimides prepared in analogy to nylon chemistry utilizing preformed monomer salts [3–5]. Although these materials were typically based on pyromellitic dianhydride, and hence resulted in insoluble polyimides, these materials were capable of being melt processed if the aliphatic diamine was composed of linear carbon chain of nine or more atoms or a branched carbon chain of at least seven atoms. Extension of these early efforts to include aromatic diamines, which of course behave quite differently from the more basic aliphatic diamines, led to the classical two-stage synthesis of aromatic polyimides and the birth of polyimide chemistry [6–10]. Much of this work, which took place at the DuPont laboratories, led to the development of the Pyralin series of soluble, polyimide precursors for wire coating applications and Kapton H-film, a free standing polyimide film. Since then, polyimide chemistry has developed into an international effort with major contributions coming from Japan, Europe, and the USSR. This led to a variety of polyimide products targeted for various applications as exemplified by Monsanto's Skybond series of polyimide precursors, Rhone-Poulenc's Kerimide and Nolimid polyimides, Gulf Chemical's Thermid system, Upjohn's polyimide 2080, Hitachi's PIQ series of polyimides, and Amoco's Torlon polyamide-imide copolymers, just to name a few. Of course, many of these systems utilized different chemistries to arrive at the final polyimide and in some cases have disappeared from the market as applications and material requirements changed.

Following this initial period of polyimide development, interest reached a steady-state and remained there until the late 1970s. During this time a major impetus to the polyimide area was provided by the aerospace industry. The need for composite matrix resins as well as structural adhesives with excellent oxidative and thermal stability appeared to be at least partially met by polyimide type resins. Ultimately, requirements of high flow and low void content in relatively thick parts directed these efforts into different directions. Another upswing occurred in the early 1980s with the potential application of polyimides as dielectric insulators and passivation coatings in the semiconductor industry. In these areas, linear polyimides derived from soluble precursors were more suitable. Here, use of relatively thin coatings (approx. 2–20 μm in thickness) allows for the escape of volatiles during curing of the precursor without significant void formation. This renewed interest in polyimides is clearly reflected by the total number of polyimide related literature references starting from their conception to the present. Of the more than 20,000 references covering this time period, more than half of these are in the period from 1982 to the present day. As shown in Fig. 1, the total number of polyimide references from 1982 to 1992 steadily increase and then sharply decrease in 1992. The magnitude of this drop in polyimide references in 1992 could be caused by a combination of several factors. For one, it may simply be an artifact of the lag in data input into the computer database. On the other hand, in 1992 a definite economic downswing started to affect the chemical, aerospace and microelectronics industry. As a result, many research efforts were redirected or cut alltogether and may be reflected in this drop of polyimide literature references. In addition, the number of patents clearly outnumber the corresponding journal references indicating potential commercial significance to much of this work. Nevertheless, this renewed interest of polyimide related activities has led to a much better and more complete understanding of polyimides, including synthetic aspects and solution behavior as well as structure-property relationships.

2 Polyimides From Poly(amic acid) Precursors

2.1 Synthetic Aspects

The classical synthetic pathway to prepare polyimides consists of a two-step scheme in which the first step involves polymerization of a soluble and thus processable poly(amic acid) intermediate, followed by a second dehydration step of this prepolymer to yield the final polyimide. This preparative pathway is representative of most of the early aromatic polyimide work and remains the most practical and widely utilized method of polyimide preparation to date. As illustrated in Scheme 4, this approach is based on the reaction of a suitable diamine with a dianhydride in a polar, aprotic solvent such as dimethyl sulfoxide (DMSO), dimethylacetamide (DMAc), dimethylformamide (DMF), or N-methylpyrrolidone (NMP), generally at ambient temperature, to yield a poly(amic acid). The poly(amic acid) is then cyclized either thermally or chemically in a subsequent step to produce the desired polyimide. This second step will be discussed in more detail in the imidization characteristics section. More specifically, step 1 in the classical two-step synthesis of polyimides

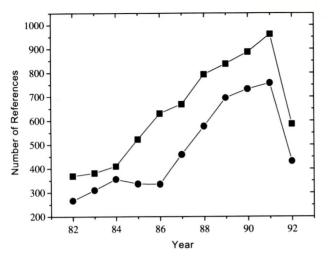

Fig. 1. Polyimide literature references over the last decade broken down into patents (■) and journal articles (●)

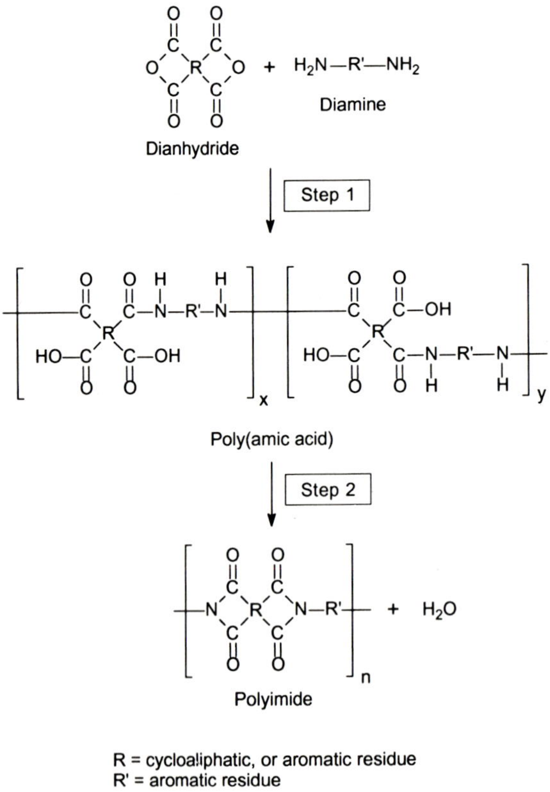

R = cycloaliphatic, or aromatic residue
R' = aromatic residue

Scheme 4

involves the hydrogen transfer polymerization of generally any aliphatic, cycloaliphatic, or aromatic dianhydride with a suitable diamine. This yields a soluble
poly(amic acid) which is composed of at least two or more structural isomers
depending on the symmetry of the starting dianhydride. The ratio of these
isomers is determined by the chemical nature of the dianhydride [11]. Unlike
the dianhydride, the nature of the diamine is of extreme importance. Thus,
highly basic diamines, e.g. aliphatic diamines, are not suited for this preparative
pathway due to their high propensity to form salts during the initial stages of the
reaction, see Scheme 5. Once sufficient dianhydride has been added to the
diamine solution, the amine groups of the excess diamine form a salt with the
generated acid groups of the amic acid linkage. The salt formation prevents
reaction of protonated amine groups with the anhydride, therefore changing the
stoichiometry of the initial monomer charge, precluding high molecular weight
formation.

Additionally, if the salt bond reacts to form an amide bond during the curing
step of the precursor, crosslinked networks can result due to diamide formation.
An elegant solution to this dilemma can be realized by utilizing silylated amines
[12]. As illustrated in Scheme 6, monosilylation of the aliphatic diamine
nitrogen atoms and subsequent reaction with the dianhydride yields a polyimide
precursor devoid of free carboxylic acid groups, a poly(amic trimethylsilyl ester)
[13]. Since no free carboxylic acid groups are involved, salt formation is
avoided. The poly(amic trimethylsilyl ester) is then readily converted to the final
polyimide by thermal or chemical imidization.

On the other hand, diamines of low basicity do not exhibit sufficient
nucleophilic character to allow them to enter into a polymer-forming reaction
with the anhydride. Ideally, the diamine should exhibit a pK_a of 4.5–6 [14].
Furthermore, depending on the exact chemical structure of the diamine, the
basicity and the reactivity of the second amine group may be significantly
influenced after reaction of the first amine-group to yield an amide linkage [14,

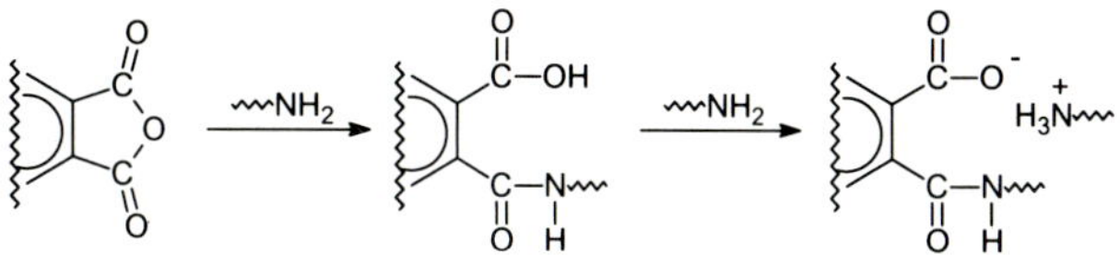

Scheme 5

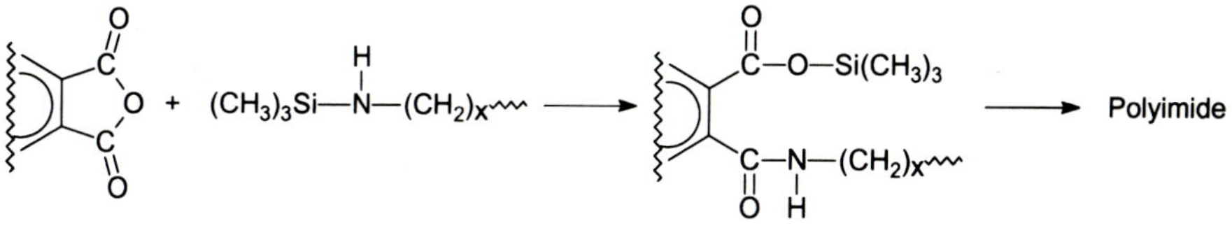

Scheme 6

15]. Keeping in mind these factors and the fact that for polymerizations of this type the absence of side reactions, monomer purity and solvent dryness are extremely critical, poly(amic acid) synthesis should be readily accomplished. As it turns out, poly(amic acid) formation is actually a very complex situation and involves some very unique features. As illustrated in Scheme 7, five additional potential reaction pathways besides the propagation reaction are possible. To

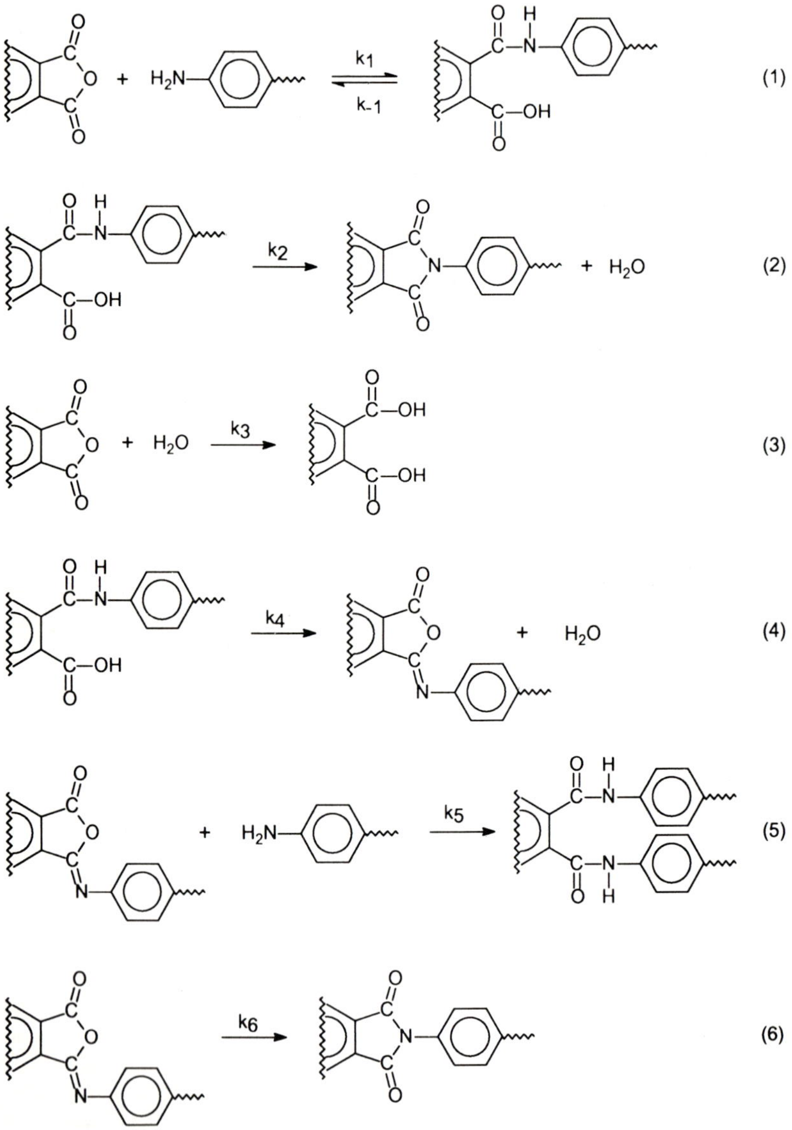

Scheme 7

complicate the situation even further, the propagation step involves an equilibrium between amic acid and starting anhydride and amine groups. The possibility of invoking an equilibrium between propagation and depolymerization was first proposed by Frost and Kesse in 1964 [16] and was based on the report by Bender and coworkers [17] that the hydrolysis of phenyl phthalamic acid involves phthalic anhydride and not the direct hydrolysis of the amide group. Many of the other possible side reactions are based on experimental observations of the polymerization and solution behavior as well as pursuant work by a number of researchers [18–20].

Compilation of the rate data extracted from various literature references [18–20], as illustrated in Table 1, suggests that propagation is the main reaction pathway. Even in the presence of water the anhydride-amide reaction is five times faster than the competing hydrolysis of anhydride by water [20]. Although this rate difference is important during the initial phase of the polymerization, during the later stages the presence of water is particularly detrimental. Thus, once the diamine concentration has reached a point where the competitive reaction with water becomes significant, hydrolysis of anhydride groups would lead to the formation of diacid moieties, which are inert toward further reaction. Of course, hydrolysis of the anhydride formed from the back reaction of amic-acid units can also occur.

Based on the rate data, it is possible to rank the relative importance of these reactions. Therefore, it follows that reactions 1 and 3 are of key importance, followed by reaction 2 and possibly reactions 4, 5 and 6. Unfortunately, little or no data exists for the latter set of reactions under the conditions employed for poly(amic acid) preparation. However, if isoimide formation were to occur to any significant degree, then reactions 5 and 6 are possible with reaction 6 being the more dominant pathway, since amine concentration due to end-groups or depolymerization would be quite low [21, 22].

Much has been written about the way in which the polymerization should be conducted. One finds references to dianhydride purification by complexing it with anisole [20, 23] or forming solid ingots of the dianhydride to insure high molecular weight formation just to name a few [24]. Although some of this

Table 1. Relative rate constants for reactions shown in Scheme 7

REACTION	RATE CONSTANT (s^{-1})
Propagation (k_1)	0.1–0.5
Depropagation (k_{-1})	10^{-5}–10^{-6}
Spontaneous Imidization (k_2)	10^{-8}–10^{-9}
Hydrolysis (k_3)	10^{-1}–10^{-2}
Isoimide Formation (k_4)	—
Diamide Formation (k_5)	—
Isomerization (k_6)	—

*Rate constants are estimated for a typical polymerization at ca. 10 wt% concentration, i.e. 0.5 M.

information was relevant and had merit at the time it was reported, the commercial availability of today's high purity aromatic dianhydrides and diamines, preclude much of the old preparative monomer work. The key to successful poly(amic acid) preparation is to utilize pure monomers and to insure absolute dryness. This is most easily achieved by sublimation of the corresponding monomers due to their inherent high thermal stability [23, 25]. Polar, aprotic solvents are generally most readily dried by vacuum distillation from phosphorous pentoxide which also aids in the removal of low levels of amine impurities present in this class of solvents. More recently, alternative solvents such as diglyme and ethyl carbitol have been utilized in cases where poly(amic acid) solubility permits their use [26]. Volksen and Cotts [27] demonstrated that monomer concentration and addition sequence had no effect on the molecular weight of the poly(amic acid) prepared from highly purified pyromellitic dianhydride (PMDA) and p, p'-oxydianiline (ODA) in freshly distilled NMP with the exclusion of atmospheric moisture, see Table 2.

2.2 Solution Behavior

The solution behavior of poly(amic acids) was until recently, probably the least understood aspect of the soluble polyimide precursor. However, the advent of sophisticated laser light scattering and size exclusion chromatography instrumentation has allowed elucidation of the solution behavior of poly(amic acids). In the early days of polyimide chemistry, when most molecular weight characterization was based on viscosity determinations, a decrease in viscosity was associated with molecular weight degradation [15, 28, 29]. Upon combination of the two monomers an increase in the viscosity to the stoichiometric equivalence point is observed, followed by a decrease in the solution viscosity as a

Table 2. Effect of synthetic variables on the molecular weight of PMDA-ODA based poly(amic acid)

PMDA/ODA	%Solids	$\bar{M}_w$ (calculated)	$\bar{M}_w$ (exp.)
0.8182	9.27	11,000	10,000
0.9804	0.31	42,000	37,000
0.9921	7.55	115,000	80,000
Addition of Solid PMDA to ODA Solution			
0.9259	9.55	10,800	9,000
0.9804	9.31	42,000	37,000
Addition of Solid ODA to PMDA Solution			
0.9259	9.58	10,000	10,500
0.9804	9.32	42,000	35,000
Additional of Solid PMDA to ODA Solution			
0.9725	18.6	30,000	29,000

function of time. Although the initial increase in solution viscosity is quite expected, the subsequent decrease in solution viscosity was generally interpreted as hydrolytic molecular weight degradation. The propensity toward molecular weight degradation, i.e. decrease in the solution viscosity, was most pronounced for polymerizations which were conducted at higher monomer concentrations, about 20 weight%. Careful examination of such systems by light scattering techniques of the resulting poly(amic acid) solutions, led to the discovery that the initially high viscosities reflected weight average molecular weights much higher than predicted by the monomer stoichiometry. These high molecular weights then decreased as a function of time and temperature to a molecular weight in accordance with the molecular weight as calculated by the Carother's equation, see Fig. 2 [30]. Additionally, it was discovered that the molecular weight distribution was skewed toward high molecular weights, with the weight average molecular weight decreasing over time and approaching a most probable molecular weight distribution [31]. The number average molecular weight remained constant throughout this time. Clearly, this re-equilibration of the molecular weight distribution can only be attributed to the presence of the monomer-polymer equilibrium, allowing for fragmentation and recombination of polymer chains to attain the theoretical molecular weight distribution. At this point one question remains. What are the underlying factors which cause this tremendous shift in the initial molecular weight distribution? The answer can be found in part by observation of polymerizations utilizing the most notorious of these monomers, namely pyromellitic dianhydride, employing high concentrations and anhydride particles of varying sizes. Whereas finely divided PMDA particles appear to dissolve quite readily, progressively larger particles lead to a polymerization characterized by PMDA particles coated by a gel-like layer, as

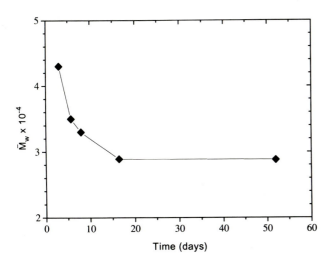

Fig. 2. Re-equilibration of the weight average molecular weight of PMDA/ODA based poly(amic acid) as a function of time in NMP at ambient temperature

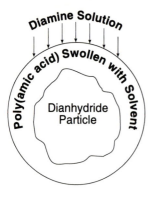

Fig. 3. Idealized scheme of the dissolution behavior of certain dianhydrides

illustrated in Fig. 3. Over time, this layer then increases in thickness while the solid dianhydride particle gradually decreases in size. Finally, the gel-particles disappear completely as the overall solution viscosity increases. These observations are in good agreement with the behavior reported by Zharkashchikov, who reports on the production of poly(amic acid) with a large molecular weight in the diffusion range, starting immediately at the surface of the dissolving particle and closely surrounding it [32]. This type of behavior where the monomer reaction is taking place at the solid-liquid interface with concomitant formation of extremely high molecular weight is reminiscent of interfacial type polymerizations. Similar observations have been reported for the preparation of poly(phenylquinoxalines) derived from diaminobenzidine (DAB), which display limited solubility in the polymerization medium and lead to molecular weight distributions skewed toward high molecular weights [33]. The resulting polymer, of course, has no means to re-equilibrate as observed for poly(amic acids). Another feature which may heavily contribute toward the interfacial behavior of certain aromatic poly(amic acid) polymerizations is the tendency of certain dianhydrides like PMDA to form 1:1 complexes with the diamine. Of course, PMDA, an electron-poor molecule, has been reported to form charge-transfer complexes with electron-rich molecules including amines [20, 34]. Such highly balanced complexes would certainly promote high molecular weight polymer formation.

2.3 Imidization Characteristics

2.3.1 Thermal Imidization

The second step of the classical polyimide synthesis, cyclodehydration or imidization, to yield the final polyimide is of equal importance to the polymer forming reaction, since ultimately the application lies in the final polyimide and its inherent properties. The final imidization step can be achieved via two different pathways. The first pathway involves the direct thermal cyclodehydration of the initial poly(amic acid) by heating the system to approximately 300 °C. The second pathway involves treating the poly(amic acid) with a chemical dehydration agent, as exemplified by a mixture of acetic anhydride/pyridine, followed by a thermal treatment to complete the imidization reaction and drive off the solvent. The straight thermal cure of poly(amic acids) represents the most practical approach to imidization, especially for commercial applications. Industrially, thermal processes are most cost-effective and can be readily instituted and controlled with high precision. Examination of a typical poly(amic acid)/NMP cure profile from ambient temperatures to 350 °C and following the evolution of volatiles as evidenced by Mass Spectrometry [35] yields the following qualitative information, see Fig. 4:

1) In the temperature regime from ambient temperature to approximately 150 °C mostly solvent is evolved.
2) In the region from 150 °C to 250 °C both solvent evolution and maximum evolution of water from imidization takes place.
3) From 250 °C and up solvent removal is virtually complete and the final imidization is completed.

Since solvent evaporation and imidization in themselves are not destructive processes, the most crucial temperature regime lies between 150 °C and 250 °C. Here solvent removal and maximum imidization occurs simultaneously causing tremendous shrinkage and the creation of maximum stress in the polymer film. At this point it is not unusual to observe cracking problems in the polymer film, depending on the inherent mechanical properties of the partially cured poly-

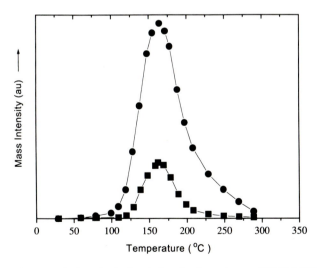

Fig. 4. Evolution of volatiles during the thermal cure of PMDA/ODA poly(amic acid). (●) Mass 18-water and (■) mass 99-NMP. Sample was predried at 90 °C for 1 hour

imide and the apparent molecular weight of the polymer. Furthermore, it is important to keep in mind that not only does reaction in this temperature range involve the maximum evolution of water, but the dynamics of the depolymerization reaction discussed earlier are also affected in such a way that the probability for hydrolysis and concomitant molecular weight breakdown is also increased. Such behavior has been clearly demonstrated for both soluble polyimides and for insoluble systems [26, 36, 37]. In all cases, the appearance and disappearance of anhydride groups as a function of temperature can be observed spectroscopically. Upon further heating and removal of water the polymer recombines to yield the expected molecular weight. Translated to insoluble polyimide systems, this means that similar behavior is to be expected. Thus, polyimides, such as poly(p, p'-oxydiphenylene pyromellitimide), PMDA/ODA, and poly(p-phenylene biphenyltetracarboximide), BPDA/PDA, which are insoluble and exhibit high glass transition temperatures would be expected to be particularly prone to molecular weight breakdown during thermal curing. Once the solvent evaporates and imidization starts to limit the mobility of the initial poly(amic acid), the system vitrifies. Further imidization pushes the glass transition temperature to its maximum value and prevents the final polyimide from regaining sufficient mobility to allow recombination reactions without significant thermal decomposition. Therefore, the advisable curing profile involves slow heating to just below 150 °C to drive off the majority of the solvent followed by a relatively rapid temperature ramp through the critical region to above 250 °C. The issue of hydrolytic molecular weight degradation during thermal curing is still somewhat controversial, although it has been demonstrated that for moderate poly(amic acid) molecular weights, the initial precursor molecular weight is retained in the final polyimide, see Table 3 [38].

Coupling these findings with the corresponding molecular weight dependence of the mechanical properties for the identical polyimide provides extremely important information necessary to optimize poly(amic acid) formulation as was recognized very early in the development of polyimides [20]. In the case of PMDA/ODA based polyimides, the elongation undergoes a rapid increase at weight average molecular weights of approximately 8000 and reaches a limiting value of about 60% for molecular weights in excess of 20000, see Fig. 5. It is

Table 3. Molecular weight data of PMDA/ODA based poly(amic acids) and their corresponding polyimides

$\bar{M}_w$ of PAA (daltons × 10³)	Cure Conditions	Age of Solution (days)	$\bar{M}_w$ of PI[a] (daltons × 10³)
10.4	300 °C	4	11
28.0	300 °C	1	22
37.0	300 °C	1	30

[a]obtained in conc. sulfuric acid

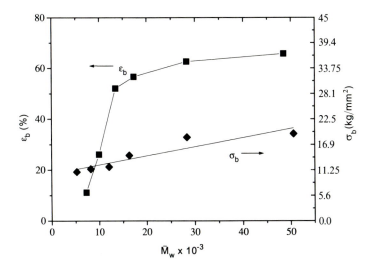

Fig. 5. Polyimide mechanical properties as a function of the weight average molecular weight of the initial poly(amic acid)

important to keep in mind that molecular weight values, here, are for the initial poly(amic acid) and are not necessarily identical to the values of the final polyimide. This molecular weight discrepancy between poly(amic acid) and polyimide becomes more pronounced for the higher molecular weight poly(amic acids). However, molecular weight ranges of the final polyimide of 20000 to 30000 are readily attainable by thermal imidization of higher molecular weight poly(amic acids). The sharp decrease in the strain-at-break at molecular weights below 10000 establishes this molecular weight as the lower limit above which excellent mechanical properties can be realized for PMDA/ODA based polyimides.

2.3.2 Chemical Imidization

The use of chemical dehydrating agents to promote ring closure reactions of maleamic and phthalamic acids is well documented [40–43]. These schemes involve reagents such as acetyl chloride, phosphorus oxychloride, acetic anhydride, N,N-dicyclohexylcarbodiimide and trifluoroacetic anhydride/triethylamine, just to name the more representative examples. With respect to poly(amic acids), the reagent most widely studied is an equimolar mixture of pyridine/acetic anhydride. In examining the conversion of benzophenone tetracarboxylic dianhydride/9, 9-fluorenedianiline based poly(amic acid) to the corresponding soluble polyimide, it was found that the cyclizing reagent is most effective when employing 4 moles per repeat unit of the poly(amic acid). Increasing the temperature from 20 °C to 100 °C decreased the reaction time from 15 h to 2 h to achieve complete imidization [41]. Spectroscopic investigations indicated that in addition to imide formation, significant amounts of isoimide linkages were also produced [44]. However, if triethylamine was used instead of pyridine, then isoimide formation was practically eliminated in addition to providing significantly faster reaction rates [45]. Based on these observations the following reaction scheme, Scheme 8, has been postulated: Amic acid linkages react with acetic anhydride to form a mixed anhydride, a reaction which is promoted by the presence of bases, such as pyridine or triethylamine. The mixed anhydride intermediate can further tautomerize from the amide to the iminol form. The amide tautomer cyclizes to the imide (pathway A), the thermodynamically favored product, whereas the iminol tautomer yields the kinetically favored isoimide (pathway B). It is also possible for the isoimide to isomerize to the more stable imide form (pathway C). Although isoimides are known to thermally isomerize to the imide, in this case, isomerization occurs via the back reaction. This back reaction is apparently initiated by nucleophilic attack of acetate ion on the isoimide [46]. Such behavior would be consistent with the fact that stronger amines, such as triethylamine, promote acetate formation and thus increase the back reaction leading to exclusive imide formation [45].

For commercial polyimides this type of imidization scheme is less attractive due to additional expense and process complexity. Nevertheless, this approach does offer advantages over the thermal imidization route by eliminating the potential of hydrolytic molecular weight degradation during cure. In this respect it is not surprising that superior mechanical properties can be realized as reported for chemically imidized PMDA/ODA based polyimide [47]. Observed elongation-at-break values which are typically three times larger than those obtainable for thermally cured specimens may provide a potential pathway to toughen otherwise brittle polyimides.

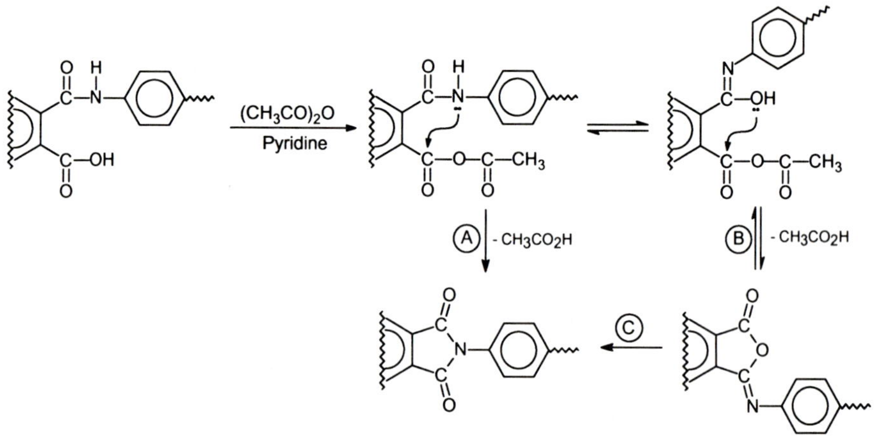

Scheme 8

3 Polyimides Via Derivatized Poly(amic acid) Precursors

3.1 Synthetic Aspects

Poly(amic acids) in which the ortho-carboxylic group has been chemically modified to either an ester- or amide moiety have been known for many years. However, their commercial significance was non-existent until very recent applications involving dielectric insulators [48] and photosensitive polyimide precursors [49, 50]. As with many synthetic pathways, there are generally several ways to arrive at the same goal. Similarly, the preparation of derivatized poly(amic acids) can be divided into two general categories:

1) Formation of the poly(amic acid) followed by derivatization of the *ortho*-carboxylic acid groups along the polymer backbone.
2) Derivatization of the monomer and subsequent activation to allow the monomer to enter a polymer forming reaction to yield the desired polymer.

In retrospect, the postulated mechanism for amic acid back reaction to anhydride and amine as the main pathway to explain hydrolytic instability of the poly(amic acid) system may have prompted the search for more stable systems in the form of derivatized poly(amic acids). Realizing that if proton transfer in the internal acid catalyzed formation of the intermediate illustrated in Scheme 9 (reaction 1) can be prevented, then the potential for the amic acid back reaction might be eliminated [51]. This, of course, can be accomplished in

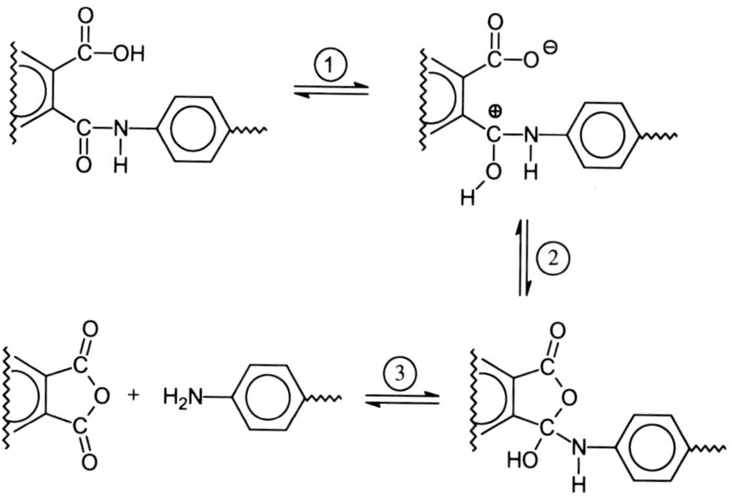

Scheme 9

several ways. The simplest approach is to quaternize the pendant carboxylic acid group with a tertiary amine and thus prevent proton transfer. The second approach is more complex and involves conversion of the free carboxylic acid group to either an amide or ester moiety. These two approaches represent the essential elements of all poly(amic acid) derivatization schemes.

3.1.1 Derivatization of Preformed Poly(amic acids)

3.1.1.1 Salt Formation

As mentioned before, the simplest way to eliminate the proton transfer step is to neutralize the pendent carboxylic acid groups with a base, such as a tertiary amine, to form a polyelectrolyte, see Scheme 10. In this case secondary and tertiary amines are preferable, since they cannot enter into a ring closure reaction, as shown in Scheme 11 (reaction 1). Primary amines, on the other hand, could enter competitive transamidation reactions during thermal curing and, if successful, cleave the polymer backbone, see Scheme 10 (reaction 2). The competitive ring closure reaction is highly probable because the aliphatic amine-nitrogen is a better nucleophile. Of course, at elevated temperatures organic salts may decompose into the individual components before amide formation

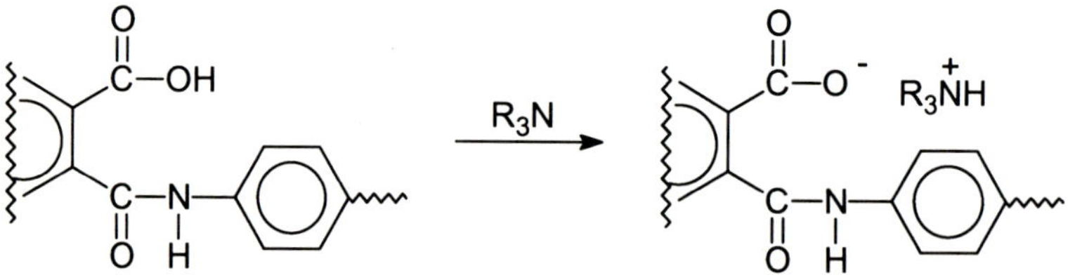

Scheme 10

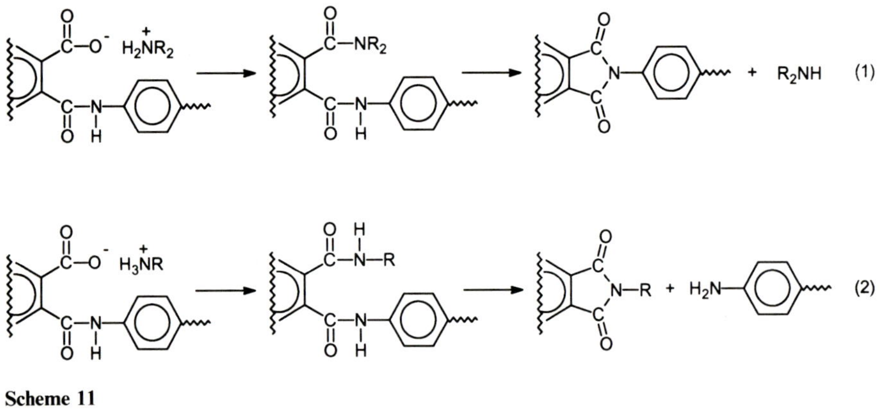

Scheme 11

takes place. In that case, the volatility of the aliphatic amine precludes amide formation and the potential for competitive ring closure. In essence, the neutralization reaction represents one of the earliest forms of poly(amic acid) derivatization [51, 52]. When 130% of the theoretical amount of a base (based on titrated free carboxylic acid groups), such as triethylamine, had been added, very stable solutions could be obtained. As a matter of fact, if the poly(amic acid) was precipitated prior to neutralization, it was readily formulated in aqueous amine solutions. However, due to the ionic nature of the polymer, the resulting solution viscosity was so high that only relatively low solids content solutions were practical. Of course, this greatly limited the utility of this approach for coating applications. Nevertheless, it has been successfully utilized in preparing photosensitive polyimide formulations in which the polyimide precursor is a poly(amic acid) quaternized with suitable unsaturated aliphatic amines [50]. Various photocrosslinkable, unsaturated amines can be easily incorporated in this fashion. Additional advantages are offered by the fact that this scheme can apply to any poly(amic acid) backbone, greatly simplifying synthetic efforts.

3.1.1.2 Reaction with Activated Halides

The preparation of esters via reaction of alkyl halides with acid salts can be useful in cases involving fairly active alkyl halides, such as benzyl or allyl halides [53]. Although these salts are generally sodium or silver salts, substituted ammonium salts are also effective [54]. A similar approach has been reported in the case of poly(amic acids) [55], as shown in Scheme 12. Thus, neutralizing a PMDA/ODA based poly(amic acid) with NaH or triethylamine and reacting the resulting polymeric salts with an excess of activated halogen compound yielded an esterified poly(amic acid). Various halogen compounds including allyl, benzyl, and phenacyl bromide, were investigated under various conditions with respect to their ability to yield the desired esterified polymer. This approach also has the potential for imide formation during the latter stages of esterification due to the presence of relatively large amounts of base and the fact that activated

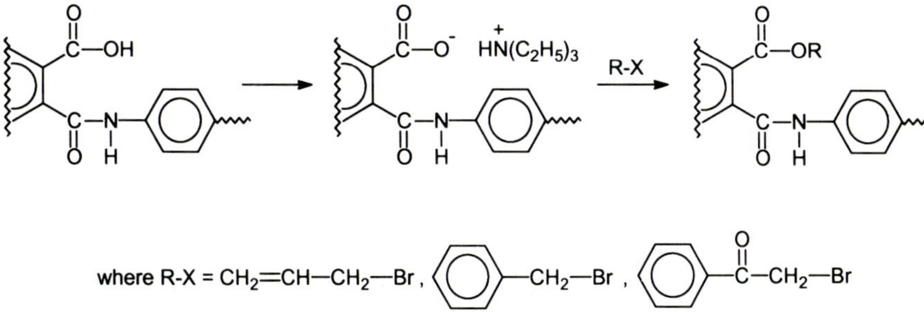

where R-X = CH_2=CH—CH_2—Br , ⟨○⟩—CH_2—Br , ⟨○⟩—$\overset{O}{\overset{\|}{C}}$—$CH_2$—Br

Scheme 12

esters are present. This base sensivity of poly(amic alkyl esters) is discussed in more detail in Sect. 3.3.2 dealing with chemical imidization.

3.1.1.3 Reactions with Poly(isoimides)

Derivatization of poly(amic acids) by methods other than neutralization and the use of the carboxylate groups as a nucleophile require transformation of this precursor into a more reactive state. This activation of the amic acid unit is realized by conversion into the corresponding isoimide derivative. Poly (isoimides) are most likely an outgrowth of the experimental work involving the chemical imidization of poly(amic acids) by anhydride/tertiary amine dehydrating mixtures. Taking advantage of much of the early work on isoimide model compounds [21, 22, 42], the use of halogenated aliphatic acid anhydrides, such as trifluoroacetic anhydride, in conjunction with triethylamine or N,N-dicyclohexylcarbodiimide (DCC) by itself were reported to form poly(isoimides) from the corresponding poly(amic acids) in high yield [56], as illustrated in Scheme 13. These poly(isoimides) are really converted to derivatized poly(amic acids) by reaction with alcohols or amines [57, 58], see Scheme 14. The conversion occurs under various conditions, but generally involves reaction of the poly(isoimide) with an excess of the derivatizing agent, i.e. alcohol or amine. The aminolysis of isomide occurs much faster than the corresponding alcoholysis reaction. However, amines with sufficient nucleophilic character can also cause the isomerization of isoimide to imide [59]. For this reason, aminolysis is much more likely to yield significant amounts of imide linkages in addition to the desired amide-amide linkages. It should also be noted that the

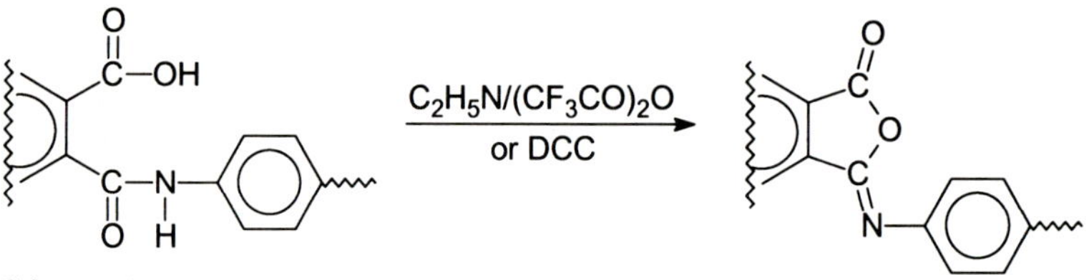

Scheme 13

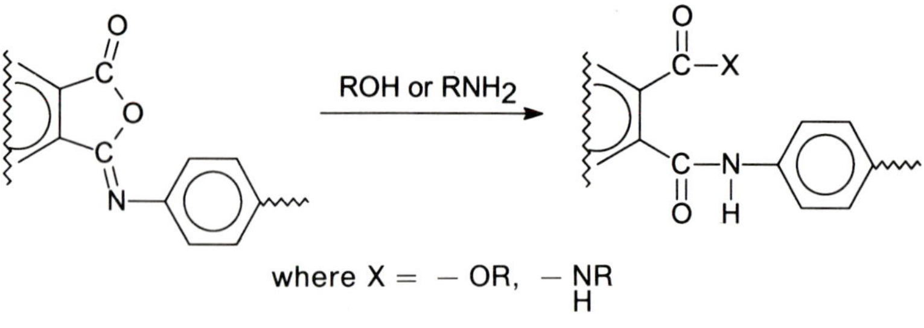

where X = − OR, − NR
                           H

Scheme 14

esterified polymer has the potential for imidization as discussed in Sect. 3.3.2. One further complication associated with this approach is in cases were the polyisoimide does not exhibit sufficient solubility to allow for aminolysis of alcoholysis in solution. In that situation, poly(amic acid) coatings have to be consolidated at intermediate temperatures, converted to the corresponding poly(isoimide) and are then immersed either in the neat derivatizing agent or a solution thereof.

3.1.2 Polyimide Precursors from Derivatized Monomers

Whenever polymers need to be derivatized, it is very difficult to obtain theoretical degrees of conversion. This difficulty is amplified as the molecular weight of the polymer increases. Additionally, if side-reactions should occur, their product may be incorporated into the polymer structure. For these reasons, it is generally accepted that modifications of the monomer prior to polymerization are preferable.

3.1.2.1 Monomer Preparation

The intial step in this synthetic approach involves derivatization of the monomer as illustrated in Scheme 15. Amidation reactions, i.e. $X = -NHR$ or $-NR_2$, proceed quite readily with most organic amines by utilizing 2 equivalents of amine to every equivalent of anhydride in an inert, low-boiling solvent such as tetrahydrofuran or ethyl acetate [60]. Of course, secondary amines are highly desirable based on arguments already elucidated for poly(amic acid) salts. The resulting salt is then neutralized with aqueous acid to yield the desired diamide diacid. In the case of alkyl esters, $X = -OR$, the dianhydride is simply refluxed in an excess of the alcohol for several hours yielding the desired diester diacid after evaporation of the excess alcohol [61]. This esterification works for all "normal" alcohols, i.e. alcohols with acidities less than methanol, and fails with more acidic alcohols such as propargyl alcohol or trifluoroethanol. The

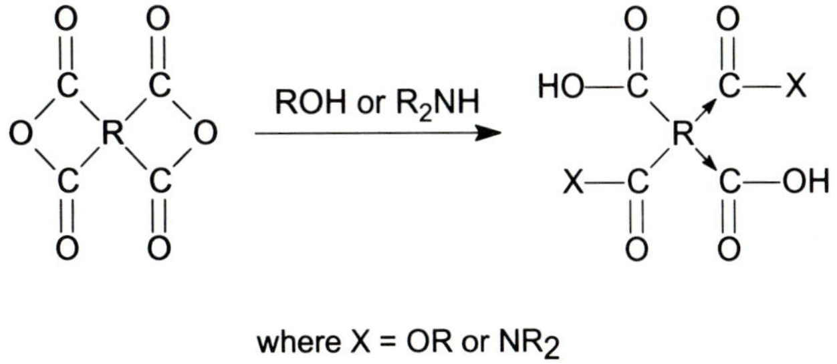

where X = OR or NR$_2$

Scheme 15

latter alcohols form so called "activated" esters because they have a propensity to revert back to the starting anhydride at elevated temperatures [62]. In this case, a modified esterification can be employed which utilizes only two equivalents of the acidic alcohol and two equivalents of a tertiary amine for every mole of dianhydride, see Scheme 16. The reaction proceeds readily at ambient temperature over a period of about 12 hours in chlorinated solvents or tetrahydrofuran. Completion of the reaction is indicated by a homogeneous solution. The solvent is then evaporated and the resulting mass is taken up in water and acidified to yield the desired diester diacid. The latter esterification scheme applies to all alcohols except tertiary-butanol, which requires even more aggressive conditions, i.e. potassium *tert*-butoxide as the base instead of a tertiary amine [63].

In addition, this esterification scheme has the distinct advantage of only requiring two equivalents of the alcohol, which greatly simplifies matters in the case of very expensive alcohols, i.e. deuterated systems, or high-boiling alcohols that would otherwise be difficult to remove.

One additional advantage provided by this approach lies in the fact that it is possible to separate structural isomers at this stage. The isomer separation permits the subsequent preparation of polymers composed of only one structural isomer. This isomer separation is readily achieved in cases of highly symmetrical molecules, such as pyromellitic dianhydride (PMDA), where only two structural isomers are possible. In the case of PMDA derived alkyl diesters, the isomer in which the carboxylic acid groups are *para*-catenated represents the less soluble isomer and generally crystallizes from a concentrated solution of the isomer mixture in the alcohol utilized in the esterification. Therefore, it is not surprising that much of the early work directed at the preparation of poly(amic alkyl esters) utilized this particular isomer, sometimes referred to as 2,5-dicarboalkoxyterephthalic acid. The corresponding isomer in which the carboxylic acid groups are *meta*-catenated, 4,6-dicarboalkoxyisophthalic acid, received no attention until very recently [64]. It turns out that the *meta*-isomer can be very readily separated, at least in the case of PMDA methyl- and ethyl esters by simple solvent extraction of the initial isomer mixture, see Table 4. Solvent extraction utilizing ethyl acetate leaves the *para*-isomer virtually insoluble while extracting the *meta*-isomer. The *meta*-isomer is then obtained in 80–90% isomeric purity by filtering off the insoluble material, representing mostly *para*-isomer, and evaporation of the ethyl acetate. Further improvement

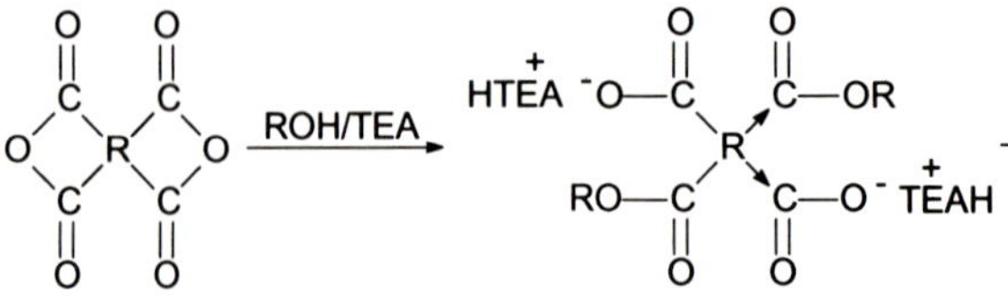

Scheme 16

Table 4. Ethyl acetate extraction data for diethyl dihydrogen pyromellitate isomer mixture

Diester Conc. (g/ml)	Soluble Fraction (wt%)	meta/para	Insoluble Fraction (wt%)
0.15	56.4	89/11	43.6
0.20	53.5	90/10	46.5
0.25	51.6	87/13	48.4
0.30	52.6	91/9	47.4
0.40	50.0	95/5	50.0
0.50	32.8*		67.2

* Crystallization occured

to isomeric purities of 99% can be achieved by further recrystallization from butyl acetate.

3.1.2.2 Polymer Preparation

Following the dianhydride derivatization step is the activation of the acid groups to enable the monomer to enter a polymer forming reaction, see Scheme 17. This activation of the dianhydride derivative can be promoted by several well-known activation processes. The most widely utilized method to accomplish this activation process is by conversion of the diacid derivative into the corresponding diacyl chloride, X = Cl[60, 65, 66, 67]. Alternatively, especially in situations where the acidic nature of the acyl chloride formation cannot be tolerated, reagents much milder in nature and providing neutral conditions are more appropriate. Among these candidates the Ueda-reagent [63, 68], N,N'-(phenylphosphino)bis[2-(3H)-benzothiazolone, and carbonyl dimidazole [69]

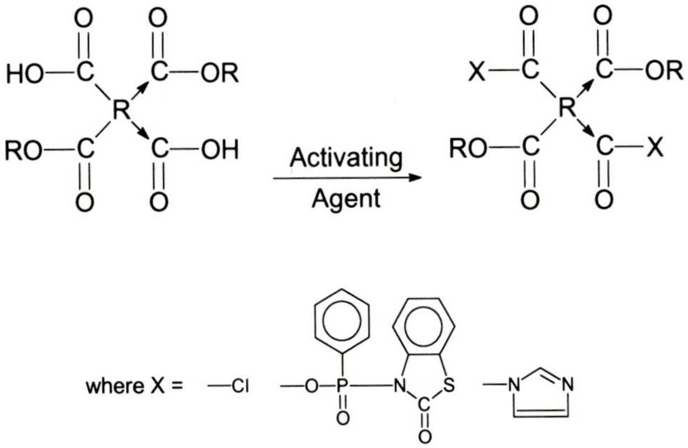

Scheme 17

have been reported as effective in sufficiently activating PMDA diester diacids to enable them to enter a polymer forming reaction with aromatic diamines under appropriate conditions. The Ueda reagent has the advantage of allowing monomer activation and polymerization in a one-pot procedure, whereas acyl chloride and azolide formation generally require some type of work-up to remove the undesirable side products, such as excess chlorinating agent and imidazole, respectively.

Once the derivatized dianhydride has been properly activated it can be polymerized with a suitable diamine as shown in Scheme 18. As already mentioned, the preferred polymerization approach is via the corresponding diacyl chlorides. In essence, at this stage the polymerization resembles the preparation of aromatic polyamide, e.g. poly(p-phenylene terephthalamide). Although polyamides can be prepared in a number of different ways as described by Morgan [70], low temperature solution polycondensation in a polar, aprotic solvent analogous to those employed for the preparation of poly(amic acids) appears most feasible. Typical the diamine is dissolved in the preferred polymerization medium, i.e. NMP or DMAC, and the diacyl chloride is then added either as a solid or dissolved in a suitable solvent. Preferred diacyl chloride solvents are tetrahydrofuran or ethyl acetate, since they can be readily obtained in a relatively dry state and they are miscible with water or methanol/water mixtures during the precipitation step. A tertiary amine is generally added as an acid scavenger to minimize protonation of the diamine and thus drive the polymerization to a maximum molecular weight. This practice, however, as mentioned earlier, can lead to premature imidization. As shown in Fig. 6, the intrinsic viscosities of PMDA/ODA based poly(amic ethyl ester) prepared in NMP with or without pyridine are virtually identical [71]. Apparently, the basic nature of the polymerization solvent is sufficient to establish an equilibrium between free and protonated amine groups to allow the polymerization to proceed to yield high molecular weight polymer. At least under these conditions the use of an acid scavenger is highly questionable. The temperature of the polymerization is preferably maintained at $0-10\,^\circ$C during this phase of the polymerization. Once monomer addition is complete, the polymerization is allowed to proceed for at least an additional 3 h, although overnight runs are quite acceptable, while being allowed to return to ambient temperature. The polyimide precursor is then precipitated in a non-solvent, such as water,

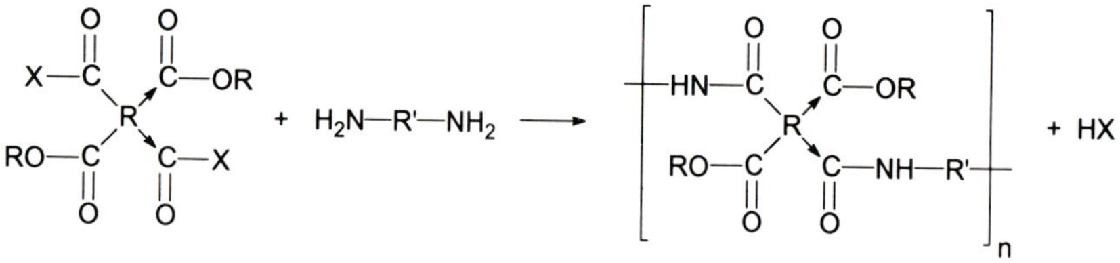

Scheme 18

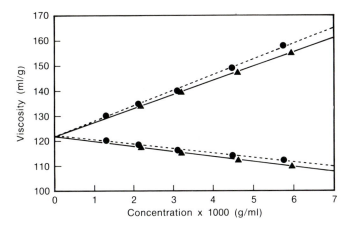

Fig. 6. Solution viscosity of *para*-PMDA/ODA poly(amic ethyl ester) in *N*-methylpyrrolidone at 27 °C. (●) with pyridine (▲) without pyridine

methanol, or mixtures of the two, filtered and vacuum dried. Since the material is in a solid form, shelf-life is virtually unlimited.

An additional modification in the above synthetic scheme is possible by introducing the aromatic diamine in the form of its trimethylsilyl derivative [72]. Monotrimethylsilyl-substituted amines are readily prepared from the free amine with hexamethyldisilazane or trimethylsilyl chloride in the presence of a tertiary amine [73, 74] whereas bis(trimethylsilyl)-substituted amines require more aggressive reagents, such as butyllithium in conjunction with trimethylsilyl chloride [75]. As illustrated in Scheme 19, monotrimethylsilyl-substituted amines react with acyl chlorides to form the corresponding amides and liberate trimethylsilyl chloride. Monotrimethylsilyl-substituted amines are reported to display increased reactivity with acyl chlorides [76]. This is of great synthetic importance since the increased reactivity allows for reaction with low basicity amines. Bis(trimethylsilyl)-substituted amines, on the other hand, react with acyl chlorides to form the corresponding *N*-trimethylsilyl amides, see Scheme 20. The *N*-trimethylsilyl amides are much more soluble in common organic solvents. However, they are hydrolytically unstable and readily convert back to the free amides.

Among the alternative polymerization schemes, the approach involving acyl imidazolides is quite unique. Although, imidazolides are generally sufficiently

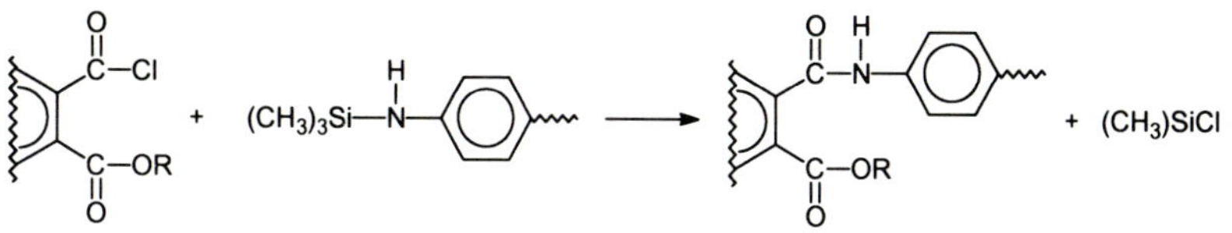

Scheme 19

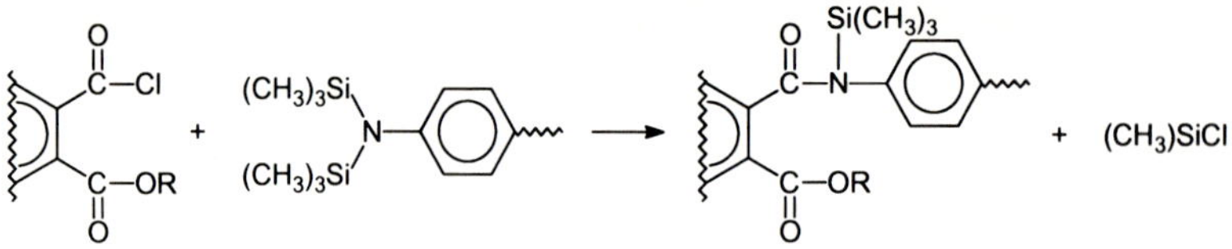

Scheme 20

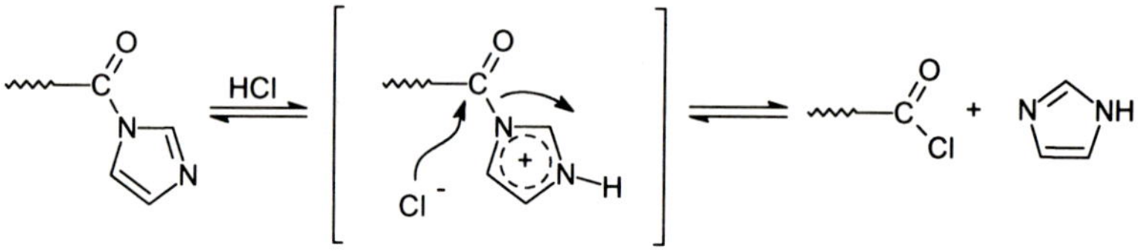

Scheme 21

activated to allow reaction with aliphatic amines, they are rather unreactive toward aromatic amines. However, if the amine is utilized in the form of its hydrochloride salt, then polymerization is possible [77]. As suggested by Staab and coworkers, nucleophilic attack of the chloride ion on the protonated acyl imidazolide to form an acyl chloride appears to be taking place under these conditions [78], see Scheme 21. Of course, the liberated imidazole, which is more basic than the aromatic amine, abstracts another proton from the aromatic amine hydrochloride. This in turn liberates the free aromatic amine and facilitates its reaction with the acyl chloride.

3.2 Solution Behavior

Poly(amic acids) in which the pendent carboxylic acid has been quaternized with an organic base, i.e. tertiary amine, behave very much like a typical poly-electrolyte. That is to say that solutions display a relatively high solution viscosity and the solution viscosity decreases with concentration until at very high dilutions the viscosity dramatically increases due to counterion diffusion and concomitant chain expansion. For comparable concentrations, quaternized poly(amic acids) display solution viscosities in aqueous base about 5 times higher than the corresponding free acid in a typical polar, aprotic solvent [51]. The nature of the organic amine exerts a pronounced effect on the aqueous solution viscosity with secondary and tertiary alkyl amines displaying viscosities approximately 4 times higher than those observed for primary alkyl amines. There appears to be no correlation with the basicity of the amine. In addition, when tertiary amines are employed to prepare the polymeric salt, solution viscosities are virtually constant as a function of time. Secondary and primary amines exhibit initial viscosities which are generally lower and also tend to display a decrease in viscosity with time.

In contrast to poly(amic acid) salts, a much larger body of data has been published for covalently derivatized poly(amic acids), such as poly(amic alkyl esters) and poly(amic alkyl amides). These polyimide precursors behave similar to the poly(amic acid) salts in that they exhibit excellent solution stability [66], see Fig. 7. They differ from the poly(amic acid) salts by having solution viscosities more comparable to poly(amic acids). This, of course, allows them to be formulated at reasonable solid contents. In contrast to poly(amic acids), which generally always show a decrease in the solution viscosity, poly(amic alkyl esters) exhibit a slightly positive slope as can be inferred from the viscosity plot. It is possible that this behavior is a result of low levels of imidization caused by amine impurities in the polar, aprotic solvent. One additional feature of the poly(amic alkyl esters) and poly(amic dialkyl amides), not available to poly(amic acids) and quaternized poly(amic acids), is the ability to prepare isomerically enriched or pure polyimide precursors. The structural difference manifests itself, among other things, in the solution behavior of the corresponding precursors [79]. As shown in Fig. 8, comparing the intrinsic viscosities of identically prepared PMDA/ODA based poly(amic ethyl ester) in its pure *meta*-and *para*-form reveals a large difference in their viscosity.

3.3 Imidization Characteristics

3.3.1 Thermal Imidization

For poly(amic acid) salts, experimental data indicates that these derivatives imidize with greater ease than the corresponding free acids. Thus, certain

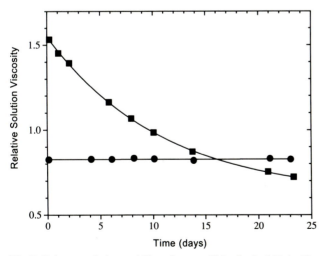

Fig. 7. Polymer solution stability of pyromellitic dianhydride/aniline phthalein based polyimide precursor in dimethylacetamide at ambient temperature. $[C] = 0.5$ wt%, (●) poly(amic methyl ester), (■) poly(amic acid) [66]

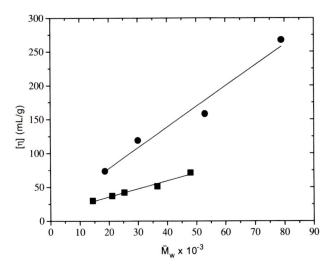

Fig. 8. Solution viscosities of *meta-* and *para-*isomers of PMDA/ODA based poly(amic ethyl ester): (●) *para-*isomer, (■) *meta-*isomer

tertiary amines promote ring closure of the polyimide precursor with rates which can be an order of magnitude faster than the free acid [52]. Unfortunately, few reliable data exist relating the structure of these polyimide precursors and their corresponding mechanical properties. Although some tensile strength data have been reported, there is no systematic comparison with the free acid of similar molecular weight, chemically imidized polymer, and/or other comparable poly(amic acid) derivatives. Poly(amic alkyl esters) and poly(amic dialkylamides) are also known to imidize thermally [65, 60]. In general, poly(amic alkyl esters) imidize at temperatures higher than the corresponding poly(amic acids), see Fig. 9. The reported imidization temperatures increase in changing from methyl- to *n*-butyl ester [65]. More specifically, the imidization temperature exhibits a direct correlation with the electron-withdrawing or donating character of the alkyl ester group [80] as well as the relative nucleophilicity of the amide-nitrogen [81]. As the electron-withdrawing character of the alkyl ester group increases, i.e. the acidity of the parent alcohol increases, the imidization temperature decreases and approaches that of the corresponding poly(amic acid). As shown in Table 5, this behavior holds true for all the esters investigated except for the t-butyl ester. One additional feature apparent from Table 5 is the variation in the overall imidization temperature regime required to achieve complete imidization as evidenced by no weight loss in the thermogravimetric weight loss curve. Although some esters start imidization at lower temperatures, i.e. trifluoroethyl, ethylglycolyl, etc., they imidize over a wide temperature range and require temperatures in excess of 300 °C to complete the imidization reaction. The isopropyl ester actually imidizes at the highest temperature but over the narrowest temperature range.

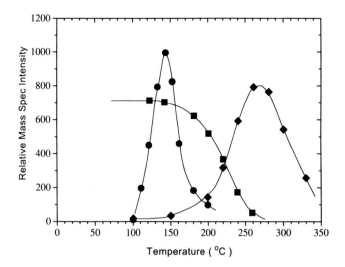

Fig. 9. Volatiles evolution profile of PMDA/ODA based poly(amic acids) compared to poly(amic ethyl ester). (●) mass 18-water, (■) mass 99-NMP, (◆) mass 46-ethanol

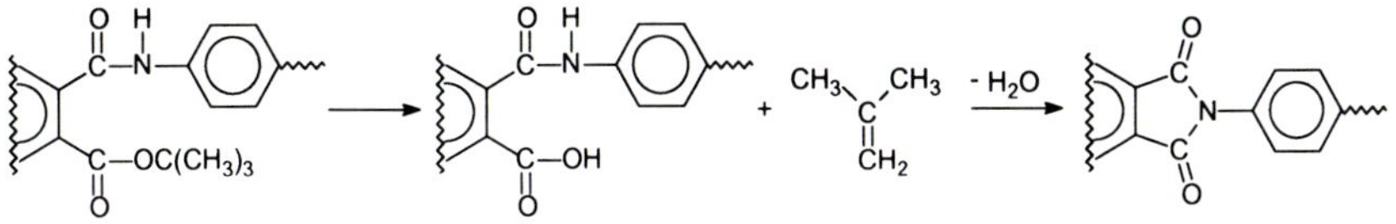

Scheme 22

Table 5. Imidization characteristics of poly(amic alkyl esters)

Poly(amic alkyl ester)	Max. Imidization Temperature (°C)	Imidization Temperature Regime (°C)
para-tert-butyl	193	170–210
para-ethylglycolyl	217	190–350
para-propargyl	224	200–300
para-ethyl	255	240–350
para-isopropyl	269	240–290

In addition to the nature of the ester group, the structural isomerism of the polymer backbone also exerts some effect on the imidization characteristics [82]. Thus, for PMDA/ODA poly(amic ethyl ester) the pure *para*-isomer was found to imidize at slightly lower temperatures as compared to the pure *meta*-isomer, see Fig. 10. This result is consistent with the report the amide groups exhibiting *para*-catenation represent the more favorable conformation for cyclization as determined for poly(amic acids) [83]. Results consistent with the

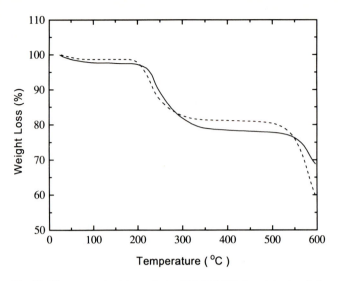

Fig. 10. Thermogravimetric analysis of PMDA/ODA poly(amic ethyl esters) in nitrogen. Heating rate = 20 °C/min, (——) *meta*-isomer, (– – – –) *para*-isomer

thermal data were also observed by forward recoil spectrometry (FRS) of deuterated precursors [84]. FRS experiments on the pure structural isomers of PMDA/ODA based poly(amic ethyl ester) allowed the determination of the extent of imidization and amount of solvent present after various thermal treatments. This was made possible by using poly(amic ethyl esters) in which the ethyl ester moiety was perdeuterated along with the use of deuterated solvents [84]. In films spun-cast from NMP, the *para*-isomers showed greater fractional conversion to the imide after a given thermal treatment than did the corresponding *meta* isomer. This imidization behavior seems to correlate with the greater solvent retention of the *para*-isomer. Subsequent release of the solvent at temperatures within the maximum imidization regime apparently plasticizes the film and facilities the ring closure reaction. Samples prepared from DMSO, do not display isomer-dependent behavior and show similar temperature-dependent behavior as the *meta* isomers prepared from NMP. Correspondingly, DMSO retained during the film formation process is released from the *meta*, *para* and mixed isomers at lower temperatures than is NMP from the *para* and mixed isomers.

Poly(amic dialkyl amides), which represent the other type of derivatized poly(amic acid) have been prepared by derivatization of poly(isoimide) [57] as well as monomer derivatization and subsequent polymerization [60]. Whereas the poly(isoimide) derivatization route has a pronounced tendency to produce poly(amic amides) with significant levels of imidization, the monomer derivatization and polymerization route reported in the literature is also not amenable to preparing well-defined poly(amic amides). The use of thionyl chloride to convert the diamide diacid to the corresponding diacyl chloride is particularly difficult due to the extreme hydrolytic sensitivity of the resulting material. In this case the use of milder activating agents which can be prepared under neutral conditions appears more applicable. Although the solution viscosity of the reported PMDA/ODA based poly(amic diethylamide) in DMAC is quite stable, addition of 1% water caused a decline in the solution viscosity. This finding is quite unexpected and contradictory to the behavior of poly(amic alkyl esters) and leads to the suspicion that the polymer may not have been as well-defined as originally intended.

While during the cure of poly(amic acids) the liberation of water can lead to hydrolytic molecular weight degradation, poly(amic alkyl esters) should not reflect such behavior. Since these polyimide precursors do not exhibit a depolymerization reaction and liberate an alcohol instead to water, higher molecular weight polyimides would be expected upon thermal cyclization. As discussed for PMDA/ODA polyimide derived from the poly(amic acid), the elongation-at-break of fully imidized samples is a good qualitative indicator of the apparent molecular weight of the final polyimide. Therefore, comparison of the elongation-at-break of polyimides derived from poly(amic acids) and poly(amic alkyl esters) of comparable molecular weights should yield higher values for the poly(amic alkyl ester) precursors. As shown in Fig. 11, this is indeed the case. Although the break in the curve occurs at similar molecular weights for both polyimide precursors, the poly(amic alkyl ester) precursor yields consistently higher elongations by almost a factor of two. This mechanical behavior is reminiscent of the data reported for chemically imidized poly(amic acid).

An additional advantage inherent to poly(amic alkyl esters) relates to their behavior at metal interfaces. When investigating the interface and adhesion of polyimide to several different metals, i.e. Au, Cr, Ni, and Cu, it was found that

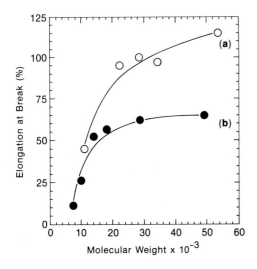

Fig. 11. Elongation-at-break as a function of precursor molecular weight for polyimides derived from (●) poly(amic acid) and (◆) poly(amic ethyl ester)

whereas the poly(amic acid) precursor gave rise to metal oxide particles in the final polyimide film, poly(amic alkyl esters) yielded metal oxide free interfaces [85, 86]. This phenomenon is particularly strong in the case of copper or copper oxide, which readily complexes with poly(amic acids). This behavior of the two different polyimide precursors has some important ramifications. Polyimides derived from poly(amic acids) exhibit greater adhesion to metal surfaces than polyimides derived from poly(amic alkyl esters). This difference can be minimized by using Cr at the interface, in which case both polyimide precursors give rise to good adhesion of the final polyimide. The formation of metal oxide particles in poly(amic acid) derived polyimide is more serious. Clearly, such behavior is to be avoided for applications requiring optimum dielectric properties. Fortunately, poly(amic alkyl esters) do not promote the metal oxide formation and offer an easy solution to this problem.

3.3.2 Base-Catalyzed Imidization

The chemical imidization of poly(amic alkyl esters) was only reported very recently [59], although reports in the literature claim chemical imidization with a traditional acetic anhydride/pyridine mixture [87]. The chemical imidization of poly(amic alkyl esters) is based on the observation that PMDA/ODA based poly(amic ethyl ester) samples, when formulated at low concentrations for size exclusion chromatography, precipitated upon standing overnight [88]. Distillation of the NMP from phosphorus pentoxide to remove low levels of methylamine, a known impurity in this particular solvent, eliminated this unusual behavior. The precipitated polymer had significant levels of imidization as evidenced by IR. Apparently, organic bases, such as alkyl amines, were able to catalyze the conversion of amic alkyl esters to the corresponding imide.

Model compound studies indicated that both the nature of the base, the nature of the alkyl ester and the amide group exerted pronounced effects on the observed imidization rates [59]. As illustrated in Table 6, the imidization rate of monomethyl p-methoxyphenyl phthalamide tracks the general basicity of the

Table 6. Effect of base strength on the imidization rate of monomethyl ester p-methoxyphenyl phthalamide at 23.3 °C.

Base	Rate Constant	pK$_a$
DBU*	5.87×10^{-4}	—
Triethylamine	9.48×10^{-6}	10.8
Piperazine	1.49×10^{-4}	9.8
N-Methylmorpholine	2.73×10^{-6}	7.4
Pyridine	0	5.2
N,N-dimethylaniline	4.38×10^{-6}	5.1

*Diluted 100-fold, 1,8-diazabicyclo [5.4.0] undec-7-ene (DBU)

various amines. The discrepancies can most likely be attributed to the fact that the tabulated pK_as were determined in water [89] and are known to change as a function of solvent medium. Furthermore, basicity may not be the only factor involved and nucleophilicity of the base may also need to be considered. This is consistent with the observation that nucleophiles such as fluoride and acetate ion are also capable of catalyzing the imidization reaction. A similar trend in the imidization rate is observed when altering the nature of the ester group and the amide-nitrogen as shown in Table 7. The susceptibility of the ester group toward imidization is related to the electro-positive character of the carbonyl group as determined by the electron-withdrawing power of the ester similar to the trend observed in the thermal imidization of poly(amic alkyl esters). The effect of the amide-nitrogen is not as clearly understood. Thus, both electron-withdrawing or electron-donating groups in the *para*-position exhibit enhanced imidization rates as compared to the unsubstituted monoalkyl aryl phthalamide. It may be a situation where both the nucleophilicity as well as the relative stability of the resulting iminolate anion exert significant effects on the imidization rate. Studying the effect of solvent for a given base, it was found that this too can significantly effect the imidization rate. Among the four solvents studied, NMP, tetrahydrofuran, chloroform, and acetonitrile, the latter two solvents yielded imidization rate constants almost two orders of magnitude higher, see Table 8.

Table 7. Imidization rates of various monoalkyl ester aryl phthalamides in N-methyl pyrollidone

Functionality	Rate constant (s^{-1})
Monoalkyl esters p-Methoxyphenyl phthalamides[a]	
Methyl	5.87×10^{-4}
Ethyl	2.85×10^{-4}
Isopropyl	4.01×10^{-5}
Monomethyl Aryl phthalamides[b]	
phenyl	5.34×10^{-6}
p-methoxyphenyl	9.48×10^{-6}
p-nitrophenyl	2.89×10^{-5}

[a] Base = DBU, [Amide Ester/Base] = 100/1, Temperaturee23.3 °C
[b] Base = Triethylamine, [Amide Ester/Base] = 1/1, Temperature = 23.3 °C

Table 8. Imidization rate of monomethyl ester p-methoxyphenyl phthalamide as a function of solvent medium

Solvent	Rate Constant (s^{-1})
N-Methylpyrrolidone	9.48×10^{-6}
Tetrahydrofuran	1.81×10^{-6}
Chloroform	8.55×10^{-5}
Acetonitrile	1.04×10^{-4}

Since the second solvent pair fall within the poor hydrogen bonding group of solvents, increased basicity of the organic base in these solvents would be consistent with the observed behavior. Based on the model compound studies, indications are that the base-catalyzed imidization process may involve a two-step mechanism, see Scheme 23. The first step corresponds to the complete or partial proton abstraction from the amide group with the formation of an iminolate anion. Since this iminolate anion has two possible tautomers, the reaction can proceed in a split reaction path to either an isoimide- or imide-type intermediate. Although isoimide model reactions indicate an extremely fast isomerization to the imide under the conditions employed for base-catalysis, all indications to date are that it is not an intermediate in the base-catalyzed imidization of amic alkyl esters.

Initial investigations of base-catalyzed imidization of polymeric systems, in particular PMDA/ODA based poly(amic alkyl esters), have been difficult due to the insolubility of the polyimide precursor at imidization levels exceeding 40%. Nevertheless, preliminary studies indicate that the base-catalyzed polymer imidization reaction appears to be significantly slower at ambient temperatures as compared to the phthalamide model compounds. It is yet unclear whether this is a direct result of the conformational aspects associated with the polymer chain or solubility considerations arising from the less soluble, partially imidized polymer chain. Since much of the initial work involved IR studies of supported

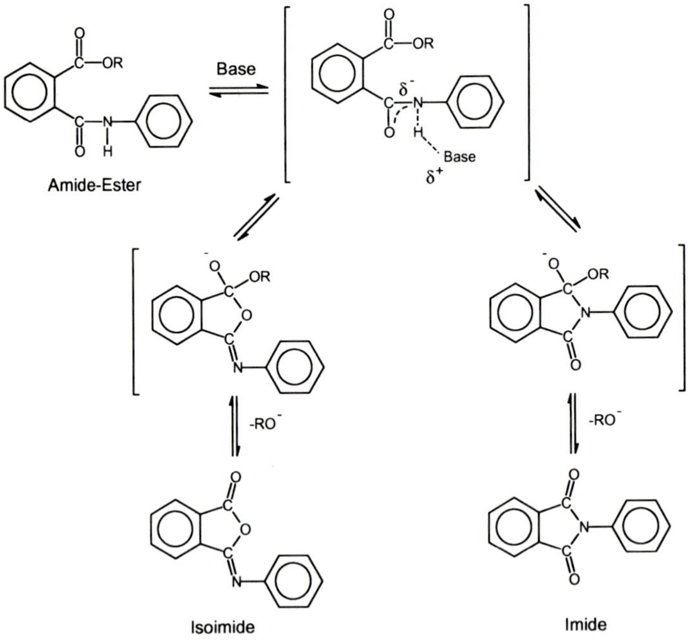

Scheme 23

polymer films with amine catalysts either added to the polymer solution prior to spin-coating or via flooding the polymer film after spin-coating, it is quite possible that much of the base is lost during curing studies at elevated temperatures. It is also possible that a skin-core effect is being observed. Since IR measurements determine the bulk imidization level, degrees of imidization at the surface could be quite high. If the evaporation of base was minimized by reapplying the neat amine prior to the post-bake step, respectable levels of imidization were achieved as illustrated in Table 9. In the case of these insoluble polyimides, base-catalyzed imidization of the precursors allows for high levels of imidization at temperatures far below those required for purely thermal cycliz-ation. The potential for utilizing this approach for a photosensitive polyimide scheme using photogenerated bases has also been reported [90].

4 Polyimides from Low Molecular Weight Precursors

The synthetic approach of preparing polyimides from monomeric reagents has been of interest since the polyimides were originally conceived [3, 91] Starting with relatively stable monomers, i.e. hydrolytically stable, which then react at elevated temperatures to form a polyimide of sufficient molecular weight offers synthetic simplicity and the possibility of preparing high solids content coating formulations. High solids content formulations exhibit attractive features in being able to provide thick polymer coatings and improved planarizing charac-teristics. This latter feature is of particular interest to microelectronic appli-cations which involve multiple layers of polyimide over irregular topographies. Among the early commercial examples which made use of this approach were the Monsanto Skybond series of polyimide precursors [92] as well as PMR-15, a resin developed by the aerospace industry [93]. The latter material is end-capped with a reactive group to provide for cross-linking of the polymer. Typically, the formulation consisted of aromatic tetracarboxylic acid diesters in conjunction with aromatic diamines. This initial formulation may have been heated at intermediate temperatures ($< 150\,°C$) to produce an oligomeric

Table 9. Degree of imidization of poly(amic ethyl esters) in the presence of diethylamine

Poly(amic ethyl ester)	Degree of Imidization (%)	
	80 °C	120 °C
PMDA/ODA	80	90
BPDA/ODA	66	80

solution (A-stage). Later, the formulation could then be cured at temperatures of approximately 300 °C to yield the final polyimide (B-stage). At that time, the reaction was believed to involve reaction of the diamine with the ester groups of the tetracarboxylic acid diester to form an intermediate oligo(amic acid) during the A-stage ($5 \leq x \leq 10$). This reaction continued during the final curing (B-stage) with concomitant imidization to increase the polymer molecular weight ($n > 50$), see Scheme 24. The tetracarboxylic acid diester and diamine pair was chosen to provide a final polyimide with a glass transition temperature, T_g, less than 400 °C and preferably greater than 300 °C. A T_g which was well within a temperature regime where the polymer was thermally stable was crucial to provide high molecular weight. The fact that molecular weight build-up or chain-extension cannot readily proceed below T_g of the system due to limited chain-end diffusion, necessitates excursions above T_g to ensure complete reaction. This characteristic has always prevented high T_g polyimides, i.e. $T_g > 350$ °C, from being prepared in this fashion. However, if chain-extension could occur at temperatures sufficiently high to allow room temperature stability, yet low enough to preclude significant levels of imidization, then such a concept could be applied to high T_g polyimides. The observation that pyromellitic dianhydride derived diesters actually revert back to the anhydride at elevated temperatures is of potential use [94], see Scheme 25. For most normal alkyl ester derivatives this temperature is too high and lies well-within the imidization regime (150–200 °C). If the ester is based on a relatively acidic alcohol so that the

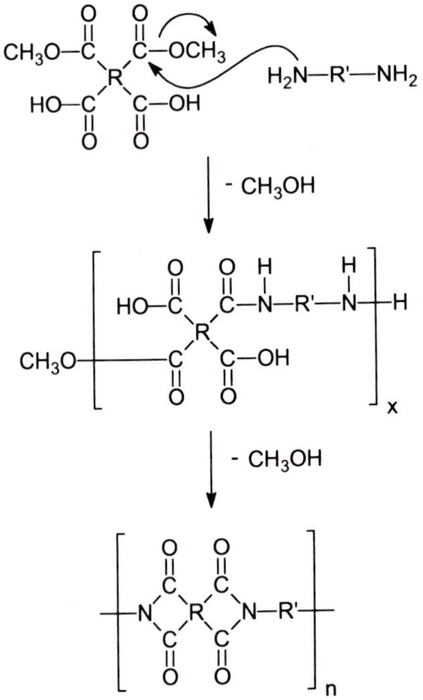

Scheme 24

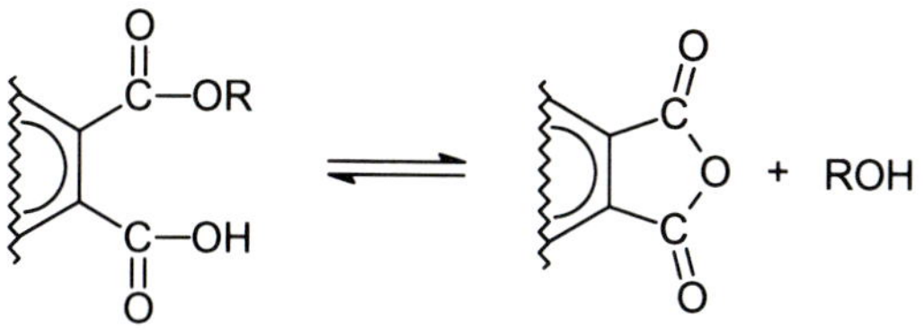

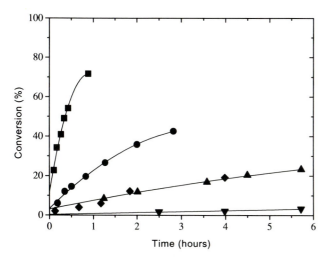

Scheme 25

Fig. 12. Reaction of monoalkyl monohydrogen phthalate with aniline. [Ester] = [Amine] = 0.25 M, T = 70 °C, Solvent = DMSO-d$_6$. (■) trifluoroethyl ester (●) 1, 3-dichloroisopropyl ester (◆) ethylglycolyl ester (▲) propargyl ester (▼) methyl ester

carbonyl is more activated toward nucleophilic attack, anhydride formation can occur at lower temperatures [62]. Therefore, the relative reactivity of PMDA based diacid diesters to revert to the anhydride at elevated temperatures was found to be directly related to the acidity of the parent alcohol, i.e. phenyl > trifluoroethyl > allyl > methyl > ethyl as evidenced by NMR studies of the reaction of aniline with various dialkyl dihydrogen pyromellitates, see Fig. 12. Combining the anhydride-forming chemistry with the ability to prepare well-defined amine-terminated oligo(amic acids) led to the development of chain-extendable oligo(amic acid) formulations [95], see Scheme 26. First, an oligo(amic acid), here R = H, with an average degree of polymerization of approximately 5 was prepared in NMP by employing a precise stoichiometric excess of p, p'-oxydianiline with pyromellitic dianhydride. Next, a sufficiently active chain-extender, i.e. dialkyl dihydrogen pyromellitate, was added to the formulation so as to readjust the overall functionality to an equimolar stoichiometry. The resulting formulations had a solids content of at least 35 wt%. Coatings were then imidized by a two-step heating program. Initially, the

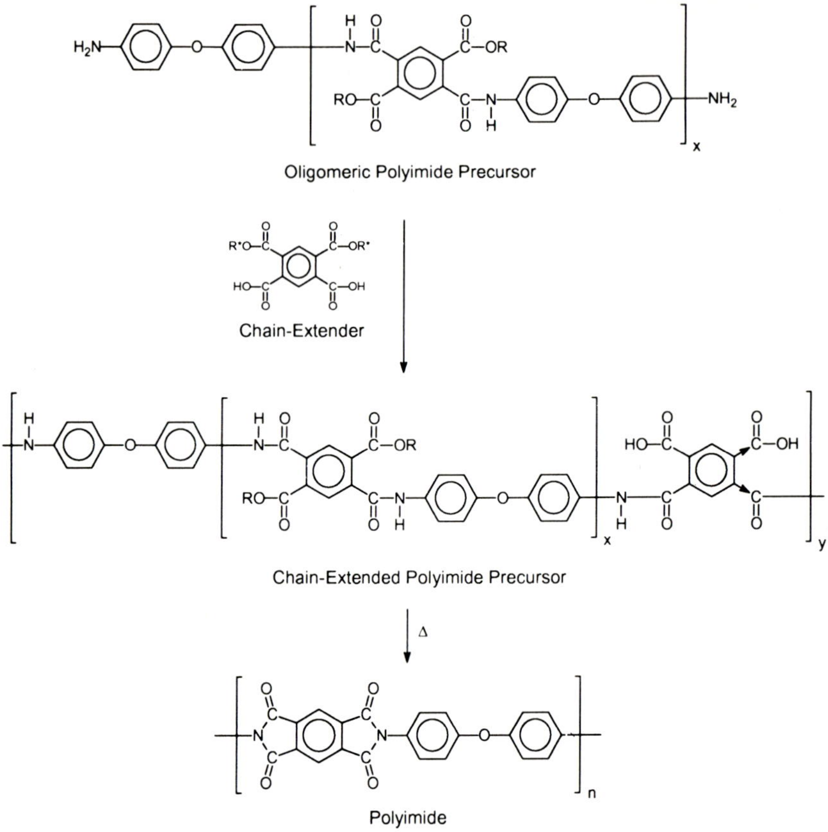

Oligomeric Polyimide Precursor

Chain-Extender

Chain-Extended Polyimide Precursor

Polyimide

where R = H, methyl, ethyl, etc.
R* = methyl, propargyl, trifluoroethyl, ethylglycolyl

Scheme 26

sample was heated slowly to a temperature not exceeding 150 °C and held at this temperature for a fixed amount of time to promote molecular weight build-up. The sample was then ramped to approximately 350 °C to fully imidize the specimen.

In light of the higher imidization temperatures associated with poly(amic alkyl esters), this approach could be extended to yield improved coating formulations [82]. Since it was now possible to balance the relative reactivity of the chain-extender with the imidization temperature of an oligo(amic alkyl ester), here R = methyl, ethyl, etc., less reactive chain-extenders could be utilized. The lower reactivity of the chain-extender would be reflected in improved solution stability and shelf-life of the formulation without sacrificing the mechanical properties of the final polyimide, see Table 10 where EGX, TFE,

Table 10. Mechanical properties of PMDA/ODA based polyimides derived from low molecular weight, chain-extendable precursors

Formulation	Tensile Strength (kg/mm^2)	Elongation (%)
Poly(amic acid)		
DP = 70	13.99	52
Oligo(amic acid)		
DP = 5, EGX	11.88	30
Oligo(amic acid)		
DP = 5, TFE	16.45	54
Oligo(amic ethyl ester)		
DP = 12, EGX	13.78	81
Oligo(amic ethyl ester)		
DP = 12, MEX	—	Brittle
Oligo(amic isopropyl ester)		
DP = 11, MEX	11.67	56

and MEX denote the chain-extenders diethylglycolyl-, ditrifluoroethyl- and dimethyl dihydrogen pyromellitate, respectively.

5. Polyimides from Dianhydrides and Diisocyanates

The preparation of imides from reaction of isocyanates with anhydrides dates back to the early days of organic chemistry [96]. With the advent of polyimide chemistry in the early 1960s, this chemistry was soon explored for the synthesis of polyimides. However, in contrast to the preparation of polyimides via their poly(amic acid) intermediate, the reaction of aromatic dianhydrides with aromatic diisocyanates is much less understood. The reaction of aromatic dianhydrides with aliphatic or aromatic diisocyanates is believed to form a cyclic seven-membered intermediate which then splits out CO_2 to form the polyimide [97], see Scheme 27. The addition of water, which has been reported to accelerate the anydride/isocyanate reaction, can result in several transformations of either the anhydride or the isocyanate reagent, see Scheme 28

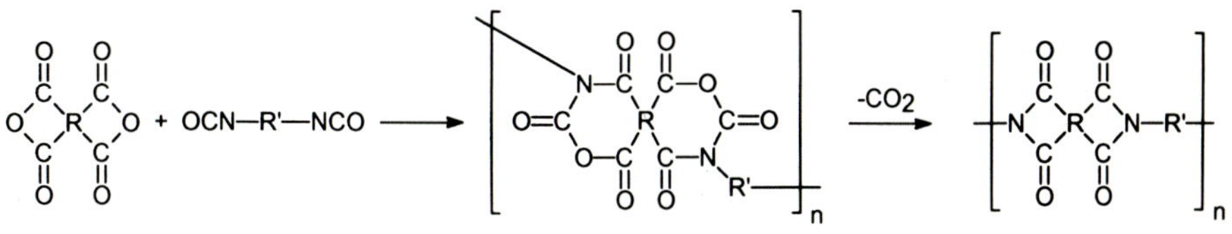

Scheme 27

[98]. Initially, water can cause the hydrolysis of the anhydride or the isocyanate, Scheme 28 (reaction 1 and 2), although the isocyanate hydrolysis has been reported to occur much more rapidly [99]. The hydrolyzed isocyanate (carbamic acid) may then react further with another isocyanate to yield a urea derivative, see Scheme 28 (reaction 3). Either hydrolysis product, carbamic acid or diacid, can then react with isocyanate to form a mixed carbamic carboxylic anhydride, see Scheme 28 (reactions 4 and 5, respectively). The mixed anhydride is believed to represent the major reaction intermediate in addition to the seven-membered cyclic intermediate, which upon heating lose CO_2 to form the desired imide. The formation of the urea derivative, Scheme 28 (reaction 3), does not constitute a molecular weight limiting side-reaction, since it too has been reported to react with anhydride to form imide [100]. These reactions, as a whole, would explain the reported reactivity of isocyanates with diesters of tetracarboxylic acids and with mixtures of anhydride as well as tetracarboxylic acid and tetracarboxylic acid diesters [101, 102]. In these cases, tertiary amines are also utilized to catalyze the reaction. Based on these reports, the overall reaction schematic of diisocyanates with tetracarboxylic acid derivatives can thus be illustrated in an idealized fashion as shown in Scheme 29.

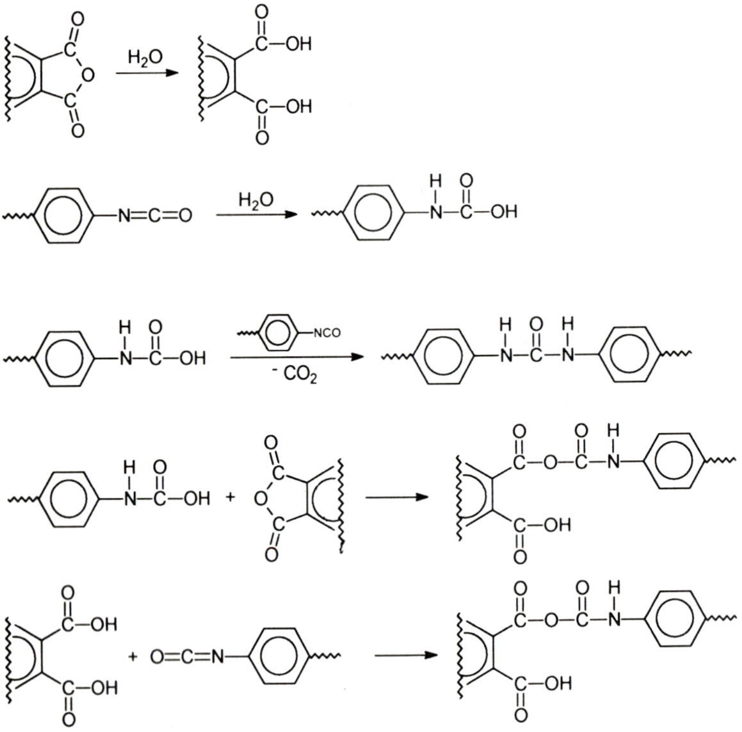

Scheme 28

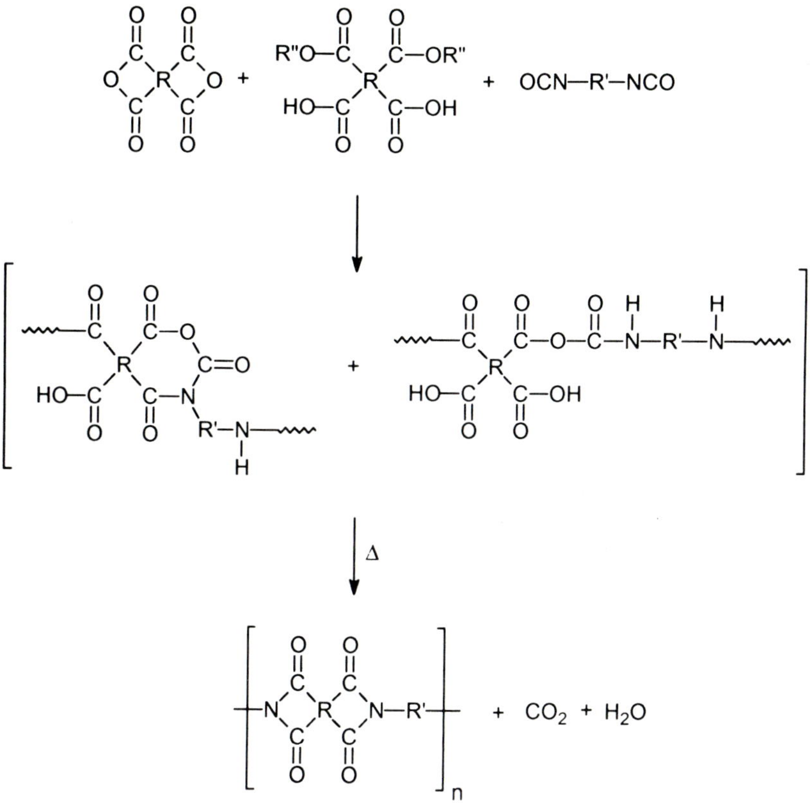

$$R''O-\overset{\overset{\displaystyle O}{\|}}{C}\quad\overset{\overset{\displaystyle O}{\|}}{C}-OR''$$

Scheme 29

In general, polyimide synthesis via the isocyanate route is conducted by allowing the initial monomer reaction to proceed at low temperatures of ca. 0–10 °C so as to minimize reactions of the diisocyanate with the solvent. Although a number of side-reactions of isocyanates with the amide-type solvents, such as dimethylacetamide and hexamethylphosporamide, have been reported [103], these side-reactions generally require temperatures of ca. 100 °C and above. The polymerization mixture is then heated to 50–100 °C with concomitant evolution of carbon dioxide and the production of soluble or insoluble low molecular weight adducts depending on the chemical nature of the monomers. In some cases, a high molecular weight polymer is produced at approximately 150 °C, however, carbon dioxide evolution is not complete until further heating to higher temperatures has occurred. Overall, this method of polyimide synthesis has not found as much application as the poly(amic acid) route. Since the evolution of carbon dioxide can be utilized as a blowing agent, however, this approach is useful for preparing polyimide foams [99].

6 Alternative Polyimide Pathways

6.1 Amine-Imide Exchange Reactions

The preparation of *N*-alkyl imides by exchange reaction of an imide with an alkyl amine was documented [104] well before the application of this chemistry to the preparation of polyimides [105], see Scheme 30. Although no experimental details are provided, the initial reaction of pyromellitimide with *p, p*-methylene dianiline in NMP takes place at reflux temperatures to apparently yield a poly(amic amide). Subsequent heating of this intermediate at elevated temperatures ($\sim 300\,^\circ$C) provides the desired polyimide with evolution of ammonia. The final polyimide is quoted to be thermally and chemically stable, however, no mechanical properties are given.

A much more elegant synthesis of polyimides by amine-imide exchange is based on the reaction of *N*-ethoxycarbonyl phthalimides with amino acids to yield phthaloyl substituted amino acids [106]. Thus, reaction of *N, N*-bis(ethoxycarboxy)pyromellitimide with *p, p*-oxydianiline in NMP at room

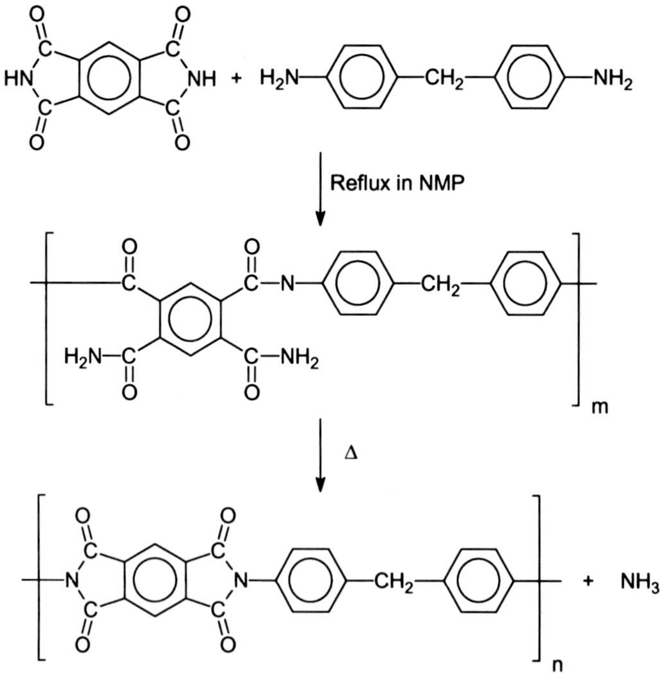

Scheme 30

temperature, see Scheme 31, yielded a soluble poly(amic N-ethoxycarboxyam-ide) [107]. After stirring for 7 days at room temperature a limiting inherent viscosity of 1.0 was reported for a 0.5 dL/g solution. Imidization of the soluble polyimide precursor, i.e. poly(amic N-ethoxycarboxyamide), at 240 °C in vacuo yielded a creasable, tough polyimide film.

Another form of amine-imide exchange reaction, also referred to as trans-imidization, involves the high temperature equilibrium of an imide in the presence of a different amine, see Scheme 32. The more basic amine, in this case methylamine, will keep the equilibrium shifted to the left. However, since methylamine is much more volatile, removal of the methylamine will gradually shift the equilibrium to the right. This approach has been particularly successful when utilizing transition metal catalysis in the preparation of poly(ether imides) [108, 109]. If the reaction is conducted at temperatures well above the T_g of the polymer, i.e. 250–300 °C for most poly(ether imides), and vacuum is applied to aid in the removal of the volatile amine, respectable molecular weights can be obtained. Improvements in this approach of poly(ether imide) preparation consist of using more electronegative amines, i.e. less basic amines, in the initial aryl diimide, such as 2-aminopyridine [110]. Use of 2-aminopyrimidine (pK_a = 3.5 at 20 °C), a less toxic amine than 2-aminopyridine (pK_a = 6.8 at 20 °C), enhances this reactivity even further as recently reported in the literature [111].

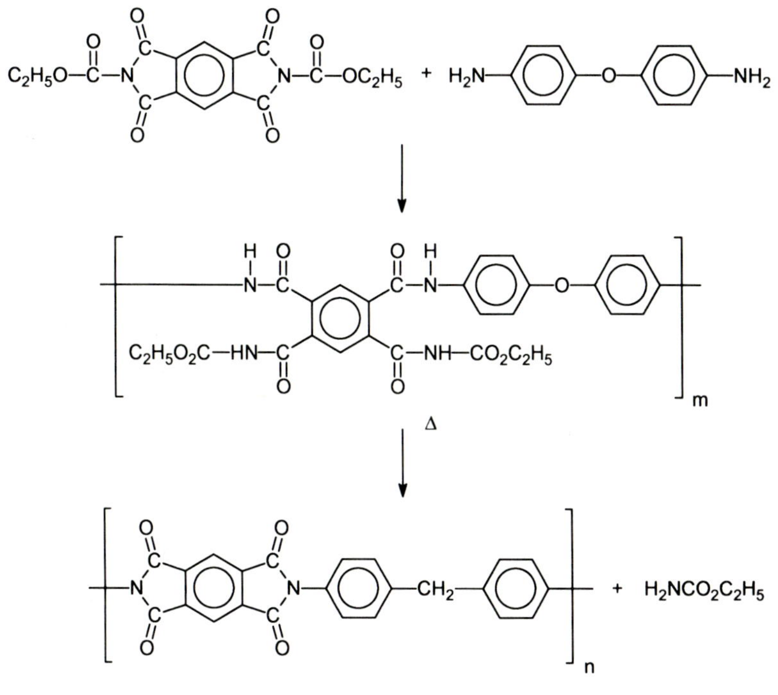

Scheme 31

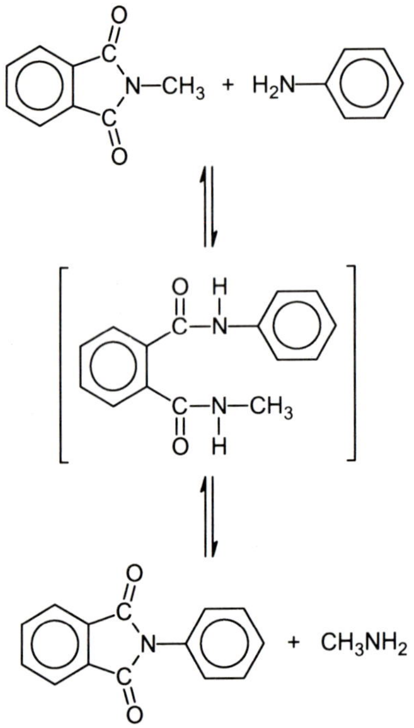

Scheme 32

Thus, *N*-pyrimidine phthalimide reacted with hexylamine at room temperature to form an amide-amide. The initial amide-amide formation proceeded more rapidly in chloroform as compared to dimethylsulfoxide (DMSO). However, the ring closure reaction to the imide was favored by the more polar, aprotic DMSO solvent, yielding the imide in nearly quantitative yield after 3 hours at 75 °C. The authors were able to utilize this synthetic approach to prepare well-defined segmented poly(imide-siloxane) block copolymers. It appears that transimidization reactions are a viable approach to preparing polyimides, given that the final polyimide has a T_g sufficiently low to allow extended excursions above the T_g to facilitate reaction without thermal decomposition. Additionally, soluble polyimides can be readily prepared by this approach. Ultimately, high T_g, insoluble polyimides are still only accessible via traditional soluble precursor routes.

6.2 Polyimides Via Phthalide Derivatives

A new synthetic, low temperature approach to polyimides has recently been reported [112]. The synthetic scheme is based on dicyanomethylidene phthalide derivatives, which are synthetic analogs to the corresponding anhydrides. As shown in Scheme 33, pyromellitic dianhydride can be readily converted to the

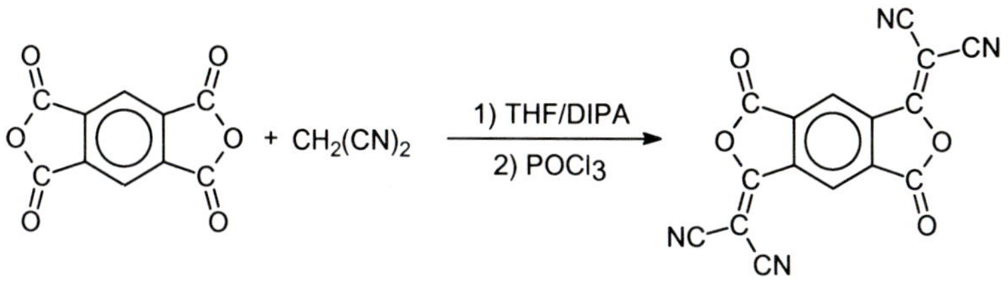

Scheme 33

bis(dicyanomethylidine) pyromellitide by reaction with malononitrile in the presence of diisopropylamine and subsequent transformation of the diisopropylamine salt to the pyromellitide with phosphorus oxychloride, $POCl_3$. Due to the fact that this pyromellitide displays analogous behavior as compared to pyromellitic dianhydride, it readily reacts with aromatic diamines to form a soluble poly(amic acid)-like intermediate, see Scheme 34. Unlike poly(amic acids), this intermediate transforms to the polyimide ($\sim 75\%$ in 3 h) in solution at ambient temperature. However, complete imidization still requires thermal

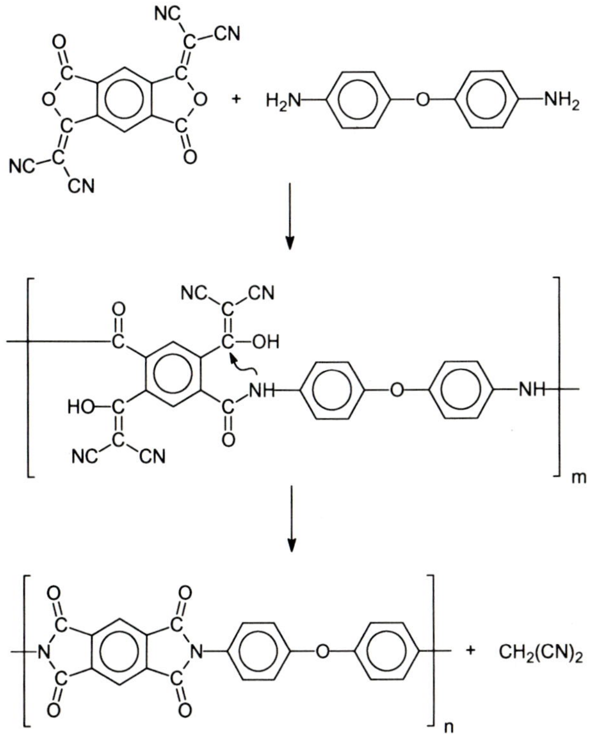

Scheme 34

excursions to elevated temperatures due to behavior analogous to that described for solid-phase thermal cyclization processes of common polyimides [113].

7 Polyimide Copolymers and Blends

The modification of polyimide properties has been an area of considerable interest prompted by the more stringent requirements of advanced polyimide applications. Unfortunately, such requirements can rarely be met by one single material. Diverging properties such as a high glass transition temperature, low thermal expansion coefficient coupled with excellent adhesion can be addressed by multicomponent systems, such as polyimide blends and copolymers. If the multicomponent system displays microphase separated morphologies, then the desired properties of each component can be realized without necessarily compromising their individual properties [114]. Furthermore, if the microphase-separated morphology is on a submicron level, then such polyimide systems would be of interest for microelectronic applications. In this regard, the better understanding of polyimide precursor behavior as well as more recent synthetic approaches have facilitated the work in this area.

7.1 Polyimide Blends

Multicomponent polyimides have to be prepared in the soluble precursor stage either by blending or synthetic modifications, i.e. block copolymer formation, due to the general insolubility and infusability of many commercial polyimides. In this respect, polymer blends represent the most practical approach to polymer modification. However, the presence of an amic acid/anhydride-amine back reaction greatly complicates such a system. The mixing of two homopoly(amic acids) in solution leads to fragmentation of both polymers and recombination of different fragments, as illustrated in Fig. 13. If this process occurs at a reasonable rate, segmented block copolymers are ultimately formed [115, 116]. In that case, a compatible system devoid of microphase-separated morphologies is generally obtained. These observations are consistent with the size exclusion chromatography and light scattering studies of BPDA/ODA poly(amic acid) mixtures. Mixing a high (DP = 150) with a low (DP = 10) molecular weight sample resulted in mixtures which eventually equilibrated to the expected most probable distribution [117]. For this particular poly(amic acid) formulation re-equilibrium required approximately 2 weeks. The re-equilibrium of the two homopoly(amic acids) is dependent on both the nature of the constituent dianhydrides and diamines as well as the solvent. Fjare [118] was able to clearly demonstrate that the back reaction, i.e. amic acid/anhydride-

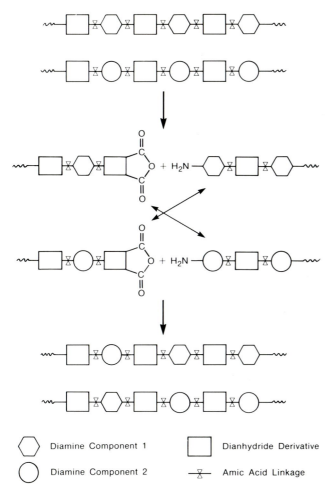

Fig. 13. Idealized scheme of the re-equilibration reaction of poly(amic acid) mixtures

amine equilibrium, is promoted by electron donating substituents on either the anhydride or amine portion of the molecule. Furthermore, less polar solvents with decreased hydrogen-bonding character, such as diglyme, also favor the back reaction.

Of course, this re-equilibration behavior of two homopoly(amic acids) can be eliminated if one of the components does not exhibit this back reaction. Recombination of different fragments is not possible and stable polyimide precursor blends are accessable [119]. For example, combining a relatively flexible poly(amic acid), hexafluoroisopropylidene diphthalic anhydride (6F)/2, 2-bis(4-aminophenoxy-4'-phenyl) propane (BDAF), with a rigid poly(amic alkyl ester), PMDA/p-phenylene diamine (PDA), produced microphase separated polyimide blends as can be interred by the persistence of the

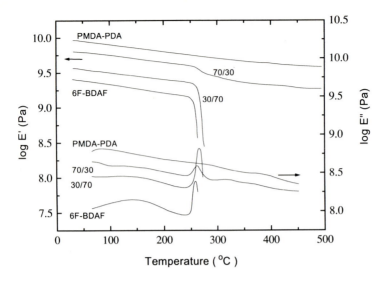

Fig. 14. Temperature dependence of the dynamic storage (E') and loss (E'') moduli at 10 Hz as a function of composition

6F-BDAF T_g for various blend compositions, see Fig. 14. The microphase-separated morphology further manifests itself in the self-adhesion behavior of polyimide films derived from such mixtures. For mixture containing at least 25 wt% of the flexible component, peel tests of polyimide bilayer samples prepared by solution casting, bulk failure of the test specimens was observed. Since the flexible component contained fluorine, the samples could be examined by X-ray photoelectron spectroscopy to determine the surface composition. At only 10% loading, the flexible component comprised 100% of the top 75 Å of the sample. The surface segregation of the flexible component is believed to be responsible for the adhesion improvements.

7.2 Polyimide Block Copolymers

In analogy to polymer blends, block copolymers offer the potential for multiphase morphologies. In general, the microphase-separated morphology obtainable with block copolymers, however, is much more controllable and often on a much smaller scale, i.e. several hundred angstrom. Polyimides, as a class of materials, have received little attention as a component in the synthesis of block and segmented copolymers [120, 121], the imide-siloxane copolymers being most prominent. The combination of excellent electrical and mechanical properties of a rigid polyimide with the self-adhesion of an engineering thermoplastic are just one example of how this approach can be utilized. Taking advantage of the poly(amic alkyl ester) route to polyimides, i.e. reaction of a diester diacyl chloride with a diamine, allows for synthetic flexibility superior to the traditional poly(amic acid) synthesis. These advantages are increased solubility of the starting monomers in solvents other than polar, aprotic solvents and the ability to work-up and purify the soluble polyimide precursor. As illustrated in Scheme 35, a suitable amine-terminated oligomer, e.g. aryl ether phenylquinoxaline (PQE), is co-reacted with a monomeric diamine and dianhydride derived dialkyl ester diacyl chloride [122]. Since the initial amine-functionality is mostly supplied by the monomeric diamine, primarily amic alkyl ester segments are formed. As the reaction approaches completion, i.e. the amine-endgroup concentration is very low, the amic alkyl ester endgroups are incorporated on a

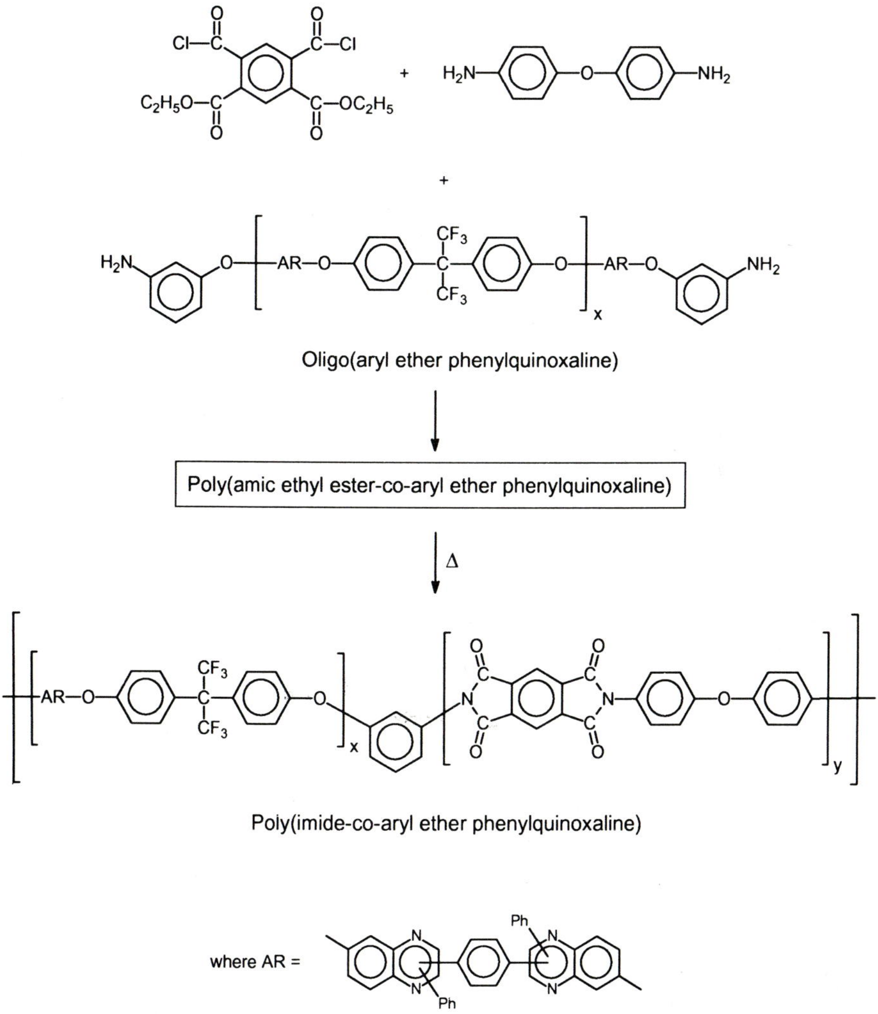

Oligo(aryl ether phenylquinoxaline)

Poly(amic ethyl ester-co-aryl ether phenylquinoxaline)

Δ

Poly(imide-co-aryl ether phenylquinoxaline)

where AR =

Scheme 35

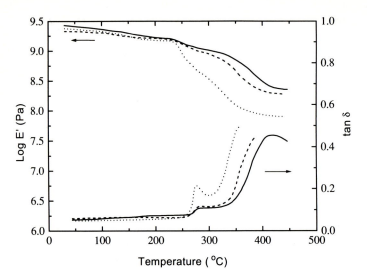

Fig. 15. Dynamic mechanical thermal behavior of poly(imide-*co*-phenyl ether phenyl quinoxaline) as a function of poly(phenyl ether phenylquinoxaline) content. (——) 16 wt%, (– – – –) 29 wt%, and (· · · ·) 55 wt%; PQE coblock molecular weight = 15 500

similar statistical basis as the PQE amine-endgroups. This results in a segmented block copolymer, i.e. poly(amic alkyl ester-*co*-aryl ether phenylquinoxaline). At this stage the copolymer is thoroughly washed to remove any homopolymer contamination, before being formulated and thermally cured. Again, the salient features of the multiphase morphology associated with such block copolymers are reflected in their dynamic mechanical behavior, see Fig. 15. All three compositions exhibited two transitions as indicated by a drop in E' and a maximum in tan δ. The first transition at approximately 250 °C reflects the PQE glass transition temperature and the second transition around 360 °C is characteristic of high T_g homopolyimides. The minimal decrease in the E' at high temperatures further indicates that the ordered morphology of the parent polyimide has been retained. Furthermore, self-adhesion measurements of the block copolymer containing at least 16 wt% of the PQE block resulted in test specimens with peel strengths which could not be measured due to bulk failure of the specimen.

8 Summary

Polyimides have experienced tremendous growth since their conception over 30 years ago. Whereas the initial goal of these materials was high thermo-oxidative stability along with excellent mechanical properties, more recent applications have focused on processability, planarization [95], low thermal expansion coefficient [123, 124], improved adhesion [125] and lower dielectric constants [126, 127]. Of course, such properties are desired in addition to and without sacrificing thermal and mechanical properties. Developments in this area were mainly driven by the emerging needs of the microelectronics industry. Since such diverse properties are very difficult to obtain with a single material, polyimide blends and segmented block copolymers have become of great interest. In addition, the multiphase morphology offered by the block copolymers is also of interest in membrane technology. Control of the polymer morphology by copolymerization can provide improved selectivity and permeability rates of gas separation membranes [128]. Another area, which has received relatively little attention in the polymer literature, excluding patent publications, concerns the use of polyimides as liquid crystal orienting layers for display devices [129]. Here, thermal requirements are not as stringent and the use of polyimides derived from cycloaliphatic dianhydrides is more prevalent [130].

Complementary to the synthetic aspects of polyimide chemistry, a wealth of information addressing structure-property relationships of a wide spectrum of polyimides can be found in the literature. Information of this nature can be readily obtained from the literature along with potential application related data [11, 131–136].

Acknowledgements. The author expresses his gratitude to J.W. Labadie, J.L. Hedrick, D. Dawson, N. Stoffel, and T. Clarke for their technical assistance and helpful suggestions.

References

1. Bogert TM, Renshaw RR (1908) J Am Chem Soc 30: 1135
2. Brandt S (1958) J prakt Chem. 4: 163
3. Edwards WM, Robinson IM (1955) US Patent 2,710,853
4. Edwards WM, Robinson IM (1959) US Patent 2,867,609
5. Edwards WM, Robinson IM (1959) US Patent 2,880,230
6. Edwards WM, (1965) US Patent 3,179,614
7. Endrey AL (1962) Can. Patent 659,328
8. Endrey AL (1965) US Patent 3,179,631
9. Endrey AL (1965) US Patent 3,179,633
10. Sroog CE, Endrey AL, Abramo SV, Berr CE, Edwards WM, Oliver KL (1985) J Polym Sci A3: 1373
11. Bessonov MI, Koton MM, Kudryavtsev VV, Laius LA (1987) In: Polyimides: Thermally stable polymers, Consultants Bureau, New York, p 23
12. Boldebuck EM, Klebe JF (1967) US Patent 3,303,157
13. Korshak VV, Vinogradova SV, Vygodskii Ye. S, Nagiev ZM, Urman YG, Alekseeva SG, Slonium IY (1983) Makromol Chem 184: 235
14. Hodgkin JH (1976) J Polym Sci: Polym Chem Ed 14: 409
15. Koton MM (1971) Poly Sci USSR A13: 1513
16. Frost LW, Kesse I (1964) J Appl Poly Sci 8: 1039
17. Bender ML, Chow YL, Chloupek FJ (1958) J Am Chem Soc 80: 5380
18. Ardashnikov AYa, Kardash IYe, Pravednikov AN (1971) Polym Sci USSR A13: 2092
19. Kolegov VI (1976) Polym Sci USSR A18: 1929

20. Dine-Hart RA, Wright WW (1967) J Appl Poly Sci 11: 609
21. Heydaya E, Hinman RL, Theodoropulos S (1966) J Org Chem 31: 1311
22. Roderick WR, Bhatia PL (1963) J Org Chem 28: 2018
23. Jones JI, Ochynski FW, Rackley FA (1962) Chemistry and Industry 1686.
24. Duran J, Viswanathan NS (1984) In: Davidson T (ed) Polymers in electronics. ACS Symp Series 242, ACS, Washington D.C. p 239
25. Wallach ML (1967) J Polym Sci A-2(5): 653
26. Fjare DE, Roginski RT (1991) Abstracts of Fourth Int Conf Polyimides, Mid-Hudson Section of Soc Plast Eng p 117
27. Volksen W, Cotts PM (1984) In: Mittal KL (ed) Polyimides: Synthesis, characterization and applications. Plenum, New York, p 163
28. Bel'nikevich NG, Adrova NA, Korzhavin LN, Koton MM, Panov YuN, Frenkel SYa (1973) Poly Sci USSR A15(8): 2057
29. Bower GM, Frost LW (1963) J Polym Sci A1: 3135
30. Cotts PM, Volksen W (1984) In: Davidson T (ed) Polymers in electronics. ACS Symposium Series 242, ACS, Washington, D.C., p 227
31. Walker CC (1988) J Polym Sci, A26: 1649
32. Zakoshchikov SA, Ignat'eva IN, Nikolayeva NV, Pomerantseva KP (1969) Polym Sci USSR A11: 2828
33. Hergenrother PM (1973) Appl Polymer Symp 22: 57
34. Tolland GW, Ferstandig LL, Heaton CD (1961) J Am Chem Soc 83: 1151
35. Volksen W (1987) In: Weber WD and Gupta MR (eds) Recent advances in polyimide science and technology, Mid-Hudson Section of Soc of Plast Eng, Poughkeepsie, New York, p 102
36. Tsapovetskii MI, Laius LA, Zhukova TI, Shibayev LA, Stepanov NG, Besonov MI, Koton MM (1988) Poly Sci USSR 30: 295
37. Kim YJ, Glass TE, Lyle GD, McGrath JE (1993) Macromolecules 26: 1344
38. Cotts PM, Volksen W (1984) In: Davidson T (ed) Polymers in electronics ACS Symp Series 242, ACS, Washington, D.C., p 227
39. Volksen W, Cotts PM, Yoon DY (1987) J Polym Sci Polym Physics, 25: 2487
40. Searl NE (1958) US Patent 2,444,536
41. Roderick WR (1957) J Am Chem Soc 79: 1710
42. Cotter RJ, Sauers CK, Whelan JM (1961) J Org Chem 26: 10
43. Vinogradova SV, Vygodskii YaS, Vorob'ev VD, Churochkina NA, Chudina LI, Spirina TN, Korshak VV (1974) Pol Sci USSR 16(3): 584
44. Koton MM, Meleshko TK, Duryavtsev VV, Nechayev PP, Kamzolkina YeV, Bogorad NN (1982) Pol Sci USSR 24: 791
45. Angelo RJ, Golike RC, Tatum WE, Kreuz JA (1987) Mid-Hudson Chapter of Soc Plas Eng., Poughkeepsie, New York p 67
46. Roderick WR. (1964) J Org Chem 29: 745
47. Wallach ML (1968), J Polym Sci A-2: 953
48. Czornyj G, Chen KR, Prade-Silva G, Arnold A, Souleotis H, Kim S, Ree M, Volksen W, Dawson D, Dipietro R (1992) Proceed 42nd Electr Comp and Techn Conference, IEEE, New York, New York, p 682
49. Ahne H, Kuehn E, Rubner R (1981) US Patent 4,366,230
50. Yoda N, Hiramoto H (1984) J Macromol Sci-Chem A21 (13&14): 1641
51. Reynolds RJW, Seddon JD (1968) J Polym Sci C (23): 45
52. Kreuz JA, Endrey AL, Gay FP, Sroog CE (1966) J Polym Sci A1 (4): 2607
53. March J (1968) Advanced organic chemistry: Reactions, mechanism and structure. McGraw-Hill, New York, p 323
54. Merker RL, Scott MJ (1961) J Org Chem 26: 5180
55. Flaim TD, Horter BL, Moss MG (1988) In: Polyimides: Synthesis, characterization and application. Proceed of 3rd Int Conf on Polyimides, Mid-Hudson Section Soc Plast Eng, Poughkeepsie, New York, p 62
56. Neth Pat App (1965) 6,413,550
57. Neth Pat Appl (1965) 6,413,551
58. Tan L, Arnold FE (1988) ACS Polym Prepr 29: 316
59. Volksen W, Pascal T, Labadie J, Sanchez M (1992) Proceed ACS Div of Polym Materials: Sci and Eng 66: 235
60. Delvigs P, Hsu LC, Serafini TT (1970) Pol Letters 8: 29
61. Bell VL, Jewell RA (1967) J Polym Sci A-1: 3043

62. Volksen W, Diller R, Yoon DY (1987) In: Weber WD and Gupta MR (eds) Recent advances in polyimide science and technology. Mid-Hudson Chapter of the Soc of Plast Eng, Poughkeepsie, New York, p 102
63. Houlihan FM, Bachman BJ, Wilkins CW, Pyrde CA (1989) Macromolecules 22: 4477
64. Diller RD, Arnold AF, Cheng YY, Cotts PM, Khojasteh M, Macy EH, Shah PR, Volksen W (1989) US Patent 4,849,501
65. Nishizaki S, Moriwaki T (1968) J Chem Soc Japan Ind Chem Sec 71: 559
66. Khar'kov SN, Krasnov PYe, Lavrova ZN, Baranova SA, Aksenova VP, Chegolya AS (1971) Polym Sci USSR A13: 940
67. Khorshak VV, Vinogradova SV, Vygodskii YaS, Geraschenko ZV (1971) Polym Sci USSR A13: 1341
68. Ueda M. Mochizuki A, Hiratsuka I, Oikawa H (1985) Bull Chem. Soc Jpn 58: 3291
69. Volksen W (1990) In: Hergenrother PM (ed) Symposium on Recent Advances in Polyimides and other High Performance Polymers, C-1
70. Morgan PW (1965) Condensation polymers by interfacial and solution methods. Interscience, New York
71. Volksen W (unpublished results)
72. Greber G, Lohmann D (1969) Angew Chem Internat Edit 8: 899
73. Oishi Y, Kakimoto M, Imai Y (1987) Macromolecules 20: 703
74. Oishi Y, Kakimoto M, Imai Y (1988) Macromolecules 21: 547
75. Ghose BN (1976) J Organomet Chem 164: 11
76. Oishi Y, Harada S, Kakimoto M, Imai Y (1992) J Polym Sci: Part A: Polym Chem. 30: 1203
77. Hayano F, Komoto H. (1972) J Polym Sci A-1: 1263
78. Staab HA, Wendell K., Datta A (1966) Liebigs Ann Chem 694: 78
79. Volksen W, Yoon DY, Hedrick JL (1992) IEEE Trans of Components, Hybrids, and Manuf Techn 15: 107
80. Cotts PM, Volksen W (1990) Polymer News 15: 106
81. Hodgkin JH (1976) J Appl Polym Sci 20: 2339
82. Volksen W, Yoon DY, Hedrick JL (1991) In: Grubb DT, Mita I, Yoon DY (eds) Materials science of high temperature polymers for microelectronics, MRS Symposium Proc, MRS, Pittsburgh, 227: 23
83. Laius LA, Tsapovetsky MI (1984) In: Mittal KL (ed) Polyimides: Synthesis, characterization, and applications, vol 1. Plenum, New York, p 295
84. Stoffel N, Kramer EJ, Volksen W, Prasad K, Russell T (1992) Bull of Am Phys Soc 37(10:511 (K34 31)
85. Kim YH, Walker GF, Kim J, Park J (1987) J Adhesion Sci Technol 1: 331
86. Kim YH, Kim J., Walker GF, Feger C, Kowalczyk SP (1988) J Adhesion Sci Technol 2: 95
87. Molodtsova YeD, Timofeyeva GI, Pavlova SA, Vygodskii YaS, Vingradova SV, Korshak VV (1977) Polym Sci USSR A19: 399
88. Kim SH, Cotts PM (1991) J Polym Sci Phys Ed 29: 109
89. Lange's Handbook of Chemistry. Dean JA, (ed) McGraw-Hill, New York (1985), p 5–14
90. McKean DR, Wallraff GM, Volksen W, Hacker NP, Sanchez MI, Labadie JW (1992) Proceed Polym Mat Sci and Eng 66: 237
91. Fryd M, Merriman BT Jr (1985) US Patent 4,533,574
92. Lavin E, Longmeadow, Markhart AH, Kass RE (1967) US Patent 3,347,808
93. Serafini TT, Delvigs P, Lightsey GR (1972) J Appl Polym Sci 16: 905
94. Johnston JC, Meador MB, Alston WB (1987) J Polym Sci Polym Chem 25: 2175
95. Economy JE, Diller RD, Volksen W, Yoon DY (1984) US Patent 4,467,000
96. Wurtz A (1854) Ann 42(3): 554
97. Meyers RA (1967) J Polym Sci A-1(7): 2757
98. Farrissey WJ, Jr., Rose JS, Carleton PS (1970) J Appl Polym Sci 14: 1093
99. Carleton PS, Farrissey WJ, Jr., Rose JS (1972) J Appl Polym Sci 16: 2893
100. D'Olieslager W, DeAguirre I (1967) Bull Soc Chim France 179
101. Alvino WM, Edelman LE (1975) J Appl Polym Sci 19: 2961
102. Alvino WM, Edelman LE (1978) J Appl Polym Sci 22: 1983
103. Ulrich H (1976) J Polym Sci: Macromol Rev 11: 93
104. Spring FS, Woods JC (1945) J Chem Soc 625
105. Neth Pat Appl (1965) 6,413,552
106. Nefkens GHL, Tesser GI, Nivard RJF (1960) Rec trav 79: 688
107. Imai Y (1970) J Polym Sci: Polym Let 8: 555

108. Takekoshi T (1974) US Patent 3,847,870
109. Takekoshi T, Kochanowski EJ (1974) US Patent 3,850,885
110. Webb JL (1988) Eur Patent 132,457
111. Rogers ME, McGrath JE (1993) Polym Preprints 34(2): 644
112. Kim JH, Moore JA (1993) Macromolecules 26: 3510
113. Numata S, Fujisaka K, Kinjo N (1984) In: Mittal KL (ed) Polyimides: Synthesis, characterization and applications. Plenum, New York, 1: 259
114. Encyclopedia of Polymer Science and Technology (1985) Wiley-Interscience, New York, 2, p 324
115. Feger C (1989) In: Lupinski JH and Moore RS (eds) Polymeric materials for electronics packaging and interconnection. ACS Symposium Series 407, ACS, Washington, D.C., p 114
116. Ree M, Yoon DY, Volksen W (1991) J Polym Sci Part B: Polym Physics 29: 1203
117. Cotts PM, Volksen W, Ferline S (1992) J Polym Sci Part B: Polym Physics 30: 373
118. Fjare D (1993) Macromolecules 26: 5143
119. Rojstaczer S, Ree M, Yoon DY, Volksen W (1992) J Polym Sci Part B: Polym Physics 30: 133
120. Jensen BJ, Hergenrother PM, Bass RG (1989) Proceed Poly Mat Sci and Eng 60: 294
121. Johnson BL, Yilgor I, McGrath JE (1984) Polymer Preprints 25: 54
122. Hedrick JL, Labadie JW, Russell TP, Palmer T (1991) Polymer 32: 950
123. Numata S, Fujisaki K, Kinfo N, Imaizuma J, Mikami Y (1987) US Patent 4,690,999
124. Numata S, Fujisaki K, Makino D, Kinjo N (1987) In: Weber WD and Gupta MR (eds) Recent advance in polyimide science and technology. Mid-Hudson Section of Soc of Plast Eng, Poughkeepsie, New York, p 164
125. Hedrick JL, Volksen W, Mohanty DK (1993) Polymer Bulletin 30: 33
126. Auman BC (1993) In: Feger C, Khojasteh MM, Htoo MS (eds) Advances in polyimide science and technology. Technomic, Lancaster, PA, p 75
127. Auman BC (1993) Macromolecules 26: 2779
128. Mecham SI, Roger ME, Kim Y, McGrath JE (1993) Polymer Preprints 34: 628
129. Ishibashi S, Hyrayama M (1992) Jpn Kokai Tokkyo Koho JP 04,110,392, CA:118: 113292
130. Sato N (1992) Jpn. Kokai Tokkyo Koho JP 04, 109,222 CA: 118: 113293
131. Adrova NA, Bessonov MI, Laius LA, Rudakov AP (eds) (1979) Polyimides: A new class of thermally stable polymers. Progress in Polymer Science Series, vol 7, Technomic, Stamford, CT
132. Feger C, Khojasteh MM, McGrath JE (eds) (1989) Polyimides: Materials, chemistry and chararcterization. Elsevier, Amsterdam
133. Wilson D, Stenzenberger HD, Hergenrother PM (eds) (1990) Polyimides. Blackie, Glasgow
134. Mittal K (ed) (1984) Polyimides: Synthesis, characterization and applications. Plenum, New York, vols 1 and 2
135. Weber WD and Gupta MR (eds) (1987) Recent advances in polyimide science and technology. Mid-Hudson Chapter of the Soc of Plast Eng, Poughkeepsie, New York
136. Feger C, Khojasteh MM, Htoo MS (1993) Advances in polyimide science and technology. Technomic, Lancaster, PA

Received December 1993

Addition Polyimides

H. D. Stenzenberger
Technochemie GMBH-Verfahrenstechnik, (A member of the Royal Dutch/Shell
Group of Companies), Gutenbergstraße 2, 69221 Dossenheim, FRG

Fiber composites and moldings based on addition poly(imides) possess an excellent balance of mechanical, thermal and electrical properties. However, currently available systems suffer from processability and long term ageing stability. This review article will examine the chemistries and concepts used to synthesize addition poly(imides) such as maleimides, ethynyl terminated imide oligomers and PMR resins. Key properties of selected resins in their cured and uncured form are presented and discussed. When appropriate, structure-property or composition-property relations are highlighted to explain and understand performance characteristics. The concept to thoughen bismaleimide resins with engineering thermoplastics is discussed extensively. Particular attention is given to more recent developments which are driven by the requirement for significantly improved long term ageing performance.

Advances in Polymer Science, Vol. 117
© Springer-Verlag Berlin Heidelberg 1994

1 Introduction

Addition poly(imide) oligomers are used as matrix resins for high performance composites based on glass-, carbon- and aramide fibers. The world wide market for advanced composites and adhesives was about $70 million in 1990. This amounted to approximately $ 30–40 million in resin sales. Currently, epoxy resins constitute over 90% of the matrix resin materials in advanced composites. The remaining 10% are unsaturated polyester and vinylester for the low temperature applications and cyanate esters and addition poly(imides) for high temperatures. More recently thermoplastics have become important and materials such as polyimides and poly(arylene ether) are becoming more competitive with addition polyimides.

The total use of advanced composites is expected to grow to around $250–300 million by 1996. The market sectors for advanced polymer composites are 40% aerospace, 30% recreational, 20% industrial, 4% transport and 6% others in 1987. The fastest growing area is aerospace which, according to actual forecasts, will use up to 65% of the total advanced polymer composites production in 1996.

The applications for addition polyimides are for electronic/electrical materials such as printed circuit boards and insulators, as matrix resins for structural composites in aircraft and as thermal insulation materials. Recently the market for polyimide based structural composites has suffered from the termination and reduction of military aircraft programs. However, the possible emergence of the High Speed Civil Transport program (HSCT) may offer an opportunity for addition polyimides (2).

Definition of addition polyimides. Addition polyimides are best defined as low molecular weight, at least difunctional monomers or prepolymers or mixtures thereof, that carry functional reactive terminations and imide functions on their backbone (Fig. 1). The reactive endgroups can undergo homo- and/or copolymerisations by thermal or catalytical means. According to the general definition, the addition polyimide can be synthesized via the classical route of reacting the tetracarboxylic dianhydride and the diamine in the presence of a monofunctional endcapper. The endcapper carries a functional group susceptible to polymerization, copolymerization or crosslinking. Accordingly, addition polyimides (thermosetting polyimides) are classified by the chemical nature of their reactive endgroups. For example, an imide oligomer containing maleimide endgroups, is therefore described as a "Maleimide-Resin". The molecular weight and the molecular weight distribution of the imide backbone can be tailored in the usual way via the stoichiometry of the tetracarboxylic dianhydride and the aromatic diamine and the synthetic conditions.

Within this chapter, a review is given on the "classical" approaches to introduce polymerizable endgroups into the (imide) oligomer backbone. The

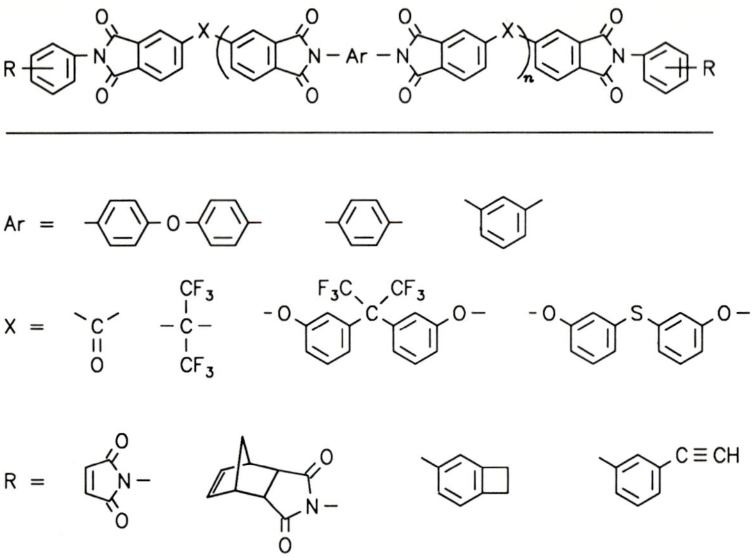

Fig. 1. Chemical structure of addition polyimide

properties of selected resins in their cured and uncured forms are presented and discussed. Particular attention is paid to more recent developments. The driving force behind the numerous developments is the need for high temperature imide polymers which provide good processability whilst maintaining good thermal properties.

2 Historical Perspectives

Addition polyimides are a relatively young class of polymers. Approximately five years after linear polyimides such as poly(4,4'-oxydiphenylene pyromellitimide) demonstrated their outstanding thermal properties, the first publication on addition polyimides appeared. In 1964, a patent was granted to Rhone Poulenc, France, for crosslinked polyimides obtained through the homo- and/or copolymerization of bismaleimides (3). Shortly afterwards a series of patents were issued on poly(amino bismaleimides) which are synthesized from bismaleimide and aromatic diamine (4). A number of bismaleimide resins based on this chemical concept became commercially available through Rhone Poulenc, France and were promoted for their application in printed circuit boards and molding compounds. Rhone Poulenc deserves the honour of having recognized the potential of bismaleimides as building blocks for temperature resistant addition poly(imides).

Following these patent disclosures, research organizations involved in air-craft/aerospace material developments recognized the potential of thermo-setting imide oligomers as resins for high temperature composites and adhesives.

In the late 1960s, the NASA Lewis Research Center sponsored the develop-ment of imide oligomers which were terminated with nadimide groups. A patent was granted to TRW-Systems in 1970 and the first publication was made in 1971 (5, 6). This new chemical approach was used by CIBA GEIGY in their first commercial polyimide product. A solution of the amide-acid prepolymer based on 4,4'-methylenedianiline (MDA), benzophenone tetracarboxylic dianhydride (BTDA) and nadic anhydride (NA) in dimethylformamide as a solvent with a molecular weight of 1300 g/mol (Mn) was sold under the trade name P13N. This varnish was promoted as a resin precursor for glass fabric laminates.

The concept of norbornene (nadic) terminated polyimides was further developed by the NASA Lewis Research Center and culminated in the so-called PMR resin concept. In 1972, Serafini published the first paper on the PMR-15 resin (7). PMR stands for Polymerization of Monomeric Reactants and the 15 for an average molecular weight of 1500 g/mol. The resin is a blend of monomers in which the benzophenone tetracarboxylic acid and the nadic acid are employed in their half ester form with MDA dissolved in a low boiling alkohol. The driving force for the development of the PMR- concept was to improve processing and to resolve problems such as resin flow and shrinkage, solution stability of amide-acid prepolymers in DMF and so forth, inherent in the P13N chemistry. In the early days of development, the main target was to achieve easy processing in combination with newly developed fibers to advanced composites. PMR-15, although an early development, is still the most widely used addition imide resin for advanced composites.

Around 1970, research was initiated under US-Air Force funding on acetyl-ene (ethynyl) terminated imide oligomers. The first publication (8) and a patent (9) appeared in 1994. Within the early development phase it was thought that curing was through ethynyl-group trimerization. The resulting aromatic ben-zene ring structure should provide a high thermal oxidative stability. One resin based on the ethynyl- cure chemistry became commercial in 1975 through the Gulf Chemicals Development Company.

Within a period of only ten years between 1964 and 1974 research in the area of curable imide oligomers produced cure concepts through maleimide-, nadim-ide- and ethynyl terminal groups. Some basic understanding on influencing the processability was developed through modification of the molecular weight, molecular weight distribution and backbone chemistry of the resins.

The following period between 1975 and 1985 was characterized by activities related to the development of prepreg systems for low pressure autoclave moulding, i.e. modification of the resin chemistries to achieve flow, tack and non-volatile cure. These requirements were dictated by the industry to meet the processing techniques already in place for epoxy and polyester resins.

Melt processable bismaleimide systems were disclosed in a US patent granted to Technochemie (now a member of the Royal Dutch/Shell Group of Com-

panies) which are specifically designed for filament winding (10). The approach
of blending selected bismaleimide monomers and further formulating them with
reactive diluents or comonomers provided formulated bismaleimide resins
which were equivalent in processing to the conventional 177 °C cured epoxy
resins.

The breakthrough with respect to application as tacky autoclaveable prepreg was achieved by US Polymeric (now British Petroleum Chemicals) with
their V378A BMI system (11). This system was used to build the delta wings of
General Dynamic's F16XL demonstrator combat aircraft (12) and found application for parts of the Mc Donnell Douglas AV8-B vertical take-off aircraft
(13).

The major concern with BMI resins has been their high inherent brittleness
due to their high crosslink density. Ciba Geigy however, demonstrated that
BMI/*o,o'*-diallyl   bisphenol-A   copolymers   are   tougher   than
tetraglycidylmethylenedianiline/4,4'-diaminodiphenyl sulfone (TGMDA/DDS)
epoxy resins (14). It is interesting to note that these BMI/DABA copolymers
were patented in Switzerland in 1975, but the significance of this invention i.e.
their toughness, was not recognized until toughness became an issue in the
aerospace industry.

Attempts have also been undertaken to improve the processability of PMR
imide resins through molecular weight adjustments and exchange of the monomers employed. LARC 160 as an example; here Jeffamine AP22, a eutectic blend
of MDA type amines, was used as a polyamine instead of the crystalline MDA.
This modification provided a "quasi" melt processable PMR resin (15). Other
modifications were studied with the aim of improving the thermal oxidative
stability by using hexafluoroisopropylidene dipthalic anhydride as a monomer
(16).

More recently PMR resin modifications were developed because MDA used in
PMR-15 is potentially hazardous. PMR-15 which is the most widely used
addition polyimide in composites, is under pressure because of the new OSHA
regulations. The target is to employ diamines which are not toxic or to prereact
the diamines to avoid the presence of free diamines.

A novel and very attractive approach to cure imide oligomers via
benzocyclobutene endgroups was first published in 1985 (17). Benzocyclobutene
endgroups are particularly attractive because they are neat hydrocarbons
(provide low dielectrics), cure thermally around 200–250 °C with no volatile
evolution and copolymerize with other functional groups, for instance, maleimide.

3 Bismaleimides

Bismaleimides are an important class of addition polyimides. They are known
for their excellent processability and balance of thermal, electrical and mechan-

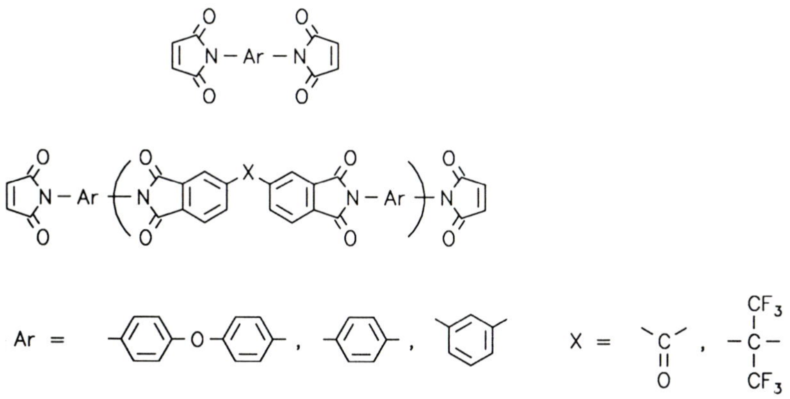

Fig. 2. Chemical structure of bismaleimide

ical properties. They are employed as resins in advanced composites and electronics such as multilayer and printed circuit boards. The general structure of a bismaleimide building block is provided in Fig. 2.

The maleimide group can undergo a variety of chemical reactions. The reactivity of the double bond is a consequence of the electron withdrawing nature of the two adjacent carbonyl groups which create a very electron-deficient double bond, and therefore is susceptible to homo- and copolymerizations. Such polymerizations may be induced by free radicals or anions. Nucleophiles such as primary and secondary amines, phenates, thiophenates, carboxylates, etc. may react via the classical Michael addition mechanism. The maleimide group furthermore is a very reactive dienophile and can therefore be employed in a variety of Diels Alder reactions. Bisdienes such as divinylbenzene, bis(vinylbenzyl) compounds, bis(propenylphenoxy) compounds and bis(benzocyclobutenes) are very attractive Diels Alder comonomers and therefore some are used as constituents for BMI resin formulations. An important chemical reaction of the maleimide group is the "ENE" reaction with allylphenyl compounds. The most attractive comonomer of this family is DABA particularly when tough bismaleimide resins are desired.

Within this section the most important chemical concepts for bismaleimide resins are discussed. Attention is paid to the classical approaches and also to more recent developments.

3.1 Building Blocks

A standard synthesis for N,N'-arylene bismaleimides is the cyclodehydration of N,N'-arylene bismaleamic acids with acetic anhydride in the presence of a catalyst (18). The yield of pure recrystallized BMI is usually 65–75% when sodium acetate is employed as a catalyst. Side reactions which lead to the

formation of isoimides, acetanilides, maleimide-acetic acid adducts and products with mixed functionalities are responsible for the relatively low yield of
pure bismaleimide. The cyclodehydration conditions such as temperature, type
and concentration of catalyst and solvent influence the yield of these byproducts
(19).

It is well documented that the isoimide is the kinetically favoured product
and that isomerization yields the thermodynamically stable imide when sodium
acetate is used as the catalyst. High catalyst concentrations provide maleimides
with low isoimide impurity. The mechanism by which the chemical imidization
is thought to occur is shown in Fig. 3. The first step in the dehydration reaction
may be formation of the acetic acid-maleamic acid mixed anhydride. This
species could lose acetic acid in one of the two ways. Path A involves
participation by the neighboring amide carbonyl oxygen to eject acetate ion
with simultaneous or subsequent loss of proton on nitrogen to form the
isoimide. Path B involves loss of acetate ion assisted by the attack of nitrogen
with simultaneous or subsequent loss of the proton on nitrogen to form the
imide. If the cyclodehydration is run in acetic anhydride in the absence of the
base catalyst, isoimide is the main reaction product.

At high temperatures with low catalyst concentration the formation of
acetanilides is favored. Maleic anhydride and acetanilides may be formed
directly from the mixed anhydride by an initial attack of the nitrogen on the
acetate carbonyl, but this process would involve a seven membered ring
transition state. Another possible route to the formation of maleic anhydride
and the acetanilides is participation by neighboring carbonyl in loosening the
amide carbon-nitrogen bond to the extent that the amine can be captured by
acetic anhydride as shown in path D.

Another base catalyzed side reaction is the formation of Michael adducts
between the maleimide and the acetic acid present in the reaction mixture. In
general, both the acetanilides and the Michael adducts are favored when high
catalyst concentrations and high temperatures are used for the synthesis.

The patent literature on the synthesis of bismaleimides is quite extensive. It
has always been the target to optimize the synthetic conditions in order to
maximize the yield of pure bismaleimide by minimizing the side reactions. Low
cyclization temperatures were found to be advantageous and could be achieved
with catalysts such as triethyl amine (20), cobalt acetate (21) and sodium
carbonate (22) or metal salts in combination with tertiary amines (23).

The various side reactions of the chemical cyclodehydration and because of
the high cost for solvents, catalyst and cyclodehydration agent, researchers have
been looking at more economic ways to manufacture bismaleimides. Efforts
have been directed towards a catalytic cyclodehydration process via azeotropic
distillation to avoid undesirable byproducts and to achieve improved yield of
pure bismaleimide. The use of Lewis acid/base salts based on p-toluene sulfonic
acid, sulfuric acid or trifluoroacetic acid and dimethylformamide (DMF), N-
methylpyrrolidone (NMP) and acetone as bases provided high yields of high
purity bismaleimide (24). In another patent dimethyldialkylammoniummethane

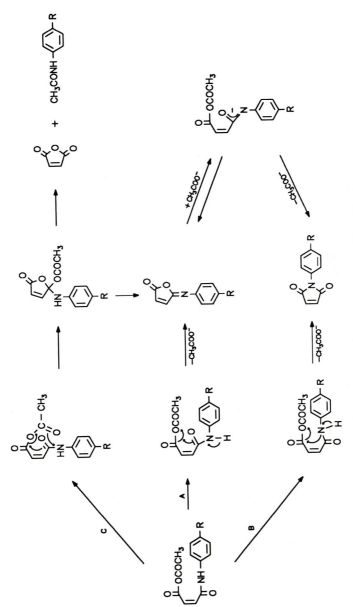

Fig. 3. Chemistry involved in the base catalysed cyclodehydraton on N,N'-bismaleamic acids

sulfonate is claimed as a useful catalyst (25). Japanese workers have developed a process for *N*-substituted maleimides which comprises heating maleic anhydride and aromatic or aliphatic primary diamine in the presence of an ion exchange resin in an organic solvent at 50–160 °C to effect cyclodehydration (26). Research in this area is still ongoing and BMI manufacturers are improving their processes in order to improve the properties of their products and competitiveness in the market place.

The most widely used BMI building block is 4,4′-bismaleimidodiphenylmethane (MDA-BMI), because the precursor diamine is readily available and cheap. However, almost every aromatic amine or diamine can be converted to the corresponding maleimide or bismaleimide respectively. Of interest to resin formulators are the melting points and polymerization characteristics of the BMI building blocks in order to tailor the flow and cure properties of their products. Low melting building blocks are desired for tacky prepreg to be processed in an autoclave at low pressure, for resins to be employed in transfer molding processes and for filament winding.

The BMI building blocks primarily used in commercial bismaleimide resins are MDA—BMI 2,4-bismaleimidotoluene, 1,3-bismaleimidobenzene and, sometimes aliphatic BMIs based on *n*-alkanes and *iso*-alkanes. However, because of toxicity problems associated with MDA and other diamines with only one or two aromatic rings, polyaromatic diamines and BMIs based on them are of increasing interest (Fig. 4). In the past, it was almost impossible to introduce novel BMI building blocks based on polyaromatic diamines because of the processing problem associated with high melting points and high melt viscosities. With new processing techniques such as powder prepregging and blending with liquid reactive diluents such as DABA, such limitations could be partially overcome. It is questionable, however, whether such expensive building blocks will find application in commercial bismaleimide resin systems.

The most widely used process for the fabrication of glass fabric prepreg involves solvent/solution techniques. The resins have to be soluble in appropriate solvents preferrably methylethylketone, acetone or other low boiling solvents. Therefore BMI building blocks with improved solubility are desired. Sometimes bismaleimides are used in epoxy resins to improve the glass transition temperature (T_g), then solubility in epoxy resin is required. It has been found that tetraalkyl- 4,4′-MDA -BMIs(27), which melt at around 150 °C and

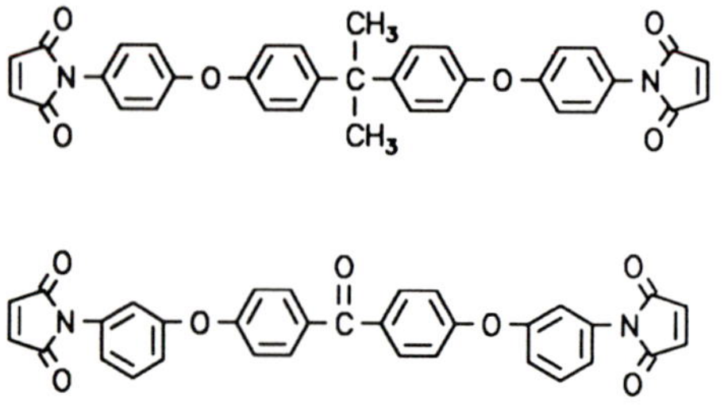

Fig. 4. Bismaleimides of polyaromatic diamines

the BMI based on the isomeric diphenylindane diamine (28) show the desired properties (Fig. 5).

At this point it has to be mentioned that the reactivity of the bismaleimide is influenced by the chemical nature of the residue between the maleimide groups. Normally, electron donating groups, such as alkyl groups, reduce the BMI reactivity, provided they are present in the phenylene ring that carries the maleimide group. This is the case for the tetra alkyl MDA-BMIs and the diphenylindane BMI. On the other hand, electron attracting groups such as SO_2, C = O, and so on, have the opposite effect. Increasing the molecular weight between the terminating maleimide groups generally provides a more latent system owing to steric effects.

The most important application for bismaleimide resin is multilayer boards. The development in this area requires resins with low dielectric constants. It is well documented in the literature that fluorine containing linear polyimides show lower dielectric constants vis a vis their non-fluorinated counterparts. Recently, Hitachi Research Laboratory, Japan, reported the thermal and dielectric behaviour of fluorine-containing bismaleimides (29). The chemical structures of the fluorinated BMIs investigated are provided in Fig. 6. The non-fluorinated four aromatic rings containing BMI, 4,4'-bis(p-maleimidophenoxyphenyl) propane, was tested in comparison.

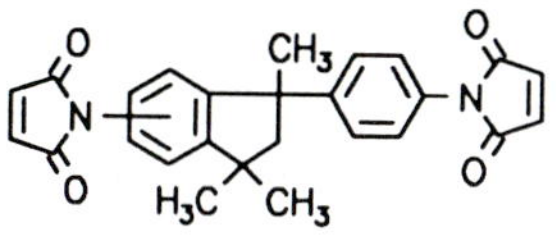

Fig. 5. Bismaleimide based on isomeric diphenylindane diamine

Mol.wt. = 815, Mp = 112°C, DC(polymer) = 2.8

Mol.wt. = 679, Mp = 136°C, DC(polymer) = 3.0

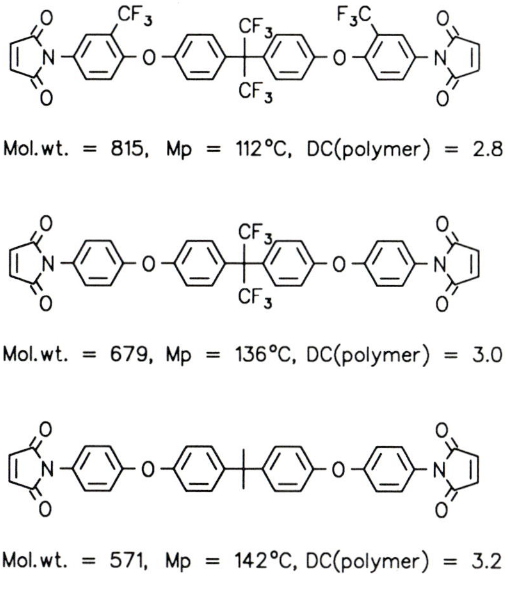

Fig. 6. Fluorine-containing bismaleimides

Mol.wt. = 571, Mp = 142°C, DC(polymer) = 3.2

It is interesting to note that the fluorine-containing BMIs are lower melting than the fluorine-free BMI. Furthermore, the expected reduction of the dielectric constants for the cured BMIs measured at 1 MHz was achieved. The values calculated by using the Clausisus-Mosotti equation are in reasonable agreement with the experimental results.

BMIs and maleimide-terminated prepolymers have been considered for systems with improved Fire, Smoke and Toxicity properties. Of particular interest are phosphorous-containing bismaleimides because they provide high Limiting Oxigen Index (LOI) values (30). 3,3'-bis(maleimidophenyl) methyl phosphine oxide is such a compound (Fig. 7).

Parker et al. (30) reported the synthesis of phosphorous containing bismaleimides and demonstrated their outstanding non-flammability characteristics. Graphite fabric composites prepared from such P-containing BMI as a matrix resin show a LOI of 100 and an anaerobic char yield of 88% at 700 °C. Other P-containing BMI have recently been synthesized (31) for example, bis(3-maleimidophenoxy-4-phenyl)phenyl phosphine oxide (Figure 8) has been tested as a composite matrix resin with the aim of improving the fibre/resin interfacial adhesion.

It is still considered desirable to synthesize new BMIs, in order to improve the processing properties of resins which they constitute and to improve the physical properties of the cured polymer. Goldfarb and coworkers (32) recently published structure-property relationships of bismaleimides containing oxyalkylene linkages (Fig. 9). By choosing the appropriate size of the main chain and pendant aliphatic groups in bismaleimides containing oxyalkylene linkages it is possible to control the breadth of the cure window without adversely affecting either thermal stability or the glass transition temperature of the cured polymer.

The most popular route to bismaleimides is the reaction of a diamine with maleic anhydride followed by cyclodehydration. The chemical structure of the

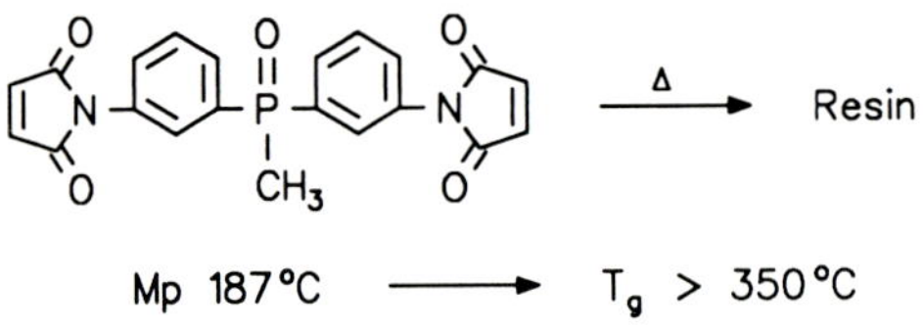

Mp 187 °C $\longrightarrow$ $T_g > 350\,°C$

Fig. 7. Chemical structure of 3,3'-bis(maleimidophenyl) methylphosphine oxide (30)

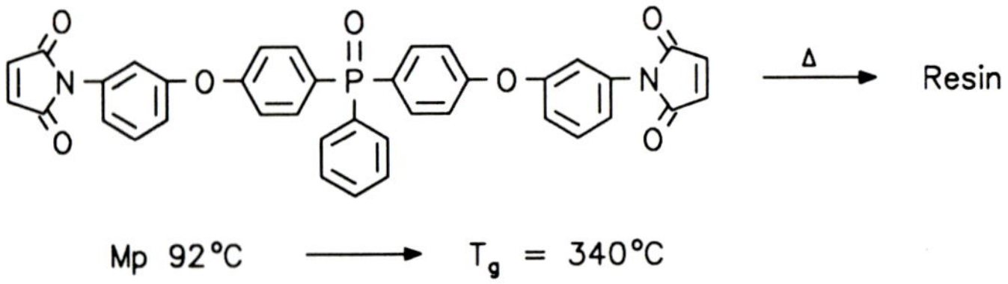

Mp 92 °C $\longrightarrow$ $T_g = 340\,°C$

Fig. 8. Bis (3-maleimidophenoxy)-4-phenyl/phenyl/phosphinooxide

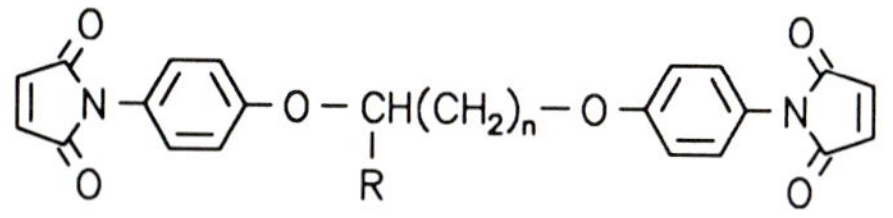

Fig. 9. Chemical structure of bismaleimides containing oxyalkylene groups

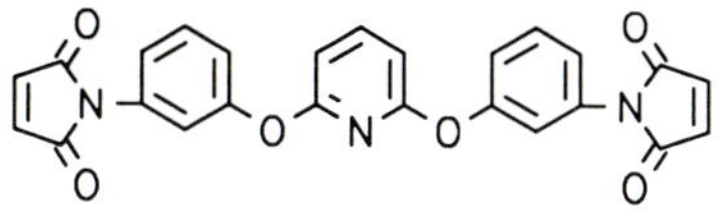

$R = H$,	$n = 1$,	Mp 215 °C	$\xrightarrow{\Delta}$	$T_g >$	400 °C
$R = H$,	$n = 2$,	Mp 122 °C	$\longrightarrow$	$T_g =$	395 °C
$R = H$,	$n = 3$,	Mp 150 °C	$\longrightarrow$	$T_g =$	390 °C
$R = H$,	$n = 4$,	Mp 50 °C	$\longrightarrow$	$T_g =$	375 °C

Mp (monomer) = 137 °C
Polymerisation Energy = 186 kJmol⁻¹
T_g (polymer) = 386 °C
Polymer Decomposition Temperature = 400 °C

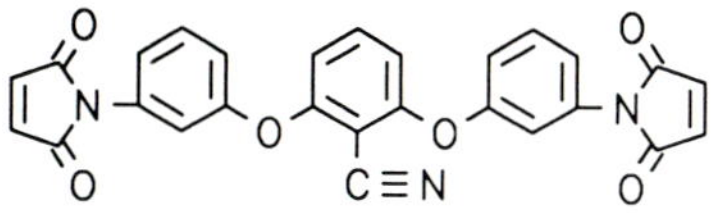

Mp (monomer) = 205 °C
Polymerisation Energy = 109 kJmol⁻¹
T_g (polymer) = 404 °C (TMA)
Polymer Decomposition Temperature = 371 °C (TGA)

Fig. 10. Structures and properties of 2,6-bis(3-maleimidophenoxy)-pyridine and 2,6-bis-(3-maleimidophenoxy)benzonitrile

diamine backbone is responsible for the properties achievable in the corresponding cured bismaleimide. A convenient synthetic route to polyaromatic diamines is the reaction of activated dihalo compounds with *m*- or *p*-amino-phenol employing a nucleophilic halodisplacement reaction. Recently, the synthesis of 3,3'-bis(maleimidophenoxy) pyridine has been published (33). The diamine, bis(*m*-aminophenoxy) pyridine, is available by reacting 2,6-dichloro pyridine with *m*-aminophenol in dimethylsulfoxide as a solvent and potassium carbonate as a catalyst at temperatures around 140 °C. Purification through recrystallization from methanol provides the high purity diamine which is converted to the corresponding BMI in the usual way. 2,6-Bis(3-maleimidophenoxy) pyridine (Fig. 10) could be of interest for use in low melting BMI resin formulations because of its relatively low melting point. The T_g of the cured homopolymer is 386 °C measured via dynamic mechanical analysis.

Another ether-type BMI, 2,6-(3-maleimidophenoxy) benzonitrile (34), is synthesized from 2,6 dichlorobenzonitrile as a precursor, again via the standard route described above. The benzonitrile structural unit in the BMI backbone is responsible for the T_g (404 °C) of the cured homopolymer.

Etherketone-type polyaromatic diamines are easy to synthesize from bis(4-fluorophenyl)ketone or bis(4-fluorobenzoyl)benzene and *m*-, or *p*-aminophenol. These have been converted to the corresponding bismaleimides (35). The melting and polymerization behavior of four isomeric ether ketone bismaleimides are provided for comparison. The "all *para*" linked 4,4'-bis-(*p*-maleimidophenoxy phenyl) benzene shows the highest melting point. The ether-ketone BMIs with the *m*-linkages in the terminating phenylene rings provide the lowest melting points. Interesting to note that the "all *para*" isomer delivers the lowest polymerization energy as measured via differential scanning calorimetry. The "*meta*" isomers although lower melting, show higher polymerization energies to a higher cure conversion. Steric effects are presumably the reason for the lower cure conversion of the "para" isomers.

Table 1. Melting and polymerization properties of isomeric ether-ketone bismaleimides

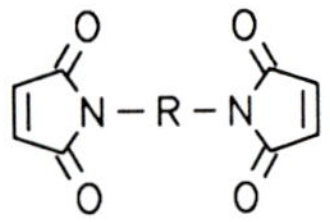

−R−	Mp, °C	T_{max}, °C	H_{pol}, Jg^{-1}
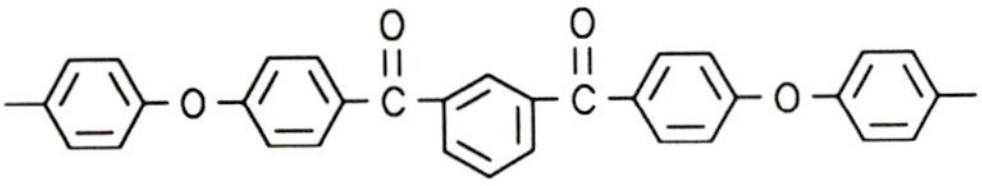	226	285	113
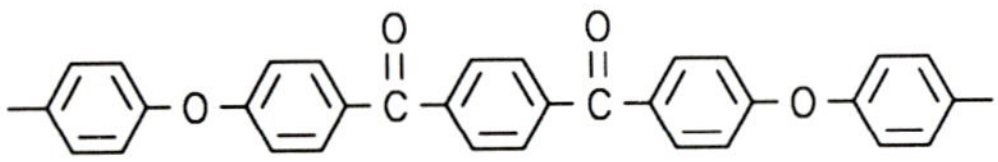	293	301	97
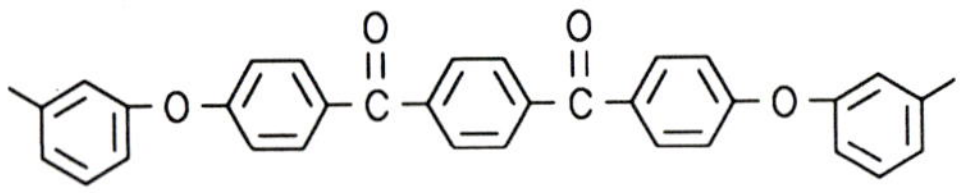	209	270	158
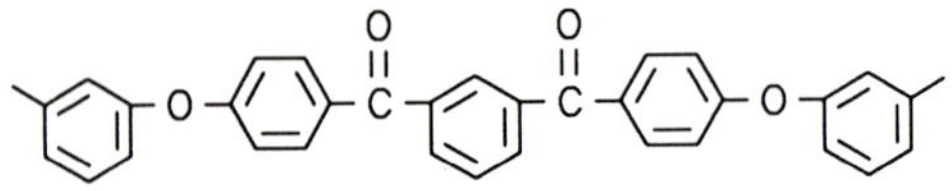	185	288	132

T = polymerization peak maximum, H = polymerization energy

3.2 Maleimide-Terminated Oligomers

Besides low molecular weight building blocks, long chain maleimide terminated oligomers have been synthesized for molding, adhesive and composite applications. The key step is the preparation of an amino terminated intermediate needed to introduce the maleimide group. For example, long-chain thermosetting arylimides synthesized from long-chain sulfone-ether diamines and maleic anhydride have been published. The resulting bismaleamic acid was cyclodehydrated in the usual way with acetic anhydride-sodium acetate (36).

The cured arylimides are heat stable systems possessing high T_gs and moderate to high modulus plateaus above their T_g, depending on the molecular weight between the (maleimide) crosslinks. The sulfone-ether backbone (see Fig. 11) alleviates problems associated with solubility of the uncured oligomer. High solids solutions can be prepared in common organic solvents.

Other backbone structures that have generated interest are the polyether ketones. An attempt was made to synthesize amino-terminated arylene ether ketones, which were subsequently converted into the corresponding maleimide-terminated oligomers (37). The aim of this approach was to obtain tough, solvent resistant, high temperature BMI thermosets.

Recently a convenient method for the synthesis of maleimide-terminated imide oligomers has been described (38). Aromatic diamine, biphenyl tetracarboxylic dianhydride and maleic anhydride are reacted in DMAc/Xylene at 50 °C to form the amic acid oligomer which was subsequently cyclodehydrated by refluxing in the presence of pyridine as a catalyst. Water is removed azeotropically over a period of three hours. The maleimide terminated imide oligomer is isolated by precipitation in water or a non-solvent. The molar ratio of the monomers can be varied widely to tailor the molecular weight and

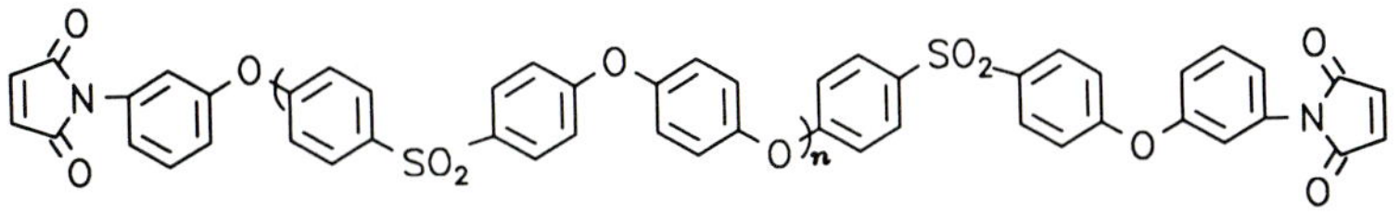

Fig. 11. Structure of maleimide terminated sulfone-ether oligomer

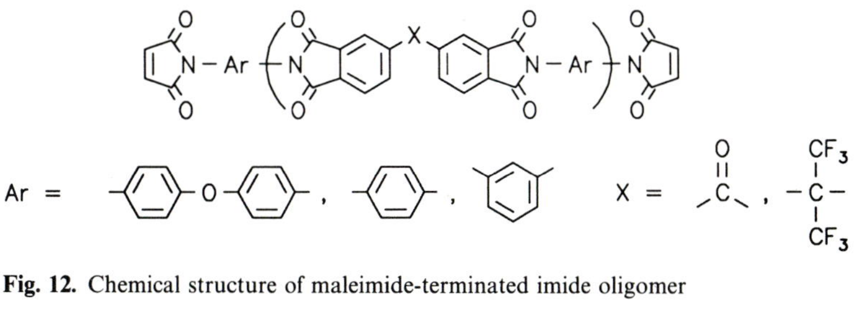

Fig. 12. Chemical structure of maleimide-terminated imide oligomer

solubility of the imide oligomer. This synthetic method is preferred when fully imidized molding powders are desired (Fig. 12).

3.3   Alternative Synthetic Routes to BMI

All BMI building blocks and maleimide-terminated oligomers discussed previously are synthesized from the corresponding polyamine precursor and maleic anhydride simply because of economic reasons. However, there are other synthetic methods available, for instance, the reaction of a functionalized monomaleimide with a polyfunctional monomer or oligomer. Such a functionalized monomaleimide is maleimido benzoic acid or its acid halide. These were used to synthesize maleimide-terminated polyamides (39, 40) or polyesters (41), respectively.

An interesting approach to maleimide-terminated phenoxy resin has recently has described (42). *para*-Maleimidobenzoic acid was reacted with diglycidylbisphenol-A epoxy resin in the presence of catalyst to provide the bismaleimide of Fig. 13. Instead of diglycidyl bisphenol-A, linear epoxy resin prepolymers can be used in this reaction to form a maleimide terminated phenoxy resin. Another suitable functionalized monomaleimide is *m*- or *p*- N-(hydroxyphenyl) maleimide which is synthesized from maleic anhydride and *m*-aminophenol in DMF as a solvent at 70 °C. The purified hydroxyphenyl maleimide was reacted with epoxy resin to form novel BMIs as outlined in Fig. 14. The new BMI and phenoxy oligomers polymerize at temperatures of 200–220 °C, but the cure temperatures can be significantly lowered when catalysts such as imidazoles or triphenylphosphine are added. The cured homopolymers show T_g of 140 and 230 °C for the n = 2 and the n = 1 polymer, respectively(43).

3.4   Bismaleimide Resin Concepts

Bismaleimide building blocks as described previously cannot be considered as resin products because they are either high melting solids or high viscosity melts above their melting points. Although building blocks usually make up 50–75% by weight of resin, other ingredients such as comonomers, reactive diluents, processing additives, elastomers and catalysts are combined with BMI so as to obtain a product suitable for the application considered. The application areas for bismaleimide resins are reinforced composites for printed circuit boards (with glass fabric), structural laminates (with glass, carbon and aramide fibres) and moldings (with short fibers and particulate fillers). In order to fulfill the processing requirements, bismaleimide building blocks have to be formulated into products that enable their use as highly concentrated solutions, molding powders or hot melts.

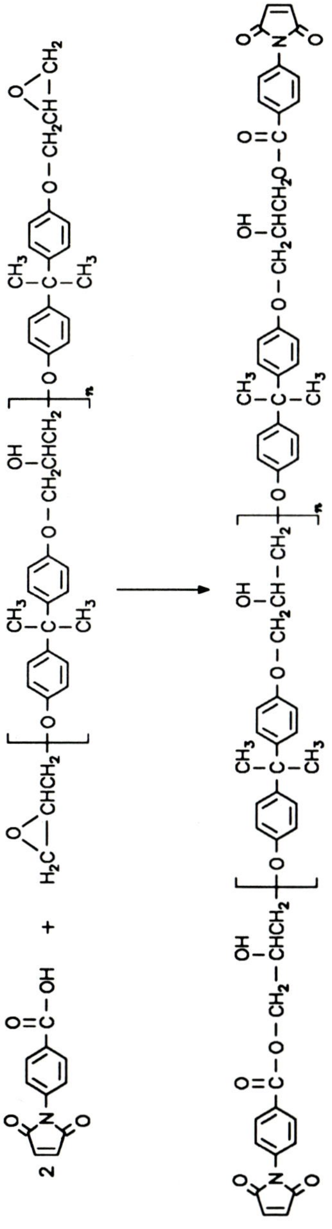

Fig. 13. Maleimide-terminated phenoxy resin obtained from 4-maleimidobenzoic acid (42)

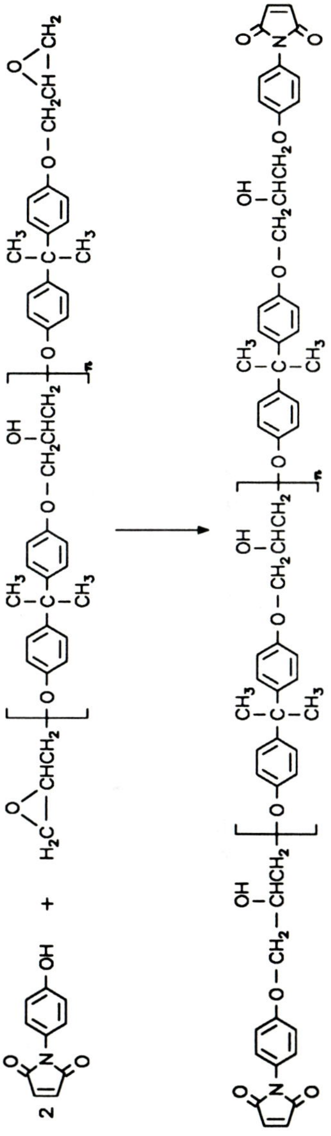

Fig. 14. Maleimide terminated phenoxy resin obtained from hydroxyphenyl maleimide (43)

3.4.1 Poly(aminobismaleimides)

The reaction of a bismaleimide with a functional nucleophile, a diamine for example, via the Michael addition reaction converts a BMI building block into a polymer. The nonstoichiometric reaction of an aromatic diamine with bismaleimide was used by Rhone Poulenc to synthesize polyaminobismaleimides (Fig. 15) (44).

The reaction product of BMI-MDA and MDA, known as Kerimide 601, is prepolymerized to such an extent that the resulting prepolymer is soluble in aprotic solvents such as NMP, DMF, and the like, and therefore can be prepregged via solution techniques. Kerimide 601 is mainly used in glass fabric laminates for electrical applications and became the industry standard for polyimide-based printed circuit boards (45).

The synthesis of BMI/diamine adducts can be performed in solution or in the melt. The molar ratio between bismaleimide and diamine can be varied widely. If excess BMI is employed, a prepolymer with mainly maleimide terminations is formed. This may be cured upon heating to temperatures around 200 °C to form the poly(aminobismaleimide). The BMI is always employed in excess to obtain a resin susceptible to crosslinking.

The BMI-MDA/MDA system has been studied extensively with respect to fracture toughness and T_g as a function of the molar ratio between the BMI and diamine (46). As expected, the fracture toughness, expressed as fracture energy (G_{1c}), increases with increasing diamine concentration. On the other hand the T_g is adversely affected (Fig. 16). For the BMI-MDA/MDA system optimum properties are achieved for a molar ratio of 2.5/1.

Recently, Enoki, Takeda and Ishii published their studies on the mechanical properties of poly(aminobismaleimides) (47). In Table 2, various properties of the MDA-BMI/MDA system as a function of the BMI/MDA molar ratio are shown. The results are consistent with the published data of Tung et al. (46) with respect to the T_g and fracture toughness. Of great interest are the data on specific gravity and water absorption showing increasing water absorption and increasing density with decreasing MDA concentration (Fig. 17). It is also interesting to note that the water absorption increases with increasing cure temperature for a given BMI/MDA molar ratio. Increased cure temperatures lead to a higher free volume in the system and therefore a higher water absorption.

In another recent publication the dielectric and thermal properties (T_g) of a series of polyaminobismaleimides are reported. Both the BMI and diamine were varied systematically with the target of identifying a system with the lowest

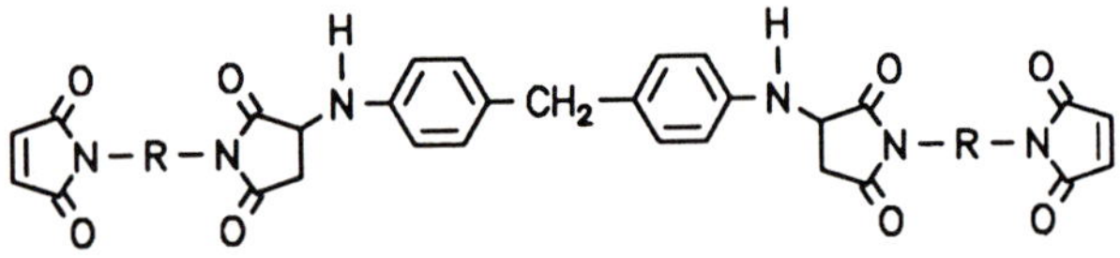

Fig. 15. Michael addition product of BMI with aromatic diamine

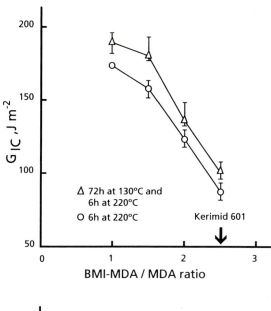

Fig. 16. Fracture energy (Glc) and glass transition temperature (T$_g$) of 4,4′ bismaleimidodiphenylmethane (MDDM) 4,4′-methylenedianiline (MDA) systems (46)

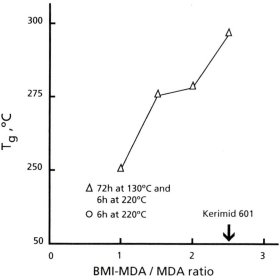

possible dielectric constant but without compromising the thermal properties (48). Table 3 lists the systems which have been screened; all BMI/diamine combinations have molar ratio of 2.5/1.

It is interesting to note that the dielectric constants (DC) varied between 2.7 and 3.7. The lowest DC value was obtained for the resin based on 1,3-(bis-4-aminophenyl)diisopropyl benzene employed both for the BMI and diamine

Table 2 Influence of molar ratio on various properties in BMI/MDA systems

BMI/molar ratio		1.2/1	1.5/1	2/1	3/1
Flexural strength	(MPa)	174	172	183	166
Flexural modulus	(GPa)	3.40	3.48	3.64	3.93
Deflection at breaking	(%)	11.3	11.2	8.00	5.35
K_{Ic}	(MN/m$^{1.5}$)	0.99	0.85	0.73	0.58
G_{Ic}	J/m^2)	242	166	128	72
T_g	(*C)	242	295	297	316

Curing condition: 175 °C/3 h + 230 °C/4 h

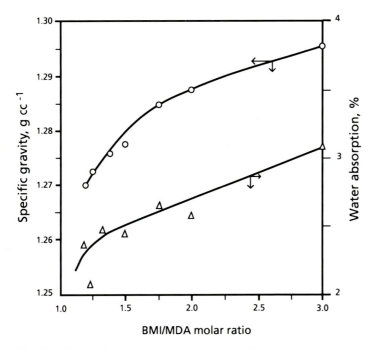

Fig. 17. Influence of BMI/MDA molar ratio on specific gravity and water absorption (47)

chain extender. DCs of 3.0 are achieved for system 6 and 13 (of Table 3) with 4,4'-bismaleimidodiphenylmethane and 1,3-bismaleimidobenzene as the BMI part and both chain extended with bis-4-aminophenyl) m-diisopropyl benzene as the diamine.

Another approach to processable bismaleimide resins via Michael addition chain extension, is the reaction of bismaleimide, or a low melting mixture of bismaleimides, with aminobenzoic hydrazide to provide a resin which is soluble in various solvents, such as acetone, methylene chloride and DMF (49). The idealized chemical structure for a 2:1 BMI-aminobenzoic hydrazide resin is

Table 3. Thermal and dielectric properties of poly(aminobisamleimides)

System No	X	Y	ε	T_g, °C
1	(structure)	(structure)	3.7	281
2	(structure)	(structure)	3.2	251
3	(structure)	(structure)	3.3	242
4	(structure)	(structure)	3.1	238
5	(structure)	(structure)	3.1	–
6	(structure)	(structure)	3.0	204
7	(structure)	(structure)	3.1	210
8	(structure)	(structure)	3.1	229
9	(structure)	(structure)	3.2	237
10	(structure)	(structure)	3.2	236
11	(structure)	(structure)	2.7	–
12	(structure)	(structure)	3.2	–

BMI/Diamine = 2.5 (molar)
Cure: 1h 200°C plus 14h 200°C

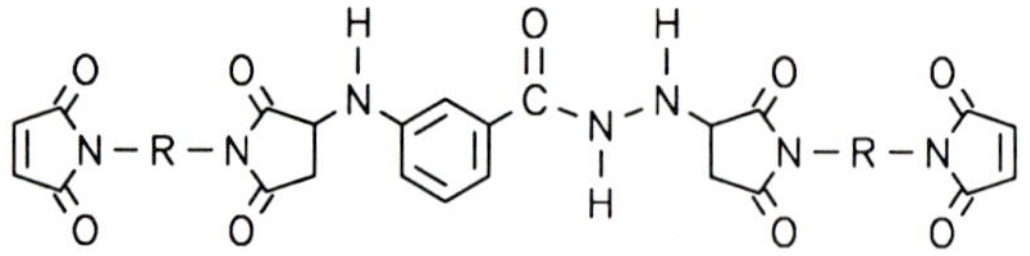

Fig. 18. Chemical structure of BMI/-aminobenzoic hydrazide 2:1 adduct

Table 4. Properties of BMI* printed circuit laminates[b]

Properties	Value
dielectric constant (1 MHz)	4.6–4.7
dielectric loss constant (1 MHz)	0.01
dielectric strength, $V/\mu m$	29.5
volume resistivity, Ω cm	10^{15}
water absorption,[c] %	< 1
peel strength of copper foil,[d] N/mm	1.2–1.4
heat stability at 287 °C, s	> 60
thermal expansion coefficient, $10^6/°C$	
x, y direction	14–16
z direction	36–38
flammability in comparison with UL 94[e]	V^0–V^1

[a] COMPIMIDE 183.
[b] Properties measured using 1–1.5 mm laminates; glass fabric U.S. 2116, resin content 45% by weight.
[c] After postcuring at 210 °C.
[d] Gould TC polyimide grade 35–μm thickness.
[e] V^0 rating for 3–mm laminates; V^1 rating for 1.5–mm laminates

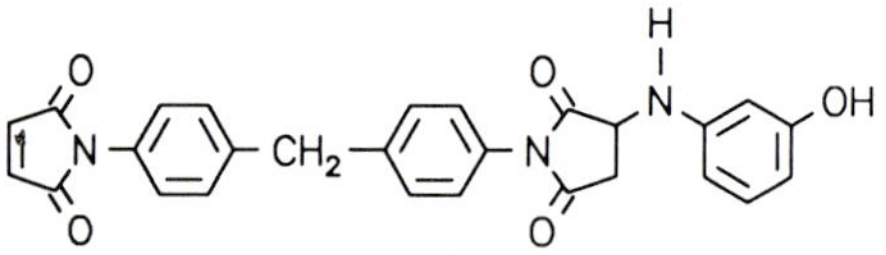

Fig. 19. Chemical structure of a BMI/*m*-aminophenol 1:1 adduct (52)

given in Fig. 18. Two resin systems based on this chemical concept are commercially available from the Shell Chemical Company/Technochemie under the COMPIMIDE trademark: Compimide 183 (50), for use in printed circuit boards, and COMPIMIDE 796 a resin suitable for low pressure autoclave molding (51). Typical properties of COMPIMIDE 183 glass fabric-PCB laminates are provided in Table 4. The Compimide 183 BMI resin offers a combination of advantageous properties, such as high T_g, low thermal expansion coefficient and flame resistance without bromine compound additives.

The Michael addition reaction has attracted many researchers as a route to convert high melting BMI building blocks into resins with improved processability as compared with the BMI building blocks. Heat resistant resin compositions have been prepared from *para*- or *meta*-aminophenol (52). The idealized structure of such a BMI/*m*-aminophenol adduct is shown in Fig. 19.

Many other polyfunctional monomers such as diphenols (53), dicarboxylic acids (54), bisthiols (55), bisthiophenols (56), secondary amines (57), biscyanamines (58) and barbituric acid (59) have been used to modify BMIs via the Michael addition chain extension reaction. These materials, however, are not commercially available.

3.4.2 Epoxy-Modification of Poly(aminobismaleimides)

BMI/amine Michael adduct resins may be further modified and blended with other thermosets or reactive diluents to achieve either specific end-use properties or processability. Epoxy resins are very suitable for the modification of BMI/primary amine adducts, because the secondary amine functionality in the aspartimide structure is a curative for the epoxy group.

A modified BMI-epoxy resin system has been introduced by Shell Chemical Company. The system is a highly reactive blend of a bismaleimide, COM-PIMIDE 1206 (55–60% by weight solution of BMI in DMF), and EPON Resin 1151, a polyfunctional epoxy resin (60). In contrast to many polyimide resins on the market, no free MDA is present in the product. This is an important feature, since MDA has been identified as an animal carcinogen and possibly a human carcinogen. This resin system has been fully evaluated for use in multilayer PCB boards (61). 2-Methylimidazole is recommended as a catalyst. However, if required, the processing window can be widened by using 2-phenylimidazole

Table 5 Properties of BMI–Epoxy[a] Electrical Laminates

Properties		Value
laminate composition, BMI–epoxy	70/30	50/50
flexural strength, 23 °C, MPa[b]	462	441
flexural modulus, 23 °C, GPa[c]	22	21
dielectric constant, 23 °C	4.32	4.48
dissipation factor, 23 °C	0.0083	0.0098
dielectric strength, $V/\mu m$	30	31
volume resistivity, $\Omega \cdot cm$	2.1×10^{16}	1.9×10^{16}
surface resistivity, $\Omega \cdot cm$	3.8×10^{16}	$> 1.9 \times 10^{16}$
flammability, UL-94	V–O	V–O
copper peel, J/m^{2d}	1208	1348
methylene chloride, mg uptake	1.1	1.7
water absorption, wt%		
103 kPa[e] steam, 1 h.	0.28	0.31
boiling, 24 h	1.32	1.24
solder shock[f]	pass	pass
etchability[g]	0.03	0.04
TGA, 5 wt% loss in air, °C	360	335

[a] COMPIMIDE 1206/EPON 1151.
[b] To convert MPa to psi, multiply by 145.
[c] To convert GPa to psi, multiply by 145,000.
[d] To convert J/m^2 to llf/in., divide by 175.
[e] 101.3 kPa = 1 atm.
[f] 20 s at 268 °C 1 h after 103 kPa[e] steam.
[g] 5 min in conc H_2SO_4

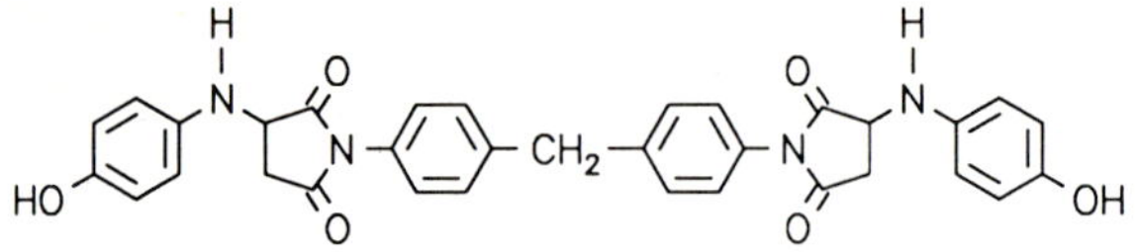

Fig. 20. BMI/*p*-aminophenol 2:1 adduct (63)

which is a more latent catalyst. There is a wide window for lamination of the system, from hot start, single pressure to dual pressure vacuum assistance. The formulation consisting of 70 parts of COMPIMIDE 1206-R60 and 30 parts of EPON Resin 1151-BH60 has an excellent property profile for the manufacture of multilayer circuit boards. Typical properties of Compimide 1206/EPON 1151 electrical laminates are compiled in Table 5.

Heat resistant resin compositions based on BMI/aminophenol-Epoxy blends are achieved by reacting a BMI/*p*-aminophenol 1:1 adduct with epoxy resin (62). Both the secondary amine and phenol functionality may react with the epoxy resin and subsequently cure through an imidazole catalyst. Imidazole catalysts promote both the epoxy/phenol reaction and the anionic maleimide crosslinking. The formation of a 1:2 BMI/aminophenol adduct, as in Fig. 20, is claimed in a patent (63). The hydroxy terminated BMI/aminophenol adduct is an advantageous curing agent for epoxy resins when high temperature performance is desired.

3.4.3 Bismaleimide/Bis(allylphenol)Resins

The copolymerization of a BMI with DABA is a resin concept that has been widely accepted by the industry because BMI/DABA blends are tacky solids at room temperature and therefore provide the desired properties in prepregs, such as drape and tack, similar to epoxies. Crystalline BMI can easily be melt-blended with DABA, which is a high viscosity fluid at room temperature. Upon heating BMI/DABA blends copolymerize via complex ENE and Diels-Alder reactions as outlined in Fig. 21. Independent NMR-studies (64, 65) show that the reactive maleimide C = C bond undergoes co-reaction with the allyl group via an "ENE" reaction to produce a propenyl-bridged intermediate which further reacts via a Diels-Alder reaction to form a 2:1 maleimide/allylphenol adduct. This 2:1 adduct, as shown in Fig. 21, may further react with maleimide to form a 3:1 maleimide/allylphenol adduct as proposed by Enoki and coworkers (65). Reportedly, DABA is an attractive comonomer for BMI because the cured copolymer is tough and temperature resistant (66). Toughness, however, is a function of the BMI/DABA ratio employed. Enoki and coworkers published the influence of the BMI/DABA molar ratio on mechanical and thermal properties as shown in Table 6.

In another study, optimized toughness properties were achieved when BMI and diallylbisphenol were employed at about 2:1 molar ratio (68). In Table 7,

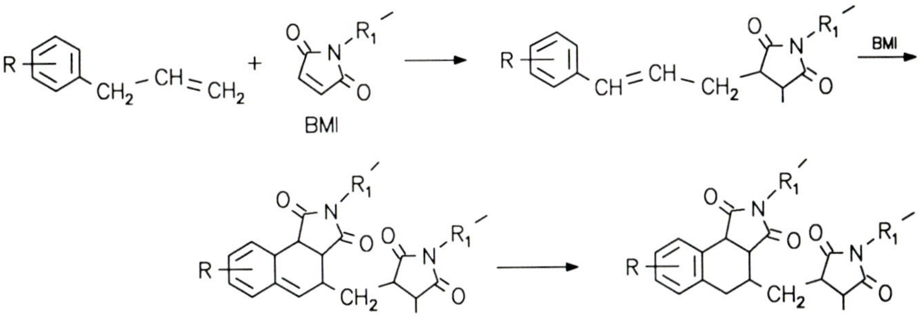

Fig. 21. Polymerization of BMI with o,o'-diallylbisphenol A via complex ENE- and Diels-Alder reactions

Table 6

BMI/DABA molar ratio		1.2/1	1.5/1	2/1	3/1
Flexural strength	(MPa)	186	188	174	131
Flexural modulus	(GPa)	4.02	3.94	4.05	4.14
Deflection at breaking	(%)	7.78	7.30	5.53	3.50
K_{Ic}	(MN/m$^{1.5}$)	0.97	0.86	0.80	0.64
G_{Ic}	(J/m^2)	197	158	133	83
T_g	(°C)	279	282	288	288

Curing condition: 175 °C/3 h + 230 °C/4 h

the mechanical properties of BMI/bis(3-allyl-4-hydroxyphenyl)-p-diisopropylbenzene resins are provided up to 250 °C. Within this system a further increase of the diallylbisphenol concentration would provide further toughness improvements, however, at the expense of the high temperature mechanical properties.

DABA is commercially available under the trademark Matrimide 5292 from Ciba Geigy. Another bisallylphenyl compound is available from the Shell Chemical Company/Technochemie under the trademark COMPIMIDE TM121. Both materials meet the processing requirements of the industry because they are honeylike fluids at room temperature and low viscosity liquids between 70–90 °C and thus act as reactive diluents and tougheners at the same time. Another family of bis(allylphenyl) compounds are the bis [3-(2-allylphenoxy)phthalimides], which are synthesized from bis(2-nitro phthalimides) and o-allylphenol (69). These allylphenoxyimides are very attractive because their copolymers with BMI are temperature resistant and tough at the same time. Comparative thermal oxidative stability data of carbon fiber laminates have recently been published(70). The chemical structure of a typical bis(allylphenoxy)phthalimide is provided in Fig. 22.

Research and development activities for allyl-type comonomers are ongoing. It has been recognized that the desired properties of cured BMI/diallylphenyl

Table 7. Mechanical properties of BMI/Bis (3–allyl–4-hy–droxy-phenyl)-p-diisopropylbenzene blends [BMI = eutectic bismaleimide blend;   BHPDB = bis(4-hydroxy-3-allyl-phenyl)-p-dii-sopropyl-benzene]

Property[1]	Composition (wt% BHPDB)			
	20	30	35	40
Flexural properties				
(RT/dry)				
Strength (MPa)	117	133	140	160
Modulus (GPa)	4.69	4.48	4.09	4.46
Elongation (%)	2.48	3.00	3.46	3.78
Flextural properties				
(177 °C/dry)				
Strength (MPa)	90	107	117	133
Modulus (GPa)	3.75	3.59	3.39	3.47
Elongation (%)	2.42	2.76	3.32	4.56
Flexural properties				
(250 °C/dry)				
Strength (MPa)	79	84	90	75
Modulus (GPa)	2.86	2.91	2.89	2.79
Elongation (%)	2.90	3.22	4.44	> 5
Fracture energy				
(G_{Ic}(J/m^2)	30	60	98	217
Moisture gain[2](%)	3.33	3.02	3.04	3.18

[1] Cure cycle: 3 h 160 °C + 4 h 210 °C + 5 h 240 °C.
[2] 1000 h, 94% RH, 70 °C

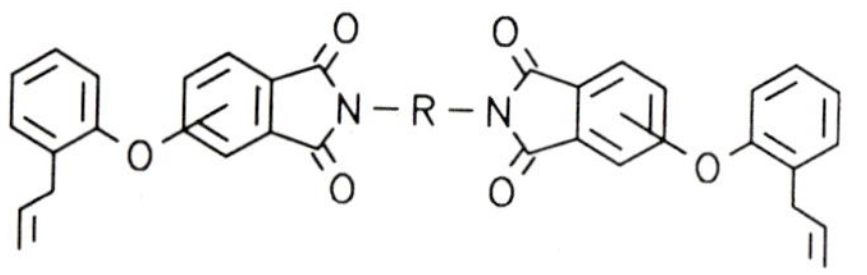

Fig. 22. Chemical structure of bis(o-allylphenoxy)phthalimide (70)

systems can be tailored by selecting the appropriate chemistry for the backbone in the diallylphenyl compound. The reaction of o-allylphenol with epoxy resin was used to synthesize a family of allylphenyl terminated phenoxy oligomers which were subsequently cured with BMI (71). Di- and polyfunctional epoxy resins may be employed in this approach. Also low and high molecular weight epoxy precursors may be selected depending on the specific end use requirements of the system. To achieve enhanced toughness properties epoxy-terminated polysiloxanes were converted into the corresponding allylphenyl terminated polysiloxane by reaction with allylphenol and then subsequently added to BMI resin to achieve polyphase toughening (72). The chemical structure of an allylphenyl-terminated polysiloxane is shown in Fig. 23.

An interesting class of ene-type BMI comonomers are the bis(allylnadimides) provided in Fig. 24. They are synthesized from allylnadic anhydride and diamine in the usual way (73). Bis(allylnadimides) can homopolymerize or copolymerize

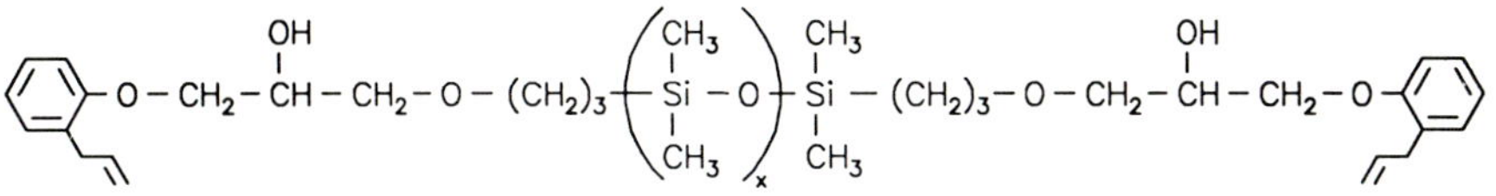

Fig. 23. Chemical structure of bis(allylphenyldimethylpolysiloxane) (72)

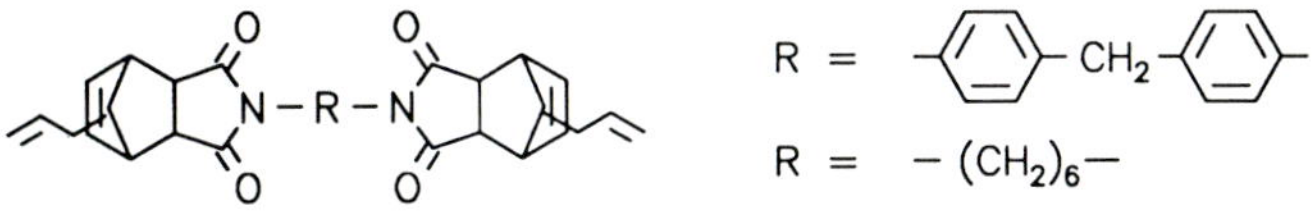

Fig. 24. Chemical structure of bis(allylnadimides)

with BMI. The key to this class of thermosets is the synthesis of the allylnadic anhydride "endcapper". Cyclopentadiene, after metallation with sodium, is reacted with allylchloride and the resulting allyl-substituted cyclopentadiene is reacted with maleic anhydride. Bis(allylnadimides) are lower melting than their maleimide counterparts and therefore provide improved processing properties. Injectable modified BMI resins have been formulated by use of allylnadic imides as reactive diluents (74).

3.4.4 Bismaleimide/Diels-Alder Copolymers

The Diels-Alder reaction can be employed to obtain thermosetting polyimides. If BMI (the bisdienophile) and the bisdiene react nonstoichiometrically, with bismaleimide in excess, a prepolymer carrying maleimide terminations is formed as an intermediate, which can be crosslinked to yield a high temperature resistant network.

Styrene can react with BMI via a complex Diels-Alder-ENE route in a 1:2 stoichiometric ratio as outlined in Fig. 25. Other vinybenzene compounds, such as propenylphenoxydiphenylsulfone (75) and bis(o-propenylphenoxy) benzophenone (76), react in a similar way with BMI. Their synthesis involves a straight forward nucleophilic halo displacement reaction. o-Allylphenol reacts with 4,4'-dichlorodiphenylsulfone or 4,4'-difluorophenylketone at 160 °C in N-methylpyrrolidone as a solvent in the presence of potassium carbonate as a catalyst. The alkaline reaction conditions are responsible for the o-allyl-o-propenyl isomerization. A wide variety of structurally similar bis(propenylphenoxy) compounds are possible by simply using p-substituted allylphenols (eugenol) or isomeric dihalophenylsulfones or dihalophenylketones in the synthesis. The bis(o-propenylphenoxy) compounds are low melting, low viscosity materials that can be melt-blended with BMI and subsequently cured at 170–230 °C. Table 8 provides the mechanical properties of a series of cured COMPIMIDE 796/bis(o-propenylphenoxy)benzophenone resins(Compimide

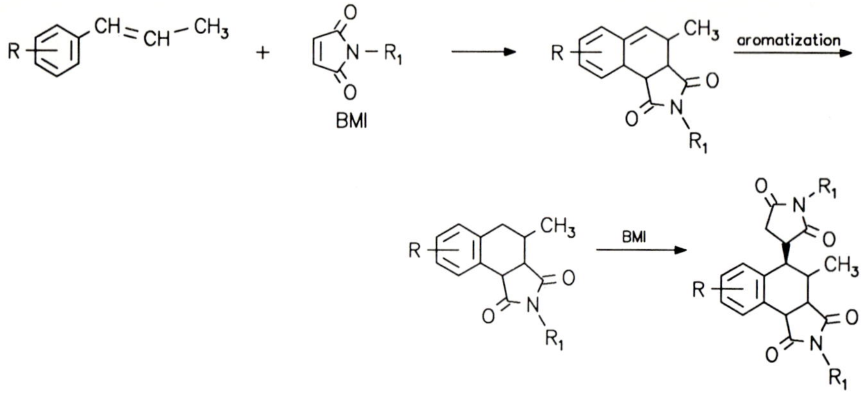

Fig. 25. Diels-Alder copolymerization of methyl styrene and bismaleimide

Table 8. Mechanical properties of bismaleimide diels-alder resins[a]

Composition BMI,[a]	toughener,[b]	Flexural strength, MPa[c]		Flexural modulus, GPa[d]		Flexural elongation, %				Water absorption,
								G_c		
%	%	23°C	250°C	23°C	250°C	23°C	250°C	J/m²[e]	T_g^f °C	%
100		76	31	4.64	3.03	1.7	1.03	63	> 300	4.30
82	18[g]	98	70	3.99	2.93	2.49	2.37	185	285	4.00
60	40[g]	114	73	3.58	2.15	3.20	4.50	267	256	2.90
80	20[h]	87	56	3.85	2.82	2.3	2.0	234	300	3.74
60	40[h]	128	83	3.49	2.38	3.9	> 5	378	277	3.63
80	20[i]	106	65	3.96	2.66	2.34	2.52	191	275(266)	3.66
60	40[i]	132	56	3.70	1.71	3.75	4.86	439	261(249)	2.59
80	20[j]	114	78	4.17	2.47	2.87	3.73	247	273(252)	3.46
60	40[j]	122	81	3.59	2.44	3.44	4.52	466	265(260)	2.90

[a] BMI = COMPIMIDE 796.
[b] Diphenylsulfones or benzophenones as indicated.
[c] To convert MPa to psi, multiply by 145.
[d] To convert GPa to psi, multiply by 145,000.
[e] To convert J/m² to ftlbf/in.², divide by 2100.
[f] By DMA analysis. Values in parens by TMA analysis.
[g] 4,4′bis(o-Propenylphenoxy) diphenylsulfone.
[h] 4,4′bis(o-Methoxy-p-propenylphenoxy)diphenylsulfone.
[i] 4,4′bis(o-Propenylphenoxy)benzophenone.
[j] 4,4′bis(o-Methoxy-p-propenylphenoxy)benzophenone

796 is a low melting BMI resin of Shell/Technochemie). These copolymer resins are attractive because they show improved toughness and reduced moisture absorption in comparison with the unmodified BMI, and also show a somewhat lowered glass transition temperature. Bis(o-propenylphenoxy)benzophenone is commercially available from Shell/Technochemie under the trademark COM-PIMIDE TM123.

The target in BMI resin technology is to improve properties such as toughness, moisture absorption and thermal stability. BMI comonomers influence such properties significantly. Diallylbisphenol A (DABA) is favoured when high toughness properties are desired. Bis(*o*-propenylphenoxy) benzophenone on the other hand provides systems with good thermal oxidative stability. Figure 26 shows the weight retention versus aging time at 275 °C for up to 1000 hours. Again, BMI structure Compimide BMI-MDA (methylene dianiline bismaleimide ex Shell/Technochemie), C796 (eutectic blend of BMIs-ex Shell/Technochemie) and the comonomer affect thermal oxidative stability.

Another important property of BMI/comonomer blends is processability. Processability includes working viscosity and cure kinetics or reactivity. A comparative study of the relative reactivity of alkenyl-functionalized modifiers for bismaleimide was recently published (78). DSC measurements suggest that propenyl-functionalized aromatic comonomers do react more readily than the allylic analogues. Recently a series of bis[3-(2-propenylphenoxy)phthalimides] was synthesized from bis(3-nitrophthalimides) and *o*-propenylphenoxy-sodium via a nucleophilic nitrodisplacement reaction (79). These propenylphenoxyimides are more reactive than their allylphenoxy analogues and TGA (thermal gravimetric analysis) data suggest a higher thermal stability for the BMI/propenylphenoxy system.

The nucleophilic halo displacement reaction was employed to synthesize 2,6-bis(*o*-propenylphenoxy)pyridine (80). This comonomer was combined with Compimide 353 to provided a system with excellent room temperature tack and therefore is suited for prepreg tape.

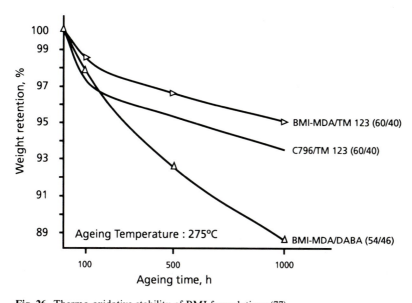

Fig. 26. Thermo-oxidative stability of BMI formulations (77)

Propenylphenoxy-terminated oligomers and their use in toughening thermosetting bismaleimide resin compositions is claimed in a recent US patent (81).
The oligomeric toughener is synthesized from epoxy resin, diphenol or poly-
phenol and o-propenylphenol. Epoxy resin is chain extended first with diphenol
and subsequently endcapped with o-propenylphenol in the presence of triphenylphosphine as a catalyst. When the glycidylether of 2,7-dihydroxynapthalene or
phenolated dicyclodipentadiene are employed as precursors, the systems are
extremely tough.

A new family of bis(dienes), the bis(vinylbenzyl) ethers, has been synthesized
and was used to modify bismaleimide (82). The chemical structure of divinylben-
zyl-bisphenol A is provided in Fig. 27.

Via DSC measurements it was shown that the new bis(vinylbenzyl)ethers can
undergo a homopolymerization and a copolymerisation with BMI. In a BMI/-
divinylbenzylether blend the Diels-Alder copolymerization is favoured over the
divinylbenzylether homopolymerisation. The T'_gs of the new copolymers are well
in excess of 270 °C when the BMI/divinylbenzylether molar ratio is 1:1 (Table
9). Isothermal weight loss studies over a period of 4000 hours indicate that the
BMI-MDA/divinylbenzylether copolymer is the most stable system of this
family.

Very interesting Diels Alder comonomers for BMI are the
bis(benzocyclobutenes). Under appropriate thermal conditions, the strained
four-membered ring of benzocyclobutene undergoes electrocylic ring opening to
generate, in situ, o-quinodimethane, which, in the presence of BMI, reacts via a
Diels-Alder reaction (83). The chemical structure of a bis(benzocyclobutene-
imide) is provided in Fig. 28. The synthesis and properties of BCB and
BMI/BCB blend systems is described in detail in chapter I of this book.

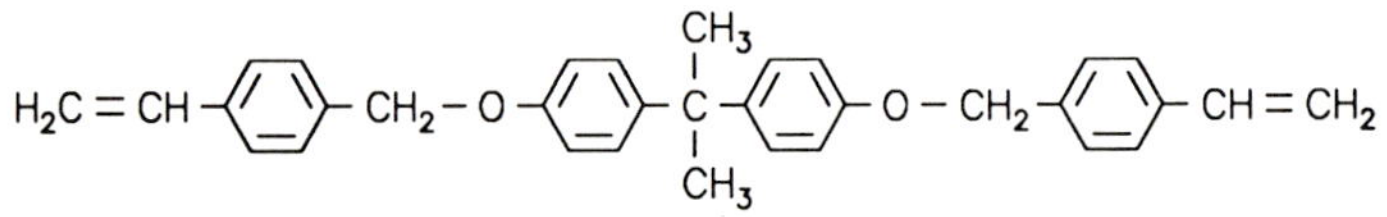

Fig. 27. Chemical structure of bis(vinylbenzyl)bisphenol A

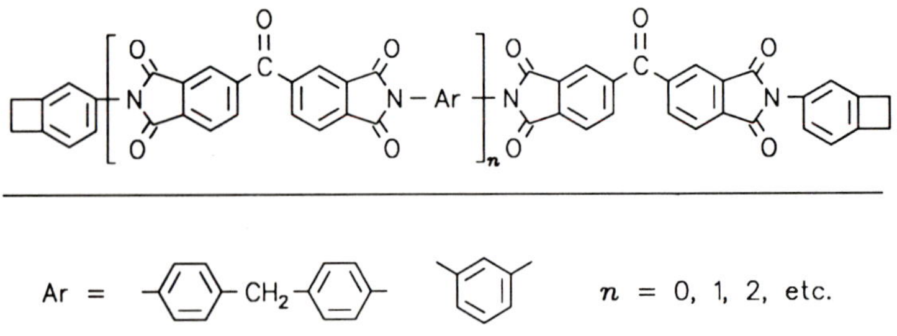

Fig. 28. Chemical structure of bis(benzocyclobutene-imide)

A widely used BMI system for carbon fibre composites is BP's V378A which is based on a low melting eutectic mixture of bismaleimides modified with 1,4-divinylbenzene as a reactive diluent and small amounts of phenoxy resin and polysulfone (84, 11). Divinylbenzene is a reactive bisdiene therefore the V378A system can be cured under standard low autoclave pressure of 7 bar and temperatures of 177 °C similar to current 177 °C curing epoxy systems. To fully develop the high temperature properties of the system, a postcure temperature of 250 °C is required. Prepreg based on V378A/T300 have been used for the manufacture of the F-16XL fighter aircraft cranked-arrow wing skins. This successful application of a bismaleimide resin was responsible for the rapidly growing interest in BMI technology for military airplane applications. The only drawback of this resin is the low fracture toughness and unpleasant odor, however, the high temperature property retention after severe humidity aging is outstanding (85).

3.4.5 Rubber-Toughened Bismaleimides

An interesting approach for toughened BMI has been reported by Kinloch and Shaw (86, 87), who copolymerized bismaleimide with liquid carboxy-terminated acrylonitrile-butadiene (CTBN) rubber. The CTBN-rubber contains C-C double bonds, both in the main and side chain, which may be employed in crosslinking reactions with BMI(Fig. 29). Kinloch used COMPIMIDE 353 (a low-melting eutectic BMI ex Shell-Technochemie) in combination with Hycar 1300X8 (ex BF Goodrich) and was able to obtain low viscosity resins at moderately elevated temperatures. In contrast to epoxy resins, Hycar 1300X8 is not compatible with the BMI at room temperature, and even at an elevated temperature (110 °C) the rubber is not soluble in the BMI. However, during cure, the rubber chemically reacts with the BMI, so that the cured system shows

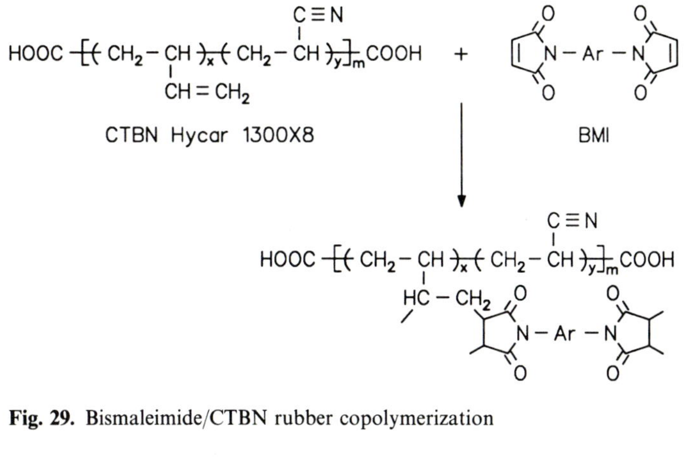

Fig. 29. Bismaleimide/CTBN rubber copolymerization

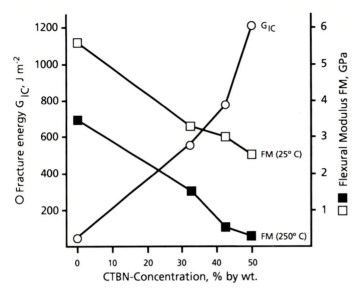

Fig. 30. BMI fracture energy as a function of CTBN-rubber concentration

a morphology similar to CTBN-rubber modified epoxy. The influence of the rubber concentration on the fracture energy, G_{Ic}, is provided in Fig. 30.

The properties of CTBN-rubber modified BMI depend on the chemical composition of the rubber (88). The acrylonitrile content influences BMI/CTBN-rubber compatability and therefore the morphology and the fracture toughness of the modified resin. High acrylonitrile rubbers are more soluble and compatible with the BMI resin and therefore provide a lower degree of phase separation and thus lower toughness. The principal drawback of CTBN-rubber toughened BMIs is the significant loss of high temperature properties, such as T_g and elastic modulus. Also, the long term aging stability at 200 °C is poor.

3.4.6 Thermoplastic-Modified Bismaleimides

Thermoplastics are increasingly used for the toughening of bismaleimide resins to improve their damage tolerance without compromising other important properties, such as T_g and modulus of elasticity. High T_g thermoplastics such as polyhydantoins (89, 90) (Resistofol N, Resistherm PH-10, Bayer AG), poly-ethersulfones (89) (P1700, Amoco Performance Products), polyimides (90) (Matrimide 5218, Ciba Geigy), polyetherimide (Ultem 1000, General Electric), and a series of poly(arylene-ethers) (91) have been tested for the toughening of BMI resins. All the thermoplastics, except the polyhydantoin PH-10, provided poly-phase morphologies, and, at approximately 20% by weight of thermoplastic, phase inverted systems with the thermoplastic as the continuous phase are obtained. The influence of the thermoplastic content on the fracture toughness of the cured BMI/thermoplastic blend system was investigated, and, as expected, higher thermoplastic levels provide increased toughness. However, the various thermoplastics show different degrees of toughening as shown in Table 9.

Many factors contribute to the toughness of a polyphase BMI/thermoplastic system, such as solubility parameters, phase adhesion, phase morphology, particle size and particle size distribution. Another important factor is the molecular weight of the thermoplastic modifier. It has been demonstrated for a particular poly(arylene-ether) backbone that high molecular weights increase the toughness of the blend system more than the low molecular weight counterparts (92).

Thermoplastic modified BMIs are mainly used as matrix resins for advanced composites; therefore it is important that the good BMI-TP toughness in neat resin form translates into the composite and thus improves damage tolerance as measured via compressive strength after impact (CAI). For a BMI/poly(etherimide) system the toughness and morphological spectrum have been investigated (93), showing that for certain thermoplastic concentrations (> 35% TP) the resin morphology is the same for the neat resin and the resin in the presence of the reinforcement. The thermoplastic is the continuous phase with very small (> 0.5 nm) BMI inclusions (94). The fibers in the composite do not influence the development of the BMI-PEI morphology and, therefore, neat resin toughness correlates with composite toughness.

The principal disadvantage of the thermoplastic approach with high molecular weight polymers is the loss of processability and the solvent sensitivity of BMI-thermoplastic blends. Therefore, attempts have been made to use reactive thermoplastics that can react into the BMI network and thus provide solvent resistance without loss of toughness. For a series of functionalized poly(arylene ethers), it has been demonstrated that a Mw (weight average molecular weight) of approximately 30 000 g/mol is required to maintain the toughness provided by the high molecular weight poly(arylene ether). However solvents do attack the system but do not completely destroy the blend system (93). Similar results have been obtained for BMI-diallylbisphenol A resin modified with amine-terminated poly(arylene ether sulfone) (95). From these studies concerning molecular weight and polymer backbone effects, it is suggested that the inherent toughness of the polymeric modifier is a major factor controlling the final network fracture toughness.

A new series of functionalized poly(arylene ether ketone)s and their use as modifiers for bismaleimide resins has recently been published (96). In contrast with the former work in this area, the functional groups are not terminal but randomly distributed along the backbone of the poly(arylene ether ketone). The synthesis involves the reaction of 4,4'-difluorobenzophenone and the potassium bisphenates of bisphenol-A and 2,2'-bis(3-allyl-4-hydroxyphenyl)-hexafluoro-isopropane or mixtures thereof in DMAc at 155 °C as outlined in Fig. 31. The concentration of the o,o'-diallylbisphenol employed in the synthesis determines the concentration of functional propenyl groups.

Table 9. Properties of thermoplastic-modified BMI* resins

Property	Control[a]	20 wt% PH-10[b]	20 wt% RN[c]	20 wt% M 5218[d]	20 wt% P 1700[e]	13 wt% Ultem[f]	20 wt% PAE-1[g]	20 wt% PAE-3[g]	20 wt% PAE-4[g]
T_g, °C	316	294	280	284	−200	220	159.7	180.3	213.4
dry flexural properties									
room temperature									
strength, MPa[b]	115	115	118		0.5	117	114	124	132
modulus, GPa[b]	3.86	3.65	3.51	3.75	3.49	3.72	3.65	3.65	3.84
elongation %	3.05	3.09	3.52		2.67	3.35	3.21	3.40	3.64
120 °C									
strength, MPa[b]	84	95	94		37	97	84	108	95
modulus, GP[b]	3.27	2.88	2.89	3.13	1.93	3.02	2.60	3.04	3.00
elongation, %	2.73	3.44	3.52		2.70	3.38	4.06	4.50	3.28
177 °C									
strength, MPa[b]		91	78			45	26	30	100
modulus, GPa[b]		2.77	2.38			1.71	0.64	0.66	3.08
elongation, %		4.52	4.74	2.64		2.99	6.61	>7.20	4.27
fracture toughness, Kq, kPa$\sqrt{m}$	573	822	1322	929	763	805	1521	1235	1025
fracture energy, G_{Ic}, J/m²	85	185	498	230	167	174	634	418	274
moisture gain[i], %	2.72	3.01		3.23			2.88	2.77	3.05

a COMPIMIDE 796–COMPIMIDE 123.
b Resisttherm PH-10 polyhydantoin.
c Resistofol N polyhydantoin.
d Matrimide 5218 polyimide.
e Polyethersulfone.
f Polyetherimide.
g Poly(arylene ether)s.
h To convert MPa to psi, multiply by 145.
i To convert J/m² to ftlbf/m²., divide by 2100

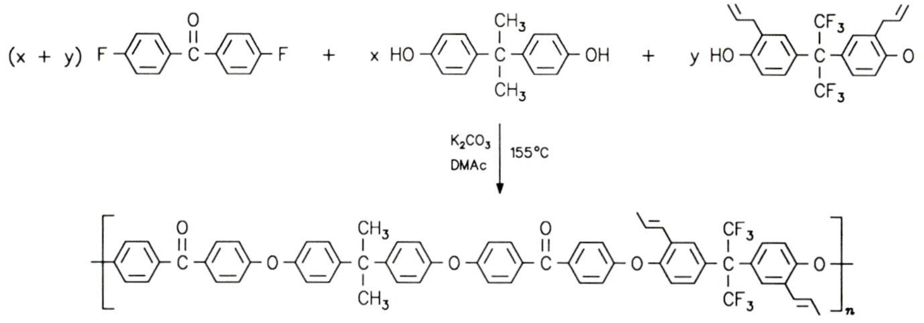

Fig. 31. Synthesis of propenyl-group-functionalized poly(arylene ether ketone)s

Table 10. Properties of BMI modified with 30 wt % of functionalized poly(arylene ether ketone) as a function of the propenyl group concentration (96)

Functionalized PAEK	MM/PG (g/PG)	Flexural		Toughness		
		Strength (MPa)	Modulus (GPa)	K_q (kN m$^{-3/2}$)	Glc (J m^{-2})	T_g (°C)
75DBA/25 6F–DABA	906	119	4.23	1558	574	151
50BA/50 6F–DABA	500	95	4.27	1365	437	142
25BA/75 6F–DABA	365	131	4.24	691	112	170
100 6F–DABA	298	130	3.99	781	153	189

PAEK = poly(arylene-ether ketone).
MM/PG = molar mass/propenyl group.
T_g = glass transition temperature.
K_q = stress intensity factor.
Glc = fracture energy.
Bismaleimide = Compimide 796/Compimide TM123 (70/30).

The properties of BMI modified with 30 wt% of functionalized poly(arylene ether ketone) are provided in Table 10. The toughness properties correlate with the propenyl group concentration of the functionalized poly(arylene ether ketone). The propenyl group concentration is expressed as the molar mass per propenyl group. If the molar mass/propenyl group is as low as 300–320 g/propenyl group, the thermoplastic behaves like a low molecular weight comonomer and then does not contribute to toughness. On the other hand, as may be expected, the solvent resistance increased with increasing crosslink density. SEM (scanning electron microscopy)-studies show a perfect interfacial adhesion for the highly functionalized system, but fracture surface morphology indicates a brittle fracture behavior. The study clearly showed that in a thermoplastic modified BMI system, the thermoplastic toughener can only be effective if the functional (crosslinking) group concentration is low enough so that the thermoplastic can develop its inherent morphology which determines the toughness.

The usual approach to manufacture tough composites based on thermosets is to blend the thermoplastic modifier with the uncured thermoset resin, either via melt blending or solvent/solution techniques and subsequently impregnate reinforcement to form the prepreg. Upon cure, the thermoplastic/thermoset blend undergoes a phase separation, the thermoplastic forms the continuous phase and can develop its characteristic morphology and thus determines the toughness of the cured thermoset/thermoplastic blend system. Another approach to toughen thermoset-based composites is "Thermoplastic-Particulate Interlayer Toughening". This technique does not require sophisticated synthesis, blending and copolymerization operations. The thermoplastic toughener is applied onto the surface of the BMI based prepreg as a fine powder, and thus after molding the system forms a ductile thermoplastic rich continuous phase between the individual fiber layers. BMI composites toughened by this method are at least as tough as the epoxy counterparts when toughness is measured via CAI-techniques.

4 PMR Polyimides

The PMR concept-PMR stands for Polymerisation of Monomeric Reactants-is a unique approach to temperature-resistant crosslinked polyimides. Although PMR resins cure via a complex crosslinking reaction, the imide backbone is synthesized in situ during processing through a condensation reaction. The most widely used resin of this family is PMR-15, developed at NASA Lewis Research Center (7). The synthesis of the resin involves dissolving the dialkylester of benzophenone tetracarboxylic acid (BTDE), MDA and the monoalkylester of 5-norbornene- 2,3 dicarboxylic acid (NE) in a low boiling alkyl alcohol and the low viscosity solution subsequently is used to impregnate fibers or fabric to provide a prepereg. The prepereg, after removal of the solvent, contains the monomeric reactants and products thereof. Upon heating to temperatures between 150–200 °C, the monomers undergo an "in situ" condensation reaction to form the norbonene-endcapped imide prepolymer. In PMR-15 the stoichiometry of the monomeric reactants is adjusted to achieve a nadimide terminated imide prepolymer with a molecular weight of 1500 g/mol. The final cure (crosslinking) is performed at temperatures between 250–320 °C. The classical PMR-15 chemistry is outlined in Fig. 32.

The PMR-15 chemistry looks very straightforward and ideal for the application as a composite matrix resin. In addition, the starting monomers are readily available and cheap. The fact that the imidization reaction, which forms the prepolymer at moderately low temperatures, could be separated from the crosslinking reaction was thought to be the key to easy laminate processing and void free laminates. However, after more than twenty years of research and development, it is known that both reactions are very complex and dependent

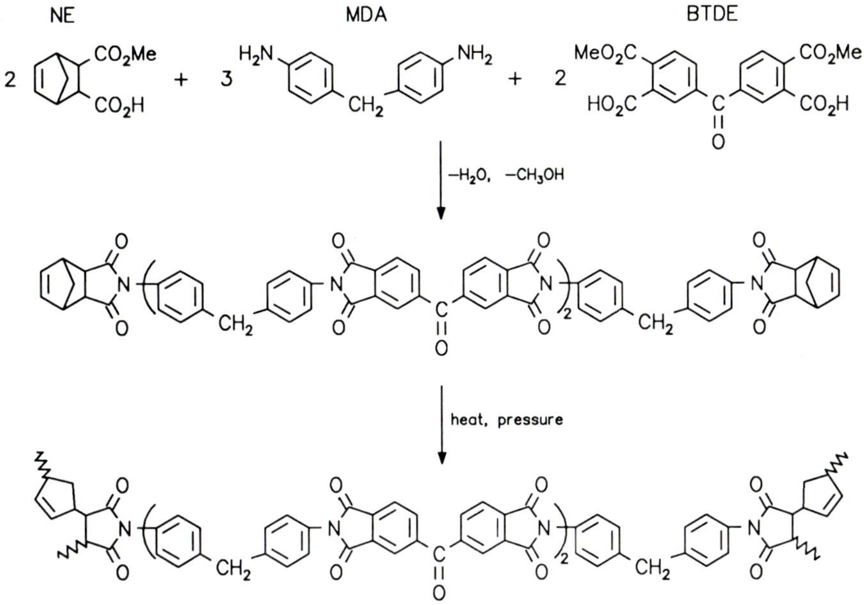

Fig. 32. PMR-15 resin chemistry

on the cure conditions employed. In particular the molecular weight and molecular weight distribution of the prepolymer can change significantly. Comparative studies of model compounds and commercial PMR-15 resin demonstrated that the temperature profile during the imidization determines the molecular weight and molecular weight distribution of the prepolymer (97, 98, 99).

At low temperatures the formation of the bis(nadimide) of MDA is favored. HPLC and spectroscopic data indicate the formation of nadic anhydride which subsequently reacts with MDA without the formation of the bis(amide acid) intermediate. The formation of the bis(nadimido diphenylmethane) offsets the stoichiometry of the PMR-15 composition leaving functional amine, ester or anhydride endgroups in the prepolymer. This may adversely affect thermal oxidative stability of the cured polymer and the physical properties of the final composite (100). Lubowitz (6), who studied the cure of bis(nadimide) model compounds, speculated from TGA studies that a reverse Diels Alder reaction which generates free cyclopentadiene and maleimide is a prerequisite for the initiation of the polymerization reaction. Meanwhile, the cure reaction of nadimides has been subject to numerous investigations (101–106). Recent work carried out at the Montanistische University, Leoben, supports the mechanism proposed by Ritchy and Wang (103) by which cyclopentadiene is definitely released during cure which subsequently may react with nadimide to form a double cyclo adduct structure.

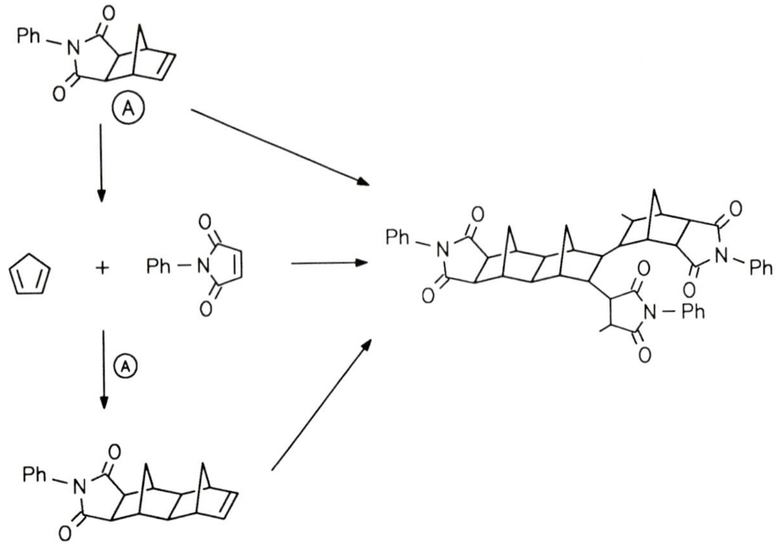

Fig. 33. N-Phenylnadimide polymerization proposed from model compound studies (106)

Wilson (106), who polymerized N-phenylnadimide under high pressure, analyzed the soluble part of the polymer obtained via H-NMR, ^{13}C-NMR-and FTIR, and confirmed the proposed mechanism given in Fig. 33 Hay et al. (107) demonstrated the complete absence of unsaturation in the polymerization product of N-phenylnadimide. Therefore the cyclopentadiene formed by the reverse Diels Alder reaction did not copolymerize but underwent a Diels Alder reaction with N-phenylnadimide to form a double cycloadduct, or it evolved as a gas. The maleimide N-phenylnadimide and the double cycloadduct are implied in a radical double bond polymerization reaction. Another aspect of the nadimide cure was published by Young (108) who proved the existence of a thermal equilibrium between geometric isomers, the *endo* and *exo* configurations of norbornene residues. Calorimetric and thermogravimetric analysis indicate that these isomers behave differently in air than in nitrogen, suggesting different mechanisms of cure depending on the atmosphere. The data are consistent with the proposed reverse Diels Alder mechanism leading to a loss of cyclopentadiene in nitrogen, and a more direct chain extension without weight loss in air.

4.1 PMR Resins with Improved Thermal Stability

The driving force for further development of the PMR concept was the need for improved high temperature stability of PMR-15 composites. New versions of PMR-15 have been developed by changing the backbone and/or endgroup chemistry. The replacement of BTDE with the dimethylester of 4,4'-(hexafluoroisopropylidene) bisphthalic ester (HFDE) and MDA with *p*-phenylene diamine (PPDA) significantly improves the thermal-oxidative stability at 316 °C as indicated by weight loss measurements performed at 316 °C (109). The chemical structure of the partially fluorinated PMR resin known as PMR-II is provided in Fig. 34.

Other members of the PMR-II family have been developed, e.i. PMR-II-30 or PMR-II-50 etc., with the aim of increasing the prepolymer molecular weight and thus reducing the concentration of aliphatic nadimide endgroups. The effect of such changes in the backbone chemistry are significant as shown in Fig. 35. Obviously the partially fluorinated backbone of PMR-II-13 provides much lower weight loss than PMR-15. Increasing the molecular weight, as in PMR-II-30, produces further improvement in thermal oxidative stability presumably through the reduction of the aliphatic endgroup content. Increasing the molecular weight, as in PMR-II-30, produces further improvement in thermal oxidative stability presumably through the reduction of the aliphatic endgroup content.

More recently a new fluorinated diamine, 2,2'-bis(trifluoromethyl)-4,4'-diaminobiphenyl (BTDB), has been combined with HFDE and NE to provide a new PMR resin, PMR-12F-71, the chemical structure of which is provided in Fig. 36. The molar ratio of the reactants was adjusted such that a formulated molecular weight of 7100 g/mol was achieved.

A further modification of the PMR-12F-71 resin comprises changing from the nadic endgroups to vinylphenyl endgroups, simply by using *p*-aminostyrene in the synthesis. This resin was called V-cap-12F-71 (see Fig. 37). The V-cap versions of PMR-II-50 (V-cap II-50) and PMR-12F-71 (V-cap-12F-71) underwent a comparative long term thermal oxidative stability testing (112). Neat resin weight loss was measured at 343 °C in air over a period of 750 hours (Fig. 38). The data clearly indicate that the 12F-PMR resins exhibit excellent thermal oxidative stability and it also shows that the NE endcap is thermally less stable than the V-cap in the PMR-II series.

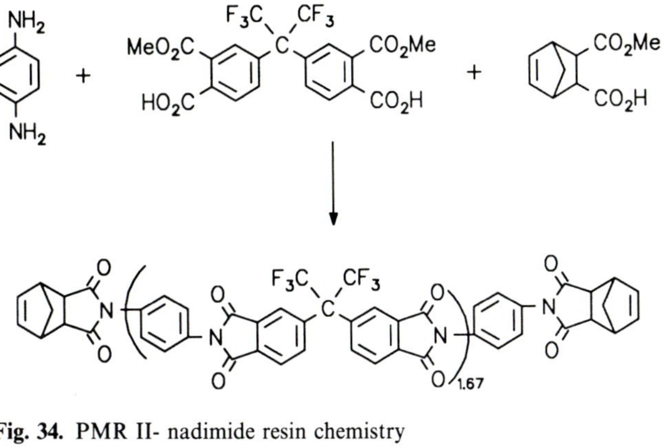

Fig. 34. PMR II- nadimide resin chemistry

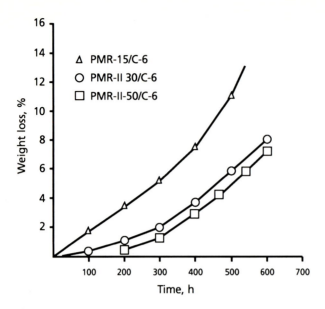

Fig. 35. Effect of backbone chemistry and molecular weight on Thermal Oxidative Stability (TOS) in air at 343 °C

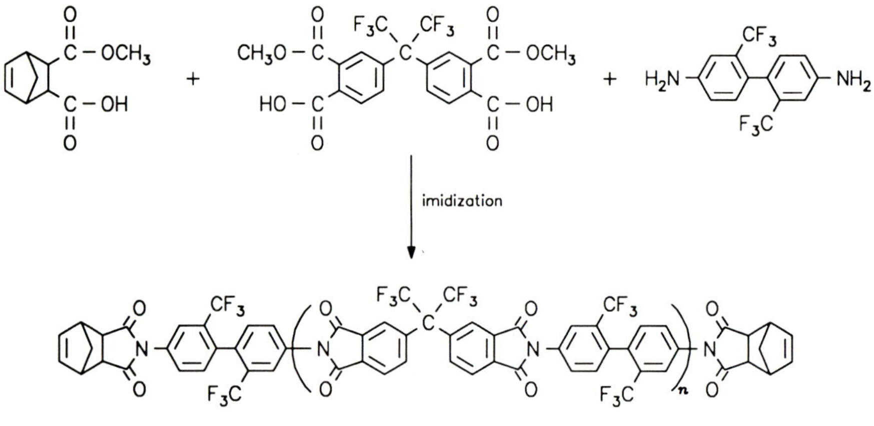

Fig. 36. Chemistry of PMR- 12F-71 resin (111)

As mentioned earlier PMR polyimide thermosts are used as matrix resins for glass- and carbon fiber composites, mainly in aeroengine applications. At this point it has to be mentioned that the thermal oxidative stability of a PMR composite is dependent on the type of fiber used (113) and the cure conditions (time/temperature/atmosphere) employed for molding. Very interesting is the observed higher thermal oxidative stability of PMR-II composites when cured/-

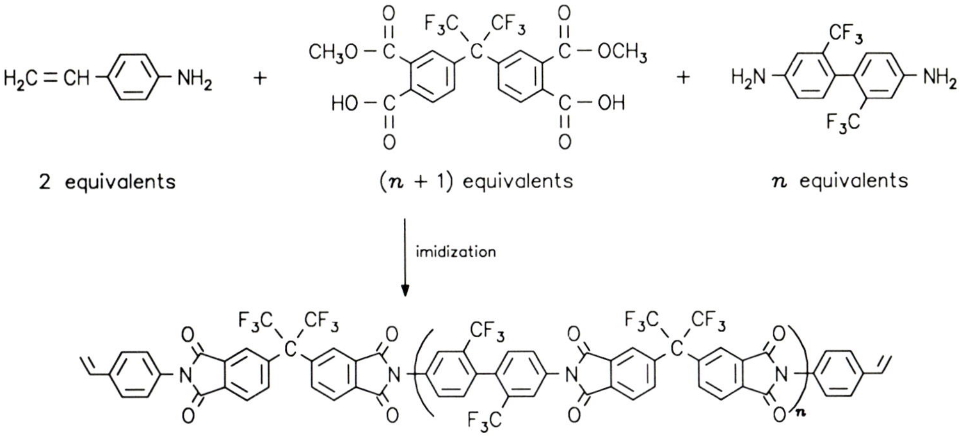

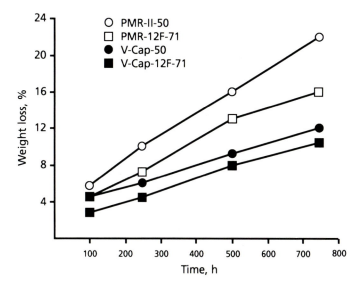

Fig. 37. Chemistry of V-cap 12F-71 PMR resin (112)

Fig. 38. Thermal oxidative stability of 12F- PMR neat resin in comparison to other PMR types in air at 343 °C

postcured in nitrogen (114). "V-CAP" PMR resin, developed by NASA Lewis Research Center, has also demonstrated its superiour performance in terms of weight loss and microcracking behaviour. General Electric fabricated an engine component (the F110 forward exhaust fairing) and performed a real application test (115).

Recently there has been reported dramatic improvements in thermal oxidative stability for PMR type resins based on p-phenlene diamine (PPDA) with

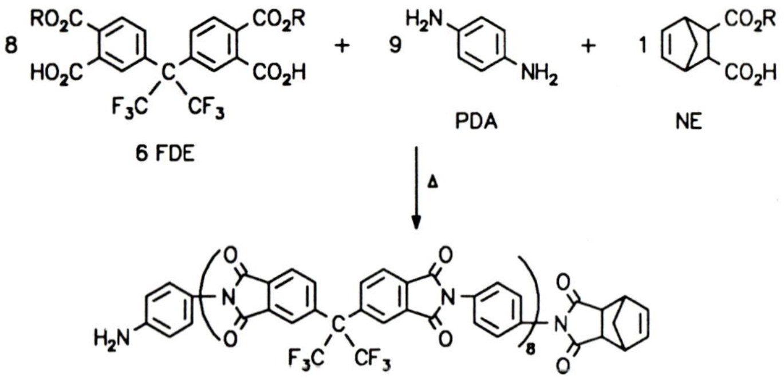

Fig. 39. AFR-700 PMR resin chemistry (116)

371 °C capability. A resin coded AFR 700, based on a NE/HFDE/PPDA backbone with a stoichiometry imbalance, providing a prepolymer with a mixture of NE and amine/or anhydride endcaps, as is shown idealistically in Fig. 39. The thermal oxidative stability improvements vis a vis PMR-15 are presumably achieved because of a reduced aliphatic (NE) endgroup concentration. Unfortunately, no publication has appeared in the open literature on the mechanical performance of AFR-700 composition.

Another PMR resin, TRW-R-8XX, is advertized by TRW; the supplier report the thermal oxidative stability properties to be superior to AFR-700. Prepregs are available from Dexter Composites INC, Cleveland, Ohio and the target applications for the material are aircraft engines, airframes, missiles etc. The weight loss is reported to be less than 3% at 371 °C after 100 hours of aging. Under the same aging conditions standard PMR-15 composites lose more than 10% weight. Because of the very high glass transition temperature of TRW-R-8XX, composites reportedly show high mechanical properties at 371 °C.

Another system, developed by NASA Langley Research Center, reportedly shows improved thermal oxidative stability as compared with PMR-15 (117). The resin is named RP-46 and its chemistry is based on BTDE, NE and 3,4'-diaminodiphenylether (3,4'-ODA), and is formulated to a molecular weight of 1500 g/mol as in PMR-15. The only difference to PMR-15 is that 3,4'-ODA replaced the MDA and therefore the improved thermal oxidative stability is attributed to this structural change.

In summary, improved thermal oxidative stability in PMR-type polyimides is achieved by tailoring a stable backbone structure through:

- the use of HFDE/PPDA as in PMR II
- the use of HFDE/BTDE as in PMR-12F-71
- the use of 3,4-ODA as in LARC-RP-46

and surprisingly by changing from NE-endcaps to styrene terminations, as in the V-CAP PMR resin systems.

4.2 PMR Resins with Improved Processability

A very important aspect in PMR resin technology is processability, i.e. impregnation of fibers to produce a prepreg and molding of the prepreg layup into a component. State of the art technology employs resin solutions to impregnate the reinforcement but other techniques such as hot melt impregnation or powder prepregging may become more important in the future because of environmental issues associated with solvent/solution techniques.

A solventless PMR resin became known under the designation LARC 160 (15), which could be processed as a hot melt. An exchange of MDA in PMR-15 with a liquid isomeric mixture of di- and trifunctional amines (Jeffamine 22) provided a mixture of monomeric reactants which was tacky at room temperature. In the presence of 3% methanol the resin could be processed via a hot melt process. Unfortunately, the cured resin was inferior with respect to thermal oxidative stability in comparison to PMR-15.

Considerable interest exists for the development of powder impregnation methods. In this technique, the resin is applied to the fiber as a dry powder or as a powder carried in a liquid slurry. However, here it is necessary to synthesize a fully imidized PMR prepolymer as a fine powder and thus the PMR concept as such is lost.

An approach for improved processing of PMR polyimide is the addition of N-phenylnadimide to the precursor solution of the monomeric reactants. After the in-situ condensation, a PMR-15 resin is obtained which is "diluted" with N-phenylnadimide in order to improve the rheological properties of the system (118). The amount of N-phenylnadimide (PN) added was in the range of 4 to 20 mol %. Just 4 mol % caused a significant and disproportionate reduction of the minimum viscosity with no concomitant loss of thermal stability.

One of the major problem areas with PMR-15 polyimide is the high final cure temperature required (188–320 °C) to fully develop the high temperature properties of the resin. Serafini and coworkers (119) showed that the use of m-aminostyrene as an endcapper instead of NE lowered the final cure temperature of the PMR polyimide from 320 to 260 °C. However, the (T_g) was lowered to 260 °C and thus limited the use temperature to 260 °C. The use of equimolar amounts of NE and p-aminostyrene in a PMR resin (120) helped to overcome this problem, however the flow properties suffered. The flow problem could again be overcome through the use of N-phenylnadimide as a reactive diluent (121). The effect increased when the *endo*-isomer of the N-phenylnadimide was used, because it melts at a lower temperature than the *exo*-isomer (122).

4.3 PMR Resins with Reduced Toxicological Hazards

It is inherent to the PMR approach that free aromatic diamine is present both in the impregnation varnish and in prepreg. In the case of PMR-15 the diamine is MDA which is a suspected human carcinogen. According to the new OSHA regulations, products that contain more than 1% of free MDA require air monitoring during processing to protect workers from exposure.

The obvious route to overcome this problem is to employ less toxic diamines. Those who actually fabricate PMR prepreg and/or components recognize that this is a tough target. The diamine has to be soluble in methanol because this is the preferred solvent in commercial prepregging operations. Reactivity problems may arise from changes in the basicity of the diamine and thus changes in the polymerization kinetics. This could give rise to changes in the polymer structure and provide inferiour cured resin properties. The substitution of MDA in PMR-15 by another diamine will definitely change the Tg of the cured network and the thermal performance. Under European community/ BRITE funding, BP Research developed a PMR resin based on 2,2-bis(4-(4-aminophenoxy)phenyl] hexafluoroisopropane (4-BDAF) which was coded B1 resin (123). The stoichiometry of the monomeric reactants was the same as in PMR-15. The formulation allowed the production of high quality laminates with a T_g of 290 °C and acceptable properties. The use of this four-ring partially fluorinated diamine was also designed to impart increased toughness to the polymer matrix, whilst its chemical nature provided moderate toxicity compared with the highly toxic MDA (4-BDAF, LD50 = 1.37 g/g, MDA, LD 50 = 0.19 g/kg (124)). The chemical structure of the imidized B1 prepolymer is provided in Fig. 40.

The major concern was the thermal oxidative stability performance of the new resin. Weight loss measurements at 250, 285 and 300 °C provided comparable low values at 250 and 285 °C. However, at 300 °C, the B1 composite exhibited a marketly lower weight loss than PMR-15. The temperature capability of B1 composite is obvious from Fig. 41, where the flexural properties of resins are plotted as a function of the ageing time at 285 °C. PMR-15 seems to be a superior resin in this test.

Another approach to overcome the MDA toxicity problem would be to employ MDA in a prereacted form. This approach would certainly provide composites with PMR-15 performance but processing may become a problem due to increased prepolymer viscosity (125).

The PMR approach inherently requires the use of monomeric reactants; therefore in one or the other processing step free diamine may be present. The

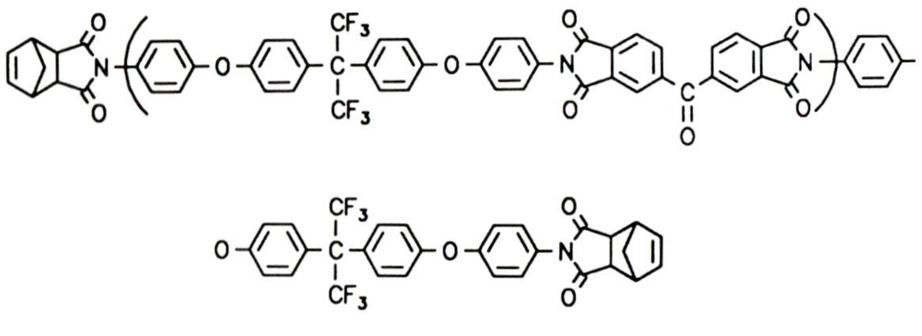

Fig. 40. Chemical structure of imidized B1 PMR resin (123)

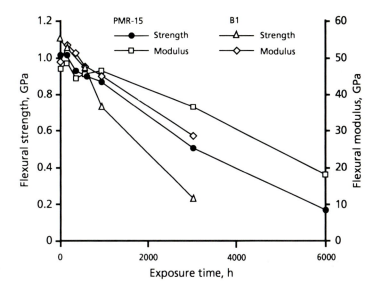

Fig. 41. Thermal oxidative stability of B1 and PMR-15 composites (aging in air at 285 °C)

possibility of employing fully imidized nadimide prepolymer powder for pre-pregging, could be the solution to overcoming the toxicity problem associated with the use of monomers.

4.4 PMR Resins with Thermally Stable Reactive Endcaps

Nadimide- and V-CAP type PMR resins use "aliphatic" endcaps (multiple carbon-carbon double bonds) to affect crosslinking. Such aliphatic moieties adversely effect the thermal oxidative stability of the cured system. Effort is therefore directed towards PMR resins which use reactive aromatic endgroups to obtain cured polymers free of aliphatic chain- or crosslinking segments.

One resin based on the BTDA/ODA backbone and 2-aminobiphenylene as an endcapper was thought to be such a resin (126). High quality laminates could be fabricated, but the T_g of the crosslinked polymer was lower than expected and therefore thermal oxidative stability was poor. The chemical structure of this thermosetting polyimide is given in Fig. 42.

Another concept has recently been reported, using aminosubstituted 2,2-paracyclophane as an endcapper in a PMR resin (127). Another version with a HFDE/PPDA backbone and the same encapper was coded N-CYCAP polyimide (amine substituted CYClophene Addition Polyimide the structure is given in Fig. 43). The resin system was tested against PMR-II. The room temperature and 343 °C properties are superior to PMR-II, but the thermal oxidative stability measured through aging at 371 °C showed superior performance for the PMR-II resin (128)

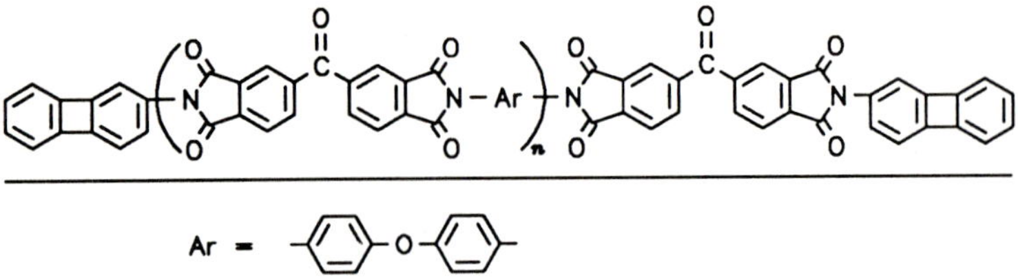

Ar =

Fig. 42. Chemical structure of biphenylene terminated polyimide

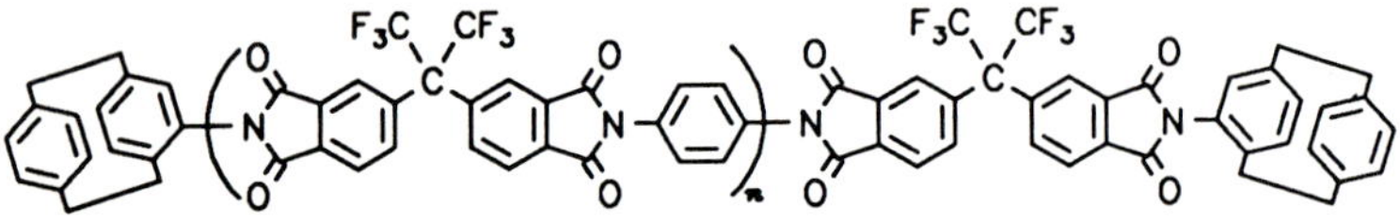

Fig. 43. N-CYCAP polyimide oligomer

5 Acetylene (Ethynyl) Functionalized Imide Oligomers

The chemical nature of the reactive endgroups of polyimide oligomers contribute significantly to the thermal oxidative stability of the crosslinked polymer. Both the aliphatic nature of the polymerizing endgroup and the degree conversion affect the thermal performance. A high concentration of aliphatic moeties in the polymer structure certainly degrades the thermal oxidative stability of the network.

The idea of synthesizing imide oligomers which carry acetylenic terminations appeared attractive because homopolymerization through acetylenic endgroups occurs without any volatile evolution and provides materials with good properties. Landis et. al (8, 9) published the synthesis of such acetylene terminated imide oligomers from benzophenone tetracarboxylic anhydride, aromatic diamine and 3-ethynylaniline via the classical route. As usual, the amide acid is formed as an intermediate which, after chemical cyclodehydration, provides the polymide. Since ethynyl-terminated polyimide is used as a matrix resin for fiber composites, processing is possible via the amide acid, which is soluble in acetone, or via the fully imidized prepolymer, which is soluble in NMP. The chemical structure of the fully imidized ethynyl-terminated polyimide is provided in Fig. 44.

As in all thermosetting polyimides, the diamine and the tetracarboxylic dianhydride employed to build the backbone can be varied. Alteration of the diamine, tetraacid or both, allow the modification of the polyimide's melting point and solubility. Of interest to the end user is the influence of chemical structure on the melting transition of the prepolymer and the T_g of the fully cured product. Lowering the uncured T_g means increasing flow and, in most

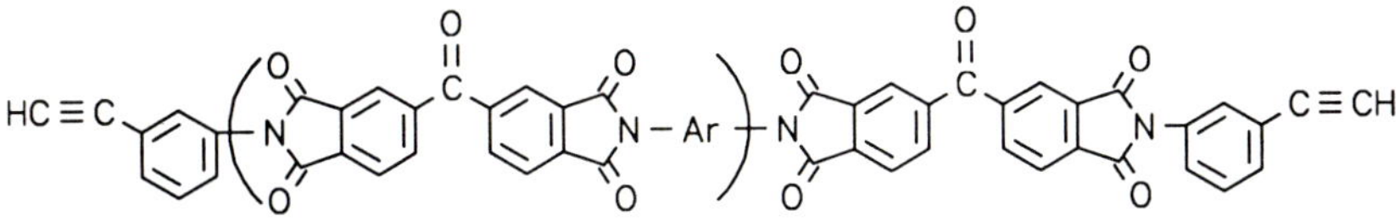

Fig. 44. Chemical structure of acetylene-terminated polyimides

Table 11. Properties of ATI resins: Uncured and cured T_gs as a function of the chemical structure (129)

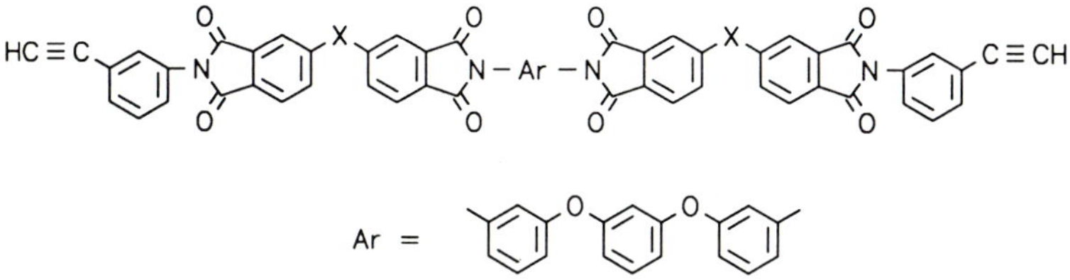

X	T_g prepolymer, °C	T_g cured polymer, °C
$\diagdown C = O$	195–200	370
O	150	253
CH_2	160	263
$F_3C \diagup \diagdown CF_3$ C	168–178	324
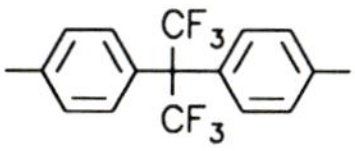 $CF_3 \diagup \diagdown CF_3$ (aromatic)	160	296
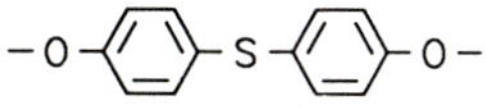 $-O-\langle\rangle-S-\langle\rangle-O-$	134	212

T_g prepolymer = Glass transition temperature of prepolymer

T_g cured polymer = after 8th postcure at 370 °C

instances, a widening of the processing window. The systematic variation of the structure of the tetraacid, provided a wide range of cured/uncured T_gs (129, see Table 11). A very attractive resin is based on hexafluoroisopropane-bisphthalic anhydride because of the low uncured and high cured T_g. This resin system became known as Thermid FA-700 and is available from National Starch Chemical Company. The resin is soluble in a variety of common solvents such as tetrahydrofurane (THF), DMF,DMAc and therefore is easy to process into composites. The first commercially available ethynyl-functionalized polyimide was Thermid 600. The resin is based on 1,3-bis(3-aminophenoxy) benzene and BTDA and is endcapped with 3-ethynylaniline. The resin has a high melting transition (198 °C), polymerizes immediately after softening and therefore has a narrow processing window.

The key to acetylene terminated polyimides is the availability of the end-capper which carries the acetylene group. Hergenrother (130) published a series of ATI resins based on 4-ethynylphthalic anhydride as endcapping agent. This approach first requires the synthesis of an amine-terminated amide acid pre-polymer, by reacting 1 mole of tetracarboxylic dianhydride with 2 moles of diamine, which subsequently is endcapped with 4-ethynylphthalic anhydride. The imide oligomer is finally obtained via chemical cyclodehydration. The properties of the ATI resin prepared via this route are not too different from those prepared from 3-ethynylaniline as an endcapper. When 1,3-bis(3-aminophenox)benzene was used as diamine, the prepolymer is completely soluble in DMAc or NMP at room temperature, whereas 4,4'-methylene dianiline and 4,4'-oxydianiline based ATIs were only partially soluble. The chemical structure of ATIs based on 4-ethynylphthalic anhydride endcapper is shown in Fig. 45.

The ethynyl terminated imide oligomers are very attractive because their cured polymers are thermally stable (131). However, improvements are required in processability. An interesting approach to this problem was the synthesis and use of ethynyl-terminated isoimide (132). If the cyclodehydration of the amide acid intermediate is performed chemically with dicyclohexylcarbodiimide, is-oimide is formed in almost quantitative yield. (Fig. 46). It is claimed that the isoimide provides better flow and solubility compared with the corresponding imide. At elevated temperatures, during cure, the isoimide rearranges into the

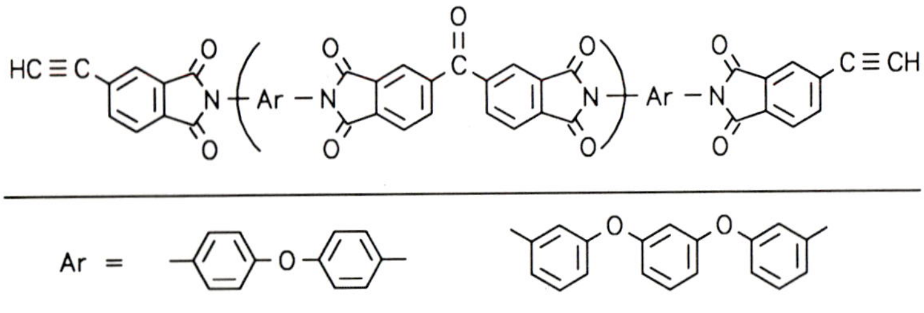

Fig. 45. Chemical structure of ATI resins based on 4-ethynylphthalic anhydride

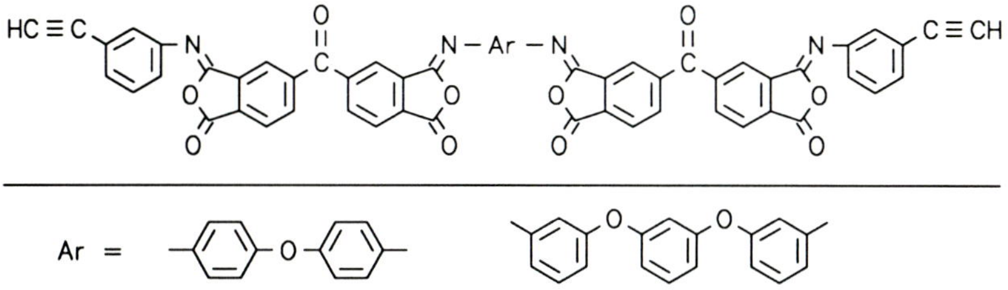

Fig. 46. Chemical structure of acetylene-terminated isoimide

Table 12. Cure onset temperatures of phenylethynyl-terminated polyimide oligomers

Endcaps	Polymerisation onset Temp., °C
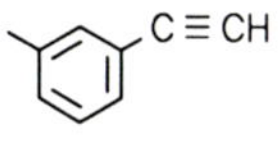 3—aminophenylacetylene	195
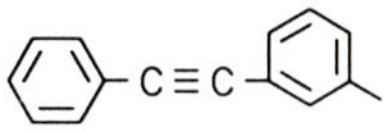 3—phenylethynylaniline	332
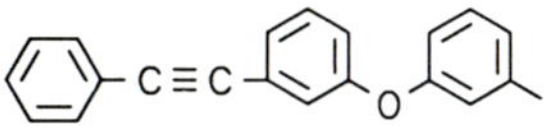 3—(3—(phenylethynyl)phenoxy)aniline	303

imide as was shown via FTIR spectroscopy (133). It is further suggested that the isoimide- imide transition and the formation of the less flexible imide backbone helps to widen the processing window. High postcure temperatures and extended postcure time is required to fully cure the acetylene functional groups.

5.1 Phenylethynyl Terminated Polyimides

The most widely used endcapper for ATI resins is 3-amino-phenylacetylene. This endcapper, however, polymerizes at relatively low temperature (200 °C) and therefore is responsible for the narrow processing window. Attempts to overcome this deficiency resulted in the use of 3-phenylethynyl aniline and 3-(3-(phenylethynyl)phenoxy)-aniline as endcappers for imide oligomers (134). The polymerization onset temperature, measured via DSC analysis, is dramatically shifted to higher temperatures, as shown in Table 12. For improved processing, the melting temperature and the cure onset temperature have to be well separated to allow melt flow or melt processing without preliminary polymerization and/or crosslinking.

5.2 Maleimide-Modified ATI Resins

Ethynylphenyl- and maleimide groups polymerize thermally in the 200–250 °C temperature range. It is therefore logical that the processability of ATIs such as Thermid 600 could be improved by blending with a low melting bismaleimide which acts as a reactive diluent. The cured copolymer is highly crosslinked and temperature resistant (135). Recently it has been reported that ethynylphenyl groups copolymerize with maleimide via a Diels-Alder reaction in the presence of ruthenium or rhodium catalysts (136). Model compound studies on the reaction of N-(3-ethynylphenyl)phthalimide with N-(4-phenoxyphenyl) maleimide in the molten state at 170 °C and in solution, show evidence for the reaction outlined in Fig. 47. This reaction takes place without a catalyst at temperatures of about 210–220 °C. This chemistry allows a wide range of resins because of the many BMIs that may be employed in this concept.

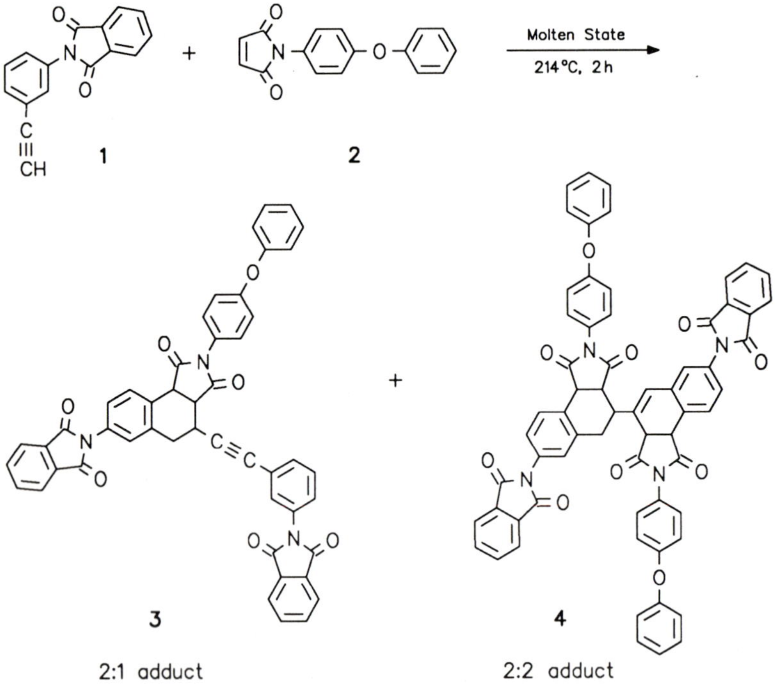

Fig. 47. Molten state reaction between ethynylphenyl compounds and maleimide

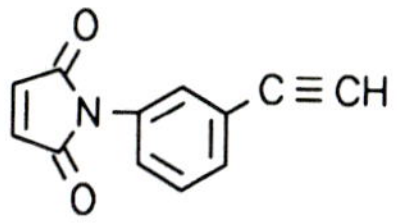

Fig. 48. Chemical structure of *N*-(3-ethynylphenyl)maleimide (137)

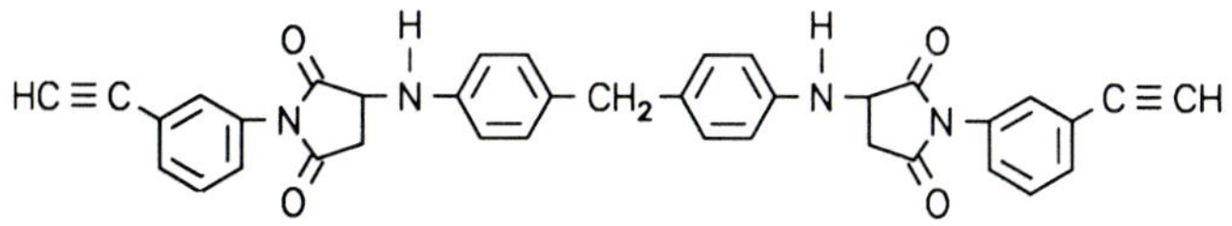

Fig. 49. Chemical structure of acetylene terminated aspartimide

Attempts to elucidate the polymerization or copolymerization kinetics of ethynyl and maleimide-functionalized monomers have been undertaken via vibrational spectroscopy (137). The thermal polymerization of *N*–(3-ethynyl-phenyl) maleimide (the structure is given in Fig. 48) was studied via IR and Raman spectroscopy. This model compound is interesting because it carries maleimide and ethynyl groups attached to the same aromatic ring. Kinetic studies indicate that both the acetylene and maleimide group react at the same rate, which strongly suggests the formation of a copolymer rather than a mixture of homopolymers.

N–(3-Ethynyl phenyl maleimide is synthesized from 3-ethynylaniline and maleic anhydride in DMAc. The cyclodehydration of the resulting amide acid is performed in the usual way with acetic anhydride and nickel acetate as a base catalyst (138). This AB-type monomer has a melting point of 129–131 °C. The polymerization/copolymerization is extremely exotherm (720 J/g) and proceeds in the 180–220 °C temperature range. Furthermore, *N*–(3-ethynyl phenyl)maleimide is an interesting building block for the the synthesis of acetylene functionalized imide resins. Acetylene terminated aspartimides (ATA) have been prepared by reacting 2 moles of *N*–(3-ethylenyl phenyl) maleimide with 1 mole of an aromatic diamine (138). The chemical structure is shown in Fig. 49. These ATA resins polymerize thermally around 200 °C and after an appropriate postcure show high T_gs. Because of their highly crosslinked nature they are considered brittle. K_{1c} values of 700–900 psi inch have been measured.

The Michael addition reaction of dimercaptodiphenylether with *N*-(3-ethynyl phenyl) maleimide allowed the synthesis of ethynyl-terminated imido-thioether as shown in Fig. 50 (139). This acetylene terminated imidothioether was blended with acetylene terminated polyarylene ether oligomers of different molecular weights and tested as composite resins (140). Blends of functionalized thermoplastics such as the acetylene terminated polyarylene ethers with brittle high-T_g imide resins are finding increased attention for tough high-T_g composites.

Of all the thermosetting imide oligomers discussed in this article, the totally aromatic acetylene terminated imides are the most promising because their

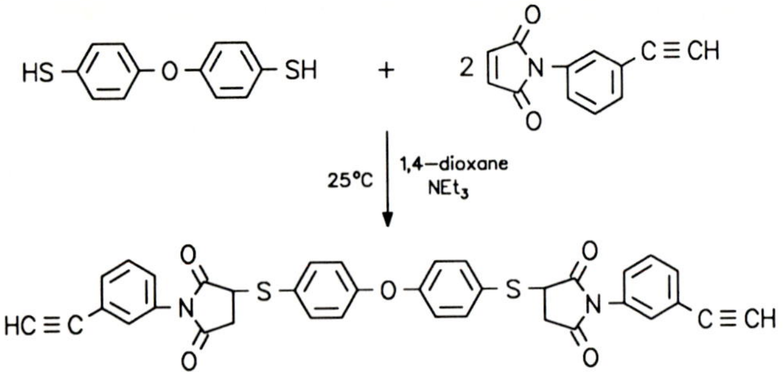

Fig. 50. Chemical structure of ethynyl-terminated imidothioether

polymers are superior in thermal oxidative stability in comparison to bismal-eimides and PMRs. For their synthesis ethynyl functionalized endcappers such as 3-ethynyl aniline or 3-ethynyl phthalic anhydride are necessary. The high cost of these endcappers has limited the broad use of this class of thermosets.

6 Benzocyclobutene-Imides

A novel cure chemistry employed for addition poly(imides) has recently been published. The successful preparation of 4-aminobenzocyclobutene allowed the synthesis of benzocyclobutene-terminated imide oligomers and bis(benzocylobutenes) (17). The benzocyclobutene group is a latent diene which isomerizes to o-guinodimethane at temperatures of about 200 °C and may homo- and/or co-polymerize for example with bismaleimide (83). Details on the benzocyclobutene chemistry are described in chapter I of this book.

7 Future Requirements

Throughout this chapter the chemical concepts employed to synthesize and cure addition poly(imides) have been discussed and their use as matrix resins for fiber composites has frequently been mentioned. The most important property of the imide backbone structure is the inherent thermal stability. The target of achieving the temperature performance of linear poly(imide) has not been reached, because of the aliphatic nature of the reactive endgroups, and because of the low molecular weight of the imide backbone required for processing. Future developments of addition polyimides will, as in the past, focus on the requirement of high thermal and thermal oxidative stability of the crosslinked polyimide and on a satisfactory processing behaviour of the uncured precursor. Further improvements of the cured resin mechanical properties are desired. A novel requirement, which takes into account the health and safety aspects, is the use of low toxicity monomers.

In summary, addition polyimides with improved properties in the areas

– thermal and thermal oxidative stability, achievable via
– the backbone chemistry of the imide prepolymer
– controlled(complete) cure

– processing
– design of the prepolymer composition and prepolymer molecular weight distribution
– low temperature cure
– cured resin mechanical properties
– high elastic modulus
– high toughness
– low moisture absorption

– toxicological hazards
– use of low toxicity monomers

are desired.

8 References

1. Fast RA, Eckert CH (1988) 33rd Int SAMPE Symp 33: 369
2. Hergenrother PM, Rogalski ME, (1992) Polym Prep 33: 334
3. Grundschober F, Sambeth J (1968) US Pat 3,380,964
4. Bargain M, Combat A, Grosjean P (1968) Brit Pat Spec 1,190,718
5. Lubowitz HR (1970) USPat 3,528,950
6. Lubowitz HR (1971) ACS Org Coat Plast Chem 31: 561
7. Serafini TT, Delvigs P, Lightsey GR (1972) J Appl Polym Sci 16: 905
8. Bilow N, Landis AL, Miller LJ (1974) US Pat 3,845,018
9. Landis AL, Bilow N, Boshan RH, Lawrence RE, Aponyi T (1974) ACS Polym Prep 15: 537
10. Stenzenberger HD (1976) US Pat 3,966,864
11. Street S (1980) 25th Nat SAMPE Symp 25: 366
12. McKague L (1983) 28th Nat SAMPE Symp 28: 640
13. Riley BL (1986) 2nd Int Conf on Fibre Reinforced Composites, Proceedings, Univ of Liverpool UK 153
14. Zahir S, Renner A (1975) Swiss Pat Appl 7988
15. St. Clair TL, Jewell RA (1976) 8th Nat SAMPE Techn Conf 8: 82
16. Vannucci RD, Alston WB (1985) NASA TXM-71682
17. Loon-Seng Tan, Arnold FE (1985) ACS Polym Prep 26: 176
18. Cole N, Gruber WF (1964) US Pat 3,127,414
19. Sauers CK (1969) J Org Chem 34: 2275
20. Haug Th, Kiefer J, Renner A (1985) Ger Pat DE 2,715,503 C2
21. Orphanides G (1979) US Pat 4, 154, 737
22. Lancaster M (1990) Europ Pat Appl 367,599
23. Orphanides G (1977) Germ Offen 27 19 903

24. Abblard J, Boudin M (1978) Germ Offen 27 51 901
25. Boudin M, Abblard J (1979) Germ Offen 28 34 919
26. Shumichi D, Yasuyuki T (1985) Europ Pat Appl 0 177,031
27. Stenzenberger HD, unpublished
28. Lee B, Chaudhari M, Gavin T (1986) 17th Nat SAMPE Techn Conf 17 172
29. Nagai A, Takahashi A, Suzuki M, Mukoh A (1992) Appl Pol Sci 44: 159
30. Varma IK, Fohlen G, Parker JA (1981) US Pat 4,276,344
31. Heisey C, Wood PA, McGrath JE, Wightman JP (1992) ACS PMSE 67: 28
32. Goldfarb IJ, Feld PA, Saikumar J (1992) ACS Polym Prep 33: 431
33. Pascal Th, Mercier R (1989) Sillion B, Polymer 30: 739
34. Pascal Th, Sillion B, Grosjean F, Grennier-Lonstalot MF, Grenier G (1990) High Performance
    Polym 2: 95
35. Stenzenberger HD, König P, unpublished
36. Kwiatkowski GT, Brode GL (1974) US Pat 4,276,377
37. Lyle GD, Senger DH, Chen DH, Kilic S, Wu SD, Mohatney, Mc Grath JE (1989) Polymer
    30: 978
38. Hirano T, Muramatsu T, Jnoue H (1989) 1st Japan Int SAMPE Symp 1
39. Holub FF, Evans ML (1971) Germ Offen 20 31 574
40. Holub FF, Evans LM (1971) Germ Offen 20 31 573
41. Kawahara H, Maikuma T (1974) Japan Kokai 74,35397 Chem Abstr 81 (1974) 170240t
42. Rao BS (1988) J Polym Sci Part C Polym Letters 26: 3
43. Park JO, Jaug S (1992) J Polym Sci Chem Ed 30: 723
44. Bargain M, Combet A, Grosjean P (1973) US Pat 3,562,223   45. N,N, Bismaleimides for
    Advanced Printed Circuit Boards Phone Poulenc
46. Tung CM, Lung CL, Liar TT (1985) ACS PMSE 52: 139
47. Enoki T, Takeda T, Ishii K (1993) J Thermosetting Plastics, Japan 14: 131
48. Hideo N, Konjiton K, Takenori N (1992) Deshi Zairyon 10: 84
49. Stenzenberger HD (1980) US Pat 4,211,861
50. Stenzenberger HD, Römer W, Herzog M, Canning M, Pierce St (Jan 1987) IPC Techn Reviews
    28
51. Stenzenberger HD (1987) 32nd Int SAMPE Symp 32: 44
52. Koyoji M, Akira S (1978) Jap Kokai 7801297
53. Forgo I, Schreiber B, Renner A, Haug Th (1975) Germ Offen 24 59 925
54. Forgo I, Renner A, Schmitter A (1974) Germ Offen 24 58 938
55. White JE, Scaia MD, Snider-Tung DA(1985) Polym Prep 26
56. Connell JW, Bassi RG, Hergenrother PM (1988) 33rd Int SAMPE Symp 33: 251
57. Asahari T, Joda N, Minami (1972) USPat 3,699,930
58. Suzuki M, Nagai K, Suzuki M, Takahashi A (1992) J Appl Polym Sci 44: 1807
59. Jing-Pin Pau, Guw-Yuh Shian, Song Shiang Lin, Ker-Ming Chen (1992) J Appl Polym Sci 45:
    103
60. Pigneri AM, Galgoci EC, Jackson RJ, Young GE (1987) 1st Int SAMPE Electr Conf. 1 657
61. Davis MJ, Sense TR (1987) IPC Fall Meeting Chicago, III
62. Kiyoji M, Katsuyuki (1978) Jap Pat Appl 78,01297
63. Kiyoji M, Tsutomu O (1977) Ger Offen 2,728,843
64. Carduner K, Chatta MS (1987) ACS PMSE 56: 660
65. Enoki T, Okubo H, Ishii K, Shibahara S (1991) Netsuku Kasei Jushi 12: 18
66. King J, Chaudhari M, Zahir S (1984) 29th Int SAME Symp 29 392
67. Enoki T, Takeda T, Ishii K (1993) J of Thermosetting Plastics Japan 14: 131
68. Stenzenberger HD, König P (1989) High Performance Polym 1: 239
69. Stenzenberger HD, König P (1989) High Performance Polym 1: 133
70. Stenzenberger HD, König P, Römer W, Herzog M, Breitigam W (1991) 36 Int SAMPE Symp
    36: 1236
71. Stenzenberger HD (1988) US Pat 4,789,704
72. Hong-Son Ryang (1989) US Pat 4,826,929
73. Renner A, Kramer A (1989) J Polym Sci Pol Chem Ed 27: 1301
74. Barret KA, Fu B, Wang A (1992) 35th Int SAMPE Symp 35
75. Stenzenberger HD, König P, Herzog M, Römer W, Canning MS, Pierce St (1986) 19th Int
    SAMPE Techn Con 19: 500
76. Stenzenberger HD, König P, Herzog M, Römer W, Pierce St, Canning MS (1987) 32nd Int
    SAMPE Sym 32: 44

77. Stenzenberger DH, unpublished
78. Barton JM, Hamerton I, Jones RJ, Stedman JC (1991) Polymer Bul 27: 163
79. Stenzenberger HD, König P (1991) High Performance Polym 3: 41
80. Eisenbarth Ph, Linden G, Altstaedt V, Peter R (1991) US Pat 5,003,017
81 Boyd JD (1990) US Pat 4,902,778
82. Ohtani K, Shinohara N, Yoshida H, Hanyuda T (1992) 13: 147
83. Loon Seng Tan, Arnold FE, Solosky (19880 J Polym Sci Polym Chem Ed 26: 3103
84. Street S (1984) In: Mittal KL (ed) Polyimides. Plenum, New York, p 77
85. McKague L (1982) in Composites for Extreme Environments, ASTM STP 768: 20
86. Kinloch AJ, Shaw ST (1983) ACS PMSE 49: 307
87. Shaw SJ, Kinloch AJ (1985) Int J Adh and Adhesives 5: 123
88. Takeda S, Kakiuchi H (1988) J Appl Pol Sci 35
89. Stenzenberger HD, Römer W, Herzog M, König P (1988) 33rd Int. SAMPE Symp 33: 1546
90. Stenzenberger HD, Römer W, Herzog M, König P, 34th Int SAMPE Symp closed session paper
91. Stenzenberger HD, Römer W, Hergenrother PM, Jensen B (1989) 34th Int SAMPE Symp 34: 2054
92. Stenzenberger HD, Römer W, Hergenrother PM, Jensen B, Breitigam W (1990) 35th Int SAMPE Symp 53: 2175
93. Rakutt D, Fitzer E, Stenzenberger HD (1990) High Performance Polymers 2: 133
94. Rakutt D, Fitzer E, Stenzenberger HD (1991) High Performance Polymers 3: 59
95. Wilkinson SP, Liptak SC, Wood PA, McGrath JE, Ward TC (1990) 36th Int SAMPE Symp 35: 482
96. Stenzenberger HD, König P (1993) High Performance Polym 5: 123
97. Dynes PJ, Liao TT, Hammermesh CL, Wutucki (1982) IN: Polyimides: synthesis, characterisation and application. Plenum, New York, p 311
98. Hay NJ, Boyle JD, James PG, Walton JR, Bare KJ, Konarski M, Wilson D (1989) High Performance Polym 1: 145
99. Grenier-Loustalot MF, Grenier PH (1991) High Performance Polym 3: 113
100. Lindenmeyer PH, Sheppard CH (1984) Characterization of PMR Polyimide Resin and Prepreg NASA Report N84–20695
101. Jones RJ, Vaughan RW, Burns EA (1972) Thermally Stable Laminating Resins, NASA CR-72984
102. Gaylord NG, Marten M (1980) ACS Polym Prep 22: 11
103. Wang AC, Ritchey MM (1981) Macromolecules 14: 825
104. Sukenik CN, Malhotra V, Varde V (1985) in Reactive Oligomers (ed Harris FW, Spinelli HJ) ACS Symp Ser 282: 53
105. Lederer et al, Paper presented at the 7th HF Mark Symposium Vienna 1988 at the Austrian Plastics Institute
106. Wilson D (1988) Brit Polym J 20: 405
107. Hay JN, Boyle JD, Parker SF, Wilson D (1989) Polymer 30: 1032
108. Young PR (1981) NASA TM-83192
109. Vannucci RD, Alston WB (1976) NASA TMX-71816
110. Wilson D (19930 ACS Symp on Recent Advances in Polyimides and other High Performance Polymers, Sparks, Nevada, Jan 18–21
111. Chuang KC, Vannucci RD, Moore BW (1992) ACS Polym Prep 33: 435
112. Chuang KC, Vannucci RD, Ansari J (1991) ACS Polym Prep 32: 197
113. Bowles KJ (1990) 35th Int SAMPE Symp 35: 147
114. Vannucci RD, Malarik D, Papadopoulos D, Waters J (1990) 22nd Int SAMPE Techn Conf 22: 175
115. McCormack WE (1992) Proc 5th NASA Lewis Research Center HITEMP Rewiew (Cleveland, Ohio, Nov 27)
116. Scola DA (1993) ACS, Symposium on Recent Advances in Polyimides and other high Performance Polymers, Sarks, Nevada, Jan 18–21
117. Pater R, NASA Langley Research Center, private communication
118. Pater R, Proc High Temp Polym Matrix Comp, NASA LeRC march 1983
119. Serafini TT, Delvigs P, Vannucci RD (1981) NASA TM-81705 120. Delvigs P (1982) NASA TM-82958
121. Delvigs P (1983) Proc High Temp Polym Matrix Comp, NASA LeRC March 23
122. Pater R (1986) NASA Report 86–11278

123. Hoyle ND, Stewart NJ, Wilson D, Baschant M, Merz H, Sikorski S, Greenwood J, Small GD (1989) High Performance Polym 1: 285
124. Mittal KL (1982) Polyimides: Synthesis, Characterization and Applications, Vol 2 117, Plenum Press, New York
125. N.N. Performance Materials 1 Oct (1990) 1
126. Droske JP, Stille JK (1983) ACS Org Coat Appl Polym Sci 48: 925
127. Baldwin LJ, Meador MAB, Meador MA (19880 ACS Polym Prep 29: 236
128. Sutter JK, Waters JF. Schuerman MA (1992) ACS Polym Prep 33: 366
129. Bilow N, Keller LB, Landis LA, Boshan RH, Castillo AA (1978) 23rd Nat SAMPE Symp 23: 791
130. Hergenrother PM (1980) ACS Polym Prep 21: 81
131. Moy TM, DePorter CD, McGrath JE (1992) ACS Polym Prep 33: 489
132. Landis AL, Naselow AB (1982) 14th Nat SAMPE Techn Conf 14: 236
133. Bott RH, Taylor LT, Ward TC (1986) ACS Polym Prep 27: 72
134. Paul Ch W, Schultz RA, Fenelli St P (1991) Fourth Int Conf on Polyimides Oct 30–Nov 1 Ellenville
135. Stenzenberger HD, unpublished
136. Soucek MD, Pater RH, Ritenour SL (1993) ACS Polym Prep 34: 530
137. Parker St F, Lander JA, Gerrard DL, Bowley HJ, Hay JN (1989) High Performance Polym 1: 311
138. Hergenrother PM, Havens SF, Connell JW (1986) ACS Polym Prep 27: 408
139. Connell JW, Bass RG (1988) Hergenrother PM, 33rd Int SAMPE Symp 33: 251
140. Connell JW, Hergenrother PM, Havens SJ (1989) High Performance Polym 1: 119

Received December 2, 1993

Thermotropic Liquid Crystalline Polymers for High Performance Applications

J. Economy and K. Goranov
University of Illinois at Urbana-Champaign, Department of Materials Science
and Engineering, 1304 W. Green Str., Urbana, IL 61801, USA

A brief review of the background to the field of liquid crystalline copolyesters (LCPs) is presented. Recent progress on interpreting the behavior of these polymers is described. In particular, the nature of the high temperature transitions of the aromatic polyesters is discussed. Reasonably definitive evidence is presented on the nature of the microstructure of these copolyesters. The tendency for randomizing or ordering of the sequence distribution on heating at elevated temperatures is discussed in terms of chemical processes involving intrachain transesterification reactions. The potential to fabricate LCPs as adhesives, protective coatings, matrix composites and structural foams is indicated.

1 Introduction

It seems very timely to put together a chapter on Thermotropic Liquid
Crystalline Polymers (LCPs) with particular emphasis on the significant progress made in the last several years. It has been 23 years since the first reports
appeared describing the commercialization of the aromatic polyesters based on
p-hydroxybenzoic acid (PHBA) [1]. Subsequently, from 1975 to 1987 a number
of companies reported on the development and/or commercialization of a
number of related copolyesters usually with PHBA as a key component (see
Fig. 1 for representative systems) [2, 3]. It is interesting to note that most of the
industrial effort has focused on synthesis and property optimization. These
trends from a world-wide point of view can be effectively assessed from a
consideration of the patents published up to 1988 (see Fig. 2) [4]. The strong
interest in these polymers can be readily understood from a brief examination of
the outstanding properties associated with this class of polymers (see Table 1).
Thus, these systems can display elastic modulus values ~ 5–10 times those for
most engineering plastics, and at least in one case the copolymer retains a
significant percentage of its mechanical properties up to 350 °C. The dielectric
constant can be as low as ~ 3.0, particularly at higher frequencies. In addition,
these polymers can display outstanding barrier characteristics to moisture and
corrosive solvents. For the most part they have been used as injection molded
parts although fibers similar to Kevlar can be prepared from the melt.

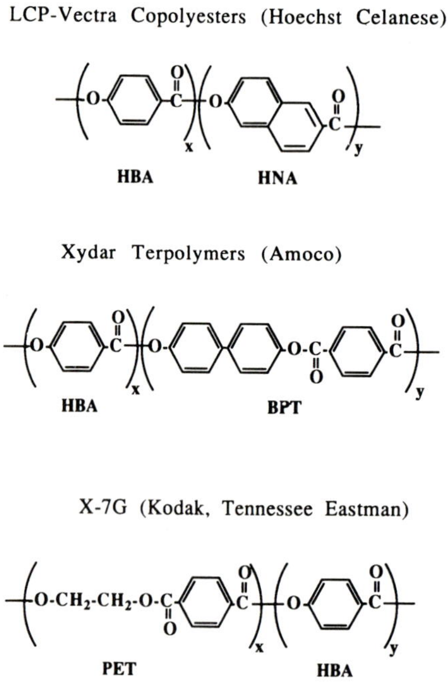

LCP-Vectra Copolyesters (Hoechst Celanese)

HBA   HNA

Xydar Terpolymers (Amoco)

HBA   BPT

X-7G (Kodak, Tennessee Eastman)

PET   HBA

Fig. 1. LC copolyester structures in commercial use

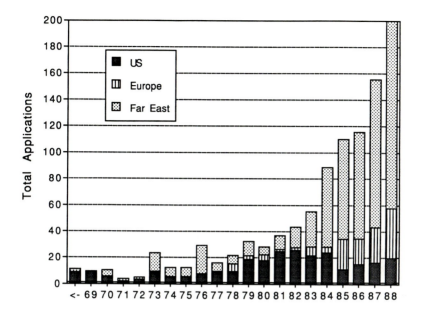

Priority Date (year)

Fig. 2. LCP patent activity by region/year – worldwide totals

Table 1. LC-polymers: profile of properties

Exceptional Inherent Stiffness
Outstanding Chemical Resistance
Excellent Flow Properties of Melt
High Heat Deformation Resistance
Low Adjustable Coefficient of Thermal Expansion
Inherent Flame Retardance
Low Notch Sensitivity
Low Dielectric Constant
Excellent Corrosion Resistance

It was first reported in the early 1970s that these melt processible polymers could best be described as thermotropic systems which usually display an nematic texture in the melt phase [5]. Subsequently, a number of additional phases have been reported ranging from discotic structures to highly ordered smectic E & G systems with three dimensional order. In the last several years an IUPAC sponsored study on nomenclature on thermotropic LPCs has been underway. A more complete set of definitions will be available shortly as a result of Recommendation No. 199 IUPAC [6].

It is important to note that only in the last few years has any meaningful progress been made on developing a more fundamental understanding of the microstructure of the copolyesters as well as the nature of the high temperature transitions and related molecular motions. Earlier workers were greatly hindered by the relative intractibility of most of the commercial systems. For example, the LCPs in Fig. 1 display limited solubility if any and usually only in highly aggressive solvents such as perfluorophenol or *m*-cresol. Consequently, indirect methods were used for the most part to study the nature of the microstructure and the potential for changes in the sequence distribution on heating near the crystal nematic transition (T_{cn}) or in the nematic melt. In the absence of more direct analytical tools it is understandable why considerable confusion developed in the published literature concerning these features [7–13].

Another area of concern relates to the rheological character of the liquid crystalline melt [14–16]. Thus, it is well recognized that these highly anisotropic melts flow very easily under shear, greatly facilitating processing such as injection molding and extrusion. On the other hand this kind of flow under shear can lead to formation of microfibrillar morphologies, skin-core effects and weld lines. To date very little progress has been made to address these drawbacks. Solid state forming, which has only been looked at briefly in the past, is examined near the end of this chapter as one possible approach to these problems.

An area where significant progress has recently been achieved is in the understanding of the role of interchain transesterification reactions at high temperatures [13]. This understanding has opened up completely new avenues for design and use of these polymers as adhesives [17, 18], protective coatings and substrates for multi-layered microelectronic packaging.

In this chapter particular emphasis is placed on recent progress in interpreting the behavior of aromatic polyesters which have been pursued commercially. The large body of literature on LCPs with spacers and the side chain LCPs are not discussed except peripherally. A more detailed discussion of these topics is available [19]. The major themes presented in this chapter are summarized as follows:

- Background
- High temperature Transitions and Molecular Motions
- Nature of the Microstructure
- Processing
- New Directions

2 Background

A brief review of the background to this field is provided here. Perhaps the first pertinent reference in the initial phases of this field was the publication of Gilkey and Caldwell in 1959 [20], where they reported the preparation of the homopolymers of PHBA and *m*-HBA by melt polymerization of the respective acetate esters. In the case of PHBA they isolated an intractable material that decomposed at 350 °C generating considerable volatiles. In 1962 we undertook to repeat this work since it seemed unreasonable that this polymer should decompose at such a low temperature. In our study we also isolated the sublimate from heating of the intractable polymer and found that it consisted almost entirely of the phenyl ester of *p*-phenoxybenzoic acid, **1**. In checking the literature we were surprised to learn that, almost 80 years earlier in 1883, Klepl had reported that heating *p*-hydroxybenzoic acid produced the dimer, trimer and a material designated as $(C_7H_4O_2)_x$ [21]. It was also reported in Beilstein that the $(C_7H_4O_2)_x$ polymer could be used as a starting material to produce substantial yields of the phenyl ester of *p*-phenoxybenzoic acid. Some 35 years later Emil Fischer undertook to synthesize the dimer and trimer of PHBA by an unambiguous route and noted that, based on similarities in the melting points, Klepl had indeed isolated these materials [22].

1

In our program it became obvious that the synthetic routes selected by Gilkey and Caldwell as well as by Klepl were undergoing some unanticipated side reactions such as high temperature acid catalyzed etherification. We reasoned that the phenyl ester of PHBA might be a more suitable starting material for polymerization since the carboxylic acid would be blocked. In fact, solution/slurry polymerization of this monomer in March 1963 yielded a very stable polymer which was shown to have the correct structure. This route with minor modification is used today in the manufacture of the homopolymer which has the trade name Ekonol® [23]. This polymer was stable at 400 °C for considerable periods of time and could be fabricated by compression sintering. By the mid 1960s it was found that solid state forming techniques such as high energy rate forging and plasma spraying could be used to fabricate solids and coatings, respectively [1].

At about that time we also successfully prepared melt processible aromatic copolyesters designed to retain properties to over 300 °C. Our intent was to synthesize random aromatic copolyesters which because of their rigid rod-like character would retain sufficient crystallinity at temperatures above 300 °C. Since we expected the T_g's of aromatic copolyesters to be well below 200 °C [24] it was important to prepare copolymers that would have a reasonable degree of crystallinity. By selecting compositions with melting points close to 400 °C, we hoped to prepare systems that could be processed at 400 °C and yet would retain their mechanical properties to well over 300 °C. After preparing a number of systems we found that the PHBA biphenol terephthalate (BPT) copolymer appeared to satisfy these goals. This system appeared to have outstanding high temperature properties, the monomers were low in cost or readily available and the copolymer did not discolor appreciably after melt processing at temperatures of 400 °C. Unwittingly, the copolymers we prepared and evaluated were also thermotropic in character. We recognized the unusual melt anisotropy and directional properties during injection molding but felt it to be more of a problem than an advantage. In our work we tended to minimize these anisotropic properties through use of fillers or by molding under modest shear conditions.

By early 1970 our program had progressed to the point that we introduced the homopolymer of PHBA as a commercial material under the tradename of Ekonol. This was followed a year or so later with two melt processible copolyesters of PHBA/BPT with compositions of 1/2 and 2/1, named Ekkcel C1000 and Ekkcel I2000, respectively. See Table 2 for a summary of properties. It is noteworthy that the I2000 (referred to later as Xydar 300) did not process easily and required temperatures in excess of 400 °C for injection molding. During that period we also heard about DuPont's development of PRD-49, a benzamide fiber [25]. The similarity between our materials and the Aramids did not escape us and within several months we had prepared compositions of PHBA-BPT which could be melt spun into fibers with properties comparable to the Aramids [26]. In the early 1970s we also established a joint venture with Sumitomo Chemical to exploit these products in the Far East marketplace. Sumitomo Chemical continued to work on the fiber and eventually made filaments available by the mid 1980s. (See Table 3 for a comparison of properties with Kevlar).

By the mid 1970s, Tennessee Eastman also announced development of a copolyester consisting of 60/40 PHBA/polyethylene terephthalate (PET) by direction reaction of acetoxybenozic acid with PET in the melt [2]. This system had the advantage of lower costs, but its use temperature was limited to 90 °C which is just above its T_g. In the early 1980s, researchers at Celanese reported

Table 2. Copolyester properties of LCP based on PHBA/BPT

Properties	1/2* Ekkcel C-1000	2/1* Ekkcel I-2000
Tensile Strength (psi)	10 000	14 000
Tensile Modulus (psi)	190 000	350 000
Elongation (%)	7–9	8
Flexutral Strength (psi):		
at 23 °C	15 000	17 000
at 260 °C	5 000	4 000
Flexutral Modulus (psi):		
at 23 °C	460 000	700 000
at 260 °C	125 000	235 000
Heat Distortion Temperature		
at 264 psi (°C)	300	293
Coefficient of Thermal		
Expansion (in/in/°F)	2.87×10^{-5}	1.60×10^{-5}
Specific Gravity (g/cc)	1.35	1.40

* describes the molar ratio of PHBA to BP and T

Table 3. High strength/modulus organic fibers

	Kevlar 49	Ekonol Fiber
Density, g/cc	1.45	1.40
Tensile strength, psi	400 000	550 000
Tensile modulus, psi	20×10^6	24×10^6
Elongation, %	2.7	3.0
Moisture absorption, %	2.0	0.01

development of an all aromatic copolyester based on PHBA and 2,6-hydroxynaphthoic acid (HNA) in a ratio of 73/27 [3]. This material could be easily injection molded or melt drawn into filaments. On the other hand the higher cost of the HNA monomer was a problem and the relatively low degree of crystallinity tended to limit the use temperature to approximately 170 °C. Hence each of these commercially available systems had certain advantages and disadvantages.

During the 1980s a number of additional companies reported similar systems including DuPont, BASF, Bayer and several Japanese companies. Up to now, however, no company has as yet introduced a copolyester where the material is easily melt processed at 350–400 °C, has a use temperature of 300 °C and a selling price of under $5.00/1b. These requirements would appear essential for a broader based market.

One other point of historical note concerns the first report of the existence of thermotropic polymers. In our work, by the mid to late 1960s, we had clearly recognized the unique shear sensitivity of these melts and the potential for highly anisotropic properties in molded samples. However, it was Professor A. Sirigu at Naples University who was the first to actually publish, in the early 1970s, on the existence of thermotropic polymers [5]. This was followed shortly thereafter by Jackson's report on the liquid crystalline nature of the PHBA/PET system [2]. During the 1980s there were a number of studies on the use of spacers between mesogenic units to lower the liquid crystalline transitions and thus permit easier study of these transition without concern for degradation reactions. As indicated earlier, this work is not included in the context of this chapter since sufficient progress has now been made in characterizing the more intractable aromatic copolyesters to permit a reasonably comprehensive interpretation of their microstructure, thermal transitions and related molecular motions.

3 Nature of High Temperature Transitions of the Aromatic Polyesters

All of the high temperature copolyesters under discussion in this chapter undergo a transition into a nematic melt. What is not obvious is the nature of the molecular motions which occur at much lower temperatures and the nature of the transitions associated with these motions. The homopolymer of PHBA has been studied in some detail over the last few years and thus provides an excellent starting point for determining the nature of the molecular motions associated with the high temperature transitions in the various copolyesters [27–29]. Because of its very ordered structure, the PHBA polymer displays additional transitions prior to forming a nematic phase.

From a DSC scan of the homopolymer of PHBA one can observe a major endotherm at $\sim 350\,°C$ and a much smaller one at 445 °C (see Fig. 3) [29]. The first transition has been examined by electron diffraction [27–30], X-ray diffraction [28] and proton and ^{13}C NMR [30]. Additional insights have been provided by synthesizing much lower molecular weight samples which permit study of these transitions at appreciably lower temperatures [29]. These low molar mass homopolymers can also go into a nematic phase under a modest shear.

With respect to the higher temperature transition at 445 °C, there are two conflicting views of this transition, namely that the phase above 445 °C is a smectic C and the other that it is nematic. Based on high temperature X-ray diffraction studies, Yoon et al. have concluded that it is a smectic C (see Fig. 4) [28]. Thus, in Fig. 4, the disappearance of the 211 peak indicates that the nematic E structure is converting to a nematic C. In our work, using polarizing optical microscopy, we have observed a nematic texture for high molar mass specimens heated rapidly to 480 °C, sheared, and then quenched. In the case of a

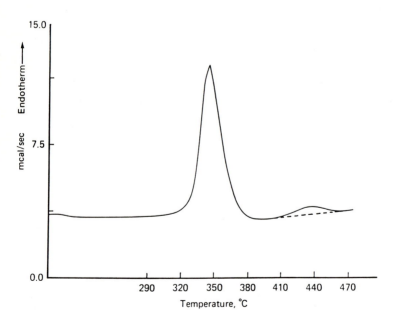

Fig. 3. DSC scan of PHBA

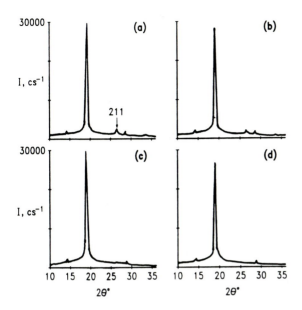

Fig. 4a–d. X-ray diffraction powder data for PHBA at: **a** 365; **b** 390; **c** 410; **d** 435 °C

low molar mass sample (DP = 15) heated to only 445 °C with no applied shear and then quenched, the sample displayed a definite nematic texture (see Fig. 5). These results can be explained on the basis that the homopolymer in a quiescent state does indeed go into a smectic C state; however, with applied shear in the case of the high molar mass sample the smectic C structure converts to a nematic phase. In the low molar mass case the molecular motions at 440 °C are such that a very modest shear is sufficient to convert the structure to the nematic phase. These studies provide for the first time a rational approach for future workers to design conditions for melt processing of the PHBA homopolymer at very high temperatures with short contact times.

Turning to the low temperature transition of the homopolymer of PHBA at 350 °C, it is generally accepted that the phase below this temperature is orthorhombic and converts to an approximate pseudohexagonal phase with a packing closely related to the orthorhombic phase (see Fig. 6) [27–29]. The fact that a number of the diffraction maxima retain the sharp definition at room temperature pattern combined with the streaking of the 006 line suggests both vertical and horizontal displacements of the chains [29]. As mentioned earlier, Yoon et al. has opted to describe the new phase as a smectic E whereas we prefer to interpret this new phase as a one dimensional plastic crystal where rotational freedom is permitted around the chain axis. This particular question is really a matter of semantics since both interpretations are correct. Perhaps the more important issue is which of these terminologies provides a more descriptive picture as to the nature of the molecular motions of the polymer above the 350 °C transition. As will be seen shortly in the case of the aromatic copolyesters, similar motions can be identified well below the crystal-nematic transition.

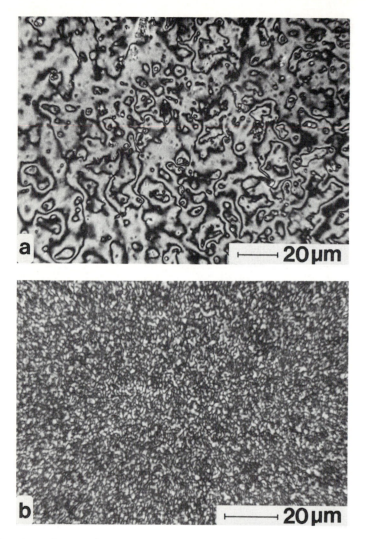

Fig. 5a–b. Nematic texture of the homopolymer PHBA: **a** DP 15, specimen quenched from 440 °C, no external shear applied; **b** DP > 100, specimen quenched from above 480 °C, external shear was applied to increase the rate at which the fluid specimen sheared between the glass surface above 480 °C

Use of Proton and ^{13}C NMR at temperatures from 27 to 400 °C provide very detailed information as to the nature of these motions [30]. Thus, it has been shown that even at 300 °C the phenylene ring displays a rapid 180° flipping motion. Above the transition temperature of 350 °C the ester unit also begins to rotate in the form of 180° flips as a result of lattice expansion (see Fig. 7). Furthermore, the entire repeat unit participates in a synchronous motion. This should be interpreted as a jumping motion rather than free or random rotation.

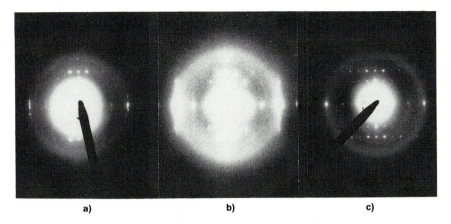

Fig. 6a–c. Electron diffraction pattern of the PHBA homopolymer (DP > 100), c-axis diffraction: **a** before heating, RT; **b** at 375 °C, furnace temperature; **c** after cooling to RT

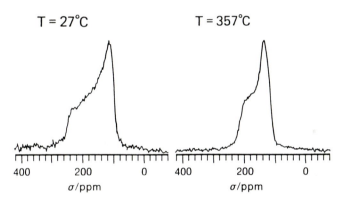

Fig. 7. ^{13}C NMR spectrum at 15.1 MHz of the carboxyl carbon in PHBA at 27 °C and 357 °C obtained by cross-polarization techniques. Note the change of width and shape of the CSA pattern

It is noteworthy that these kinds of motions are consistent with discrete flips observed in many plastic crystals.

One other point of interest is that the low molar mass PHBA oligomers display much lower transition temperatures, e.g., at a DP ~ 15 the value is 290 °C versus 350 °C for the high molar mass polymer (see Table 4). Such oligomers are usually isolated as single crystals from solution/slurry polymerization. The availability of such single crystals permits one to follow both structural and morphological changes on heating above the 290 °C transition [29]. Thus, on heating above 300 °C followed by quenching, the oligomer retains its single crystal morphology. On the other hand, when shear is applied at 300 °C, a well defined nematic texture is observed (see Fig. 8). At a higher molar mass (DP = 39) the sample is far more difficult to convert by an applied shear

Table 4. Transition enthalpies and entropies of PHBA as a function of increasing molecular weight

Material	DP	T_{cn}, °C	ΔH^a	ΔS^b
Dimer		230	2,53	5.03
Tetramer		262	2,30	4.31
Polymer 200	13	278	1,52	2.75
Polymer 250	29	303	1,18	2.04
Polymer 300	92	318	1,00	1.69
Polymer 350	> 166	348	1,25	2.02

[a] cal/mol, oxybenzoyl unit
[b] cal/Kmol, oxybenzoyl unit

into a nematic melt. From these results it would appear that at low molar mass the stability of the smectic E (plastic crystal) structure is relatively low and the packing can be easily disrupted by shear to produce a nematic phase.

Work to characterize the molecular motions in the PHBA/HNA copolyesters by NMR has been carried out in some detail by Davies et al. [31] Typically he observes onset of motions of the phenylene unit at significantly lower temperatures than that of the naphthalene ring but in both cases well below the crystal-nematic transition. In some recent work on the PHBA/BPT copolyesters we have also been able to identify the onset of motions of the various units with increasing temperature by selectively deuterating each of the rings [32]. As shown in Fig. 9, each of the rings begins to display the same kind of flips as observed with PHBA, but at distinctly different temperatures. On the other hand, the ester units presumably would not begin to spin until one approaches the crystal nematic transition at much higher temperatures.

Fahie et al. have recently reported on the ability of LCP chains to orient in an applied magnetic field [33]. Thus, working with a selectively deuterated 50/50 PHBA/HNA system, as well ^{13}C NMR for 30/70 PHBA/HNA, he was able to demonstrate ordering in a magnetic field by turning the sample by 90° in the magnetic field to reorient the samples (see Fig. 10).

4 Nature of the Microstructure in the Aromatic Copolyesters

This topic has been the subject of considerable debate over the past decade. In fact, with respect to the three best known structures indicated in Fig. 1, there are almost as many reports indicating that the sequence distributions are blocky as there are claiming that they are random [7–13]. A similar confusion has surrounded possible changes in the microstructure on heating these copolyesters in the nematic melt. In this latter case, some workers have argued that these systems are stable in the melt [7], while others claim that they undergo ordering

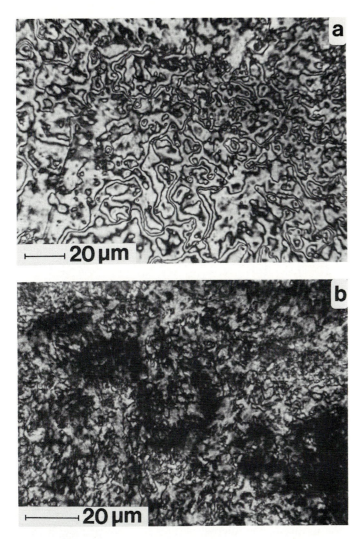

Fig. 8a,b. Nematic structure induced by moderate external shear in lower M_n samples: **a** DP 15, specimen quenched from 300 °C; **b** DP 39, specimen quenched from 330 °C

[8] and finally we have suggested that the dominant process is randomization through interchain transesterification reactions [13]. So, in a sense, all the possible permutations have been reported, but clearly only one should be allowable. Undoubtedly most of the problems associated with these reports arose from the difficulties in characterizing these relatively intractable copolymers. As noted earlier they are insoluble or only soluble in relatively small concentrations in very aggressive solvents. In this section, progress on interpreting the microstructure of the three systems indicated in Fig. 1 is reviewed.

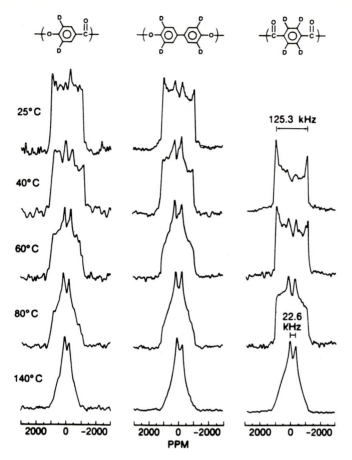

Fig. 9. PHBA/BPT – Xydar, molecular motion observed in ^{2}H quadrupole echo spectra (61.4 MHz)

To understand the nature of the microstructures that are present in these systems it is important to start by examining the role of the synthetic route in preparing these copolyesters. In the case of the PHBA/PET copolyester, this system presents a degree of complexity which is unusual. In the mid seventies, Jackson and Kuhfus reported that this system was random, but they did not comment on the possibility of compositional variations [2]. Fortunately, because of the modest solubility of this system, most of the confusion has recently been eliminated through detailed NMR studies [9]. Thus, it has been shown that the 60/40 PHBA/PET copolymers originally available from Tennessee Eastman actually consisted of two distinct compositions, namely 44/56 and 62/38 PHBA/PET (see Fig. 11). Furthermore, the soluble PET rich fraction ($\sim 20\%$) was shown to be blocky while the insoluble PHBA rich fraction (80%) was more random (see Fig. 12). These observations can be explained from a

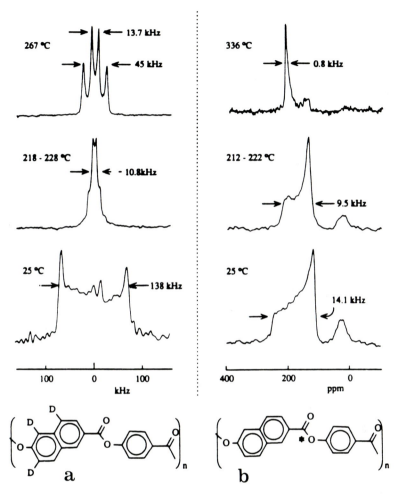

Fig. 10a. Deuterium NMR spectra of PHBA/PHNA 52/48 at the temperatures indicated. **b** ^{13}C NMR spectra of PHBA (^{13}C)/PHNA 30/70 at the temperatures indicated

consideration of the polymerization process. In this process acetoxybenzoic acid supposedly reacts in the melt with PET via ester interchange reactions to produce the random copolyester with a 60/40 ratio. A more likely interpretation is that during the reaction as the ratio of PHBA in the copolymer increases above a critical range, and phase separation occurs into a nematic and isotropic melt with distinctly different compositions. Because of the relative immiscibility of the two phases during the later stages of polymerization, there is little chance for interchain transesterification between the two phases which would homogenize the two distinct compositions. To test this hypothesis, we looked at the possibility of trying to homogenize the nonuniform composition by heating the 60/40 system at elevated temperatures under shear. Unfortunately, we were

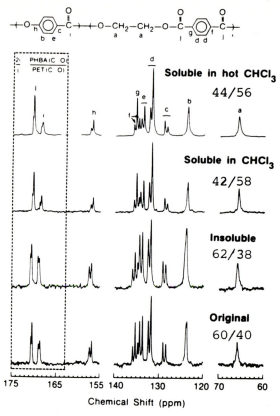

Fig. 11. ^{13}C NMR spectra of PHBA/PET fractions with differing solubilities in the PET/60PHBA copolyester (X-7G, Kodak). Numbers indicate the molar ratio of PET/PHBA calculated from NMR

unable to observe any measurable changes, since, at the higher temperatures necessary for rapid ester interchange, degradation also began to occur.

The HBA/HNA system provides a more suitable system for study, since it is prepared by melt polymerization of the two monomers and is far more stable at elevated temperatures compared to the PHBA/PET. The HBA/HNA copolymers are soluble in pentafluorophenol permitting use of NMR techniques to characterize diad sequences. In Fig. 13b,c the ^{13}C NMR spectrum of the carboxyl carbon region of the HBA/HNA copolyesters of the 73/27 and 48/52 systems is shown [34]. Also shown in Fig. 13a,d are the spectra of ^{13}C enriched HBA and HNA containing copolymers permitting unique identification of the diad sequences. As a result of this technique it was possible to determine the reactivity ratios of the two monomers by analyzing the 50/50 copolymer after polymerization to a molar mass value of 2000 [35]. Examination of the copolymer by ^{13}C NMR showed the same ratio of monomers as in the starting

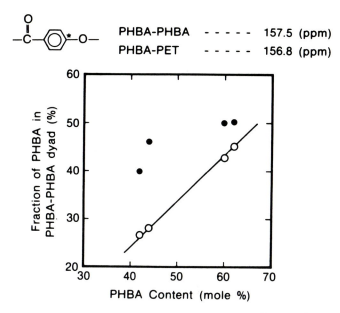

PHBA-PHBA - - - - - 157.5 (ppm)

PHBA-PET   - - - - - 156.8 (ppm)

Fig. 12. PHBA dyad sequences for PHBA/PET copolyesters; ● – experimental values, ○ – calculated values for totally random copolyesters

mixture. Furthermore, analysis of the diad sequences indicated a distribution of the four possible diads identical to what one might predict for a random copolymer.

From these results one can conclude that the reactivities of the two monomers are about the same and that there is no tendency to form blocky units. However, one could argue that, just as with the interchain trans-esterification reactions of PET with AcBA, such processes could occur in the HBA/HNA even more rapidly than the polymerization reaction tending to randomize the structure. A unique experiment has been carried out which permits one to distinguish between polymerization by transesterification reactions and interchain transesterification reactions at a temperature of 245 °C [35]. As shown in Fig. 14, ^{13}C labeled carbonyl in a AcBA monomer (B*) was reacted with the dimer of AcBA-HNA(BN). The only resonances in the carboxyl region of the spectrum will arise from the ^{13}C enriched carboxylic unit. In the absence of interchain transesterification one should observe only B*-B diads. The fact that B*N diads are observed at a concentration of 14% indicates that interchain transesterification does occur but it is relatively small. Blackwell, et al. have also examined the nature of the microstructure of the HBA/HNA using X-ray diffraction techniques [36]. From his study, which depends on modeling of the diffraction pattern, he concluded that the sequence distribution is random. It should be noted that Windle has suggested that this random sequence could be described as a periodic layered structure where some chain to

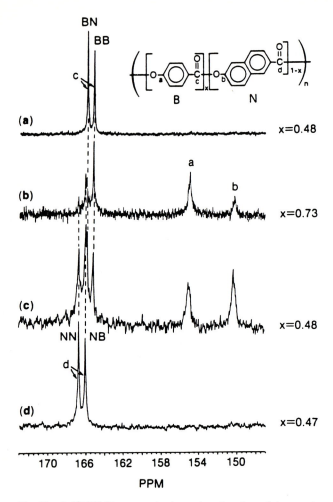

Fig. 13a–d. ^{13}C NMR spectra in the carboxyl region of the copolyesters HBA/HNA in PFP at 80 °C, x is the fraction of HBA units in the copolyester: **a** synthesized from ^{13}C labeled HBA; **b** the commercially available 73/27 HBA/HNA Vectra, Hoechst Celanese Co.; **c** synthesized from the 48/52 HBA/HNA mixture; **d** synthesized from ^{13}C labeled HNA

chain ordering exists [37]. At present, it does not appear necessary to invoke such a concept to interpret the crystallinity present in these systems.

There has been some question as to the effect on the microstructure of annealing the 73/27 HBA/HNA copolymer at a temperature 70 °C below T_{CN}. As shown in Fig. 15, distinct changes can be detected in the DSC scan as a result of annealing. However, examination of the two specimens by ^{13}C NMR shows no changes in the diad sequences (see Fig. 16).

Turning to the Xydar system (HBA/BPT), there has also been some confusion as to the interpretation of the microstructure of this copolymer. In the past

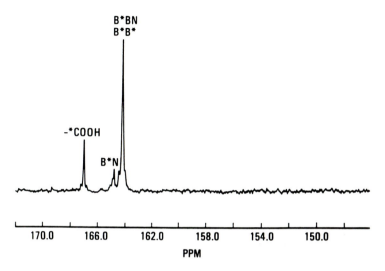

Fig. 14. Evidence for interchain transesterification from ^{13}C NMR of oligomers (n = 5) in PFP using 99% carboxyl ^{13}C enriched in the benzoyl unit

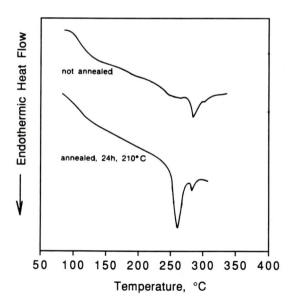

Fig. 15. DSC scan of PHBA/ PHNA (73/27)

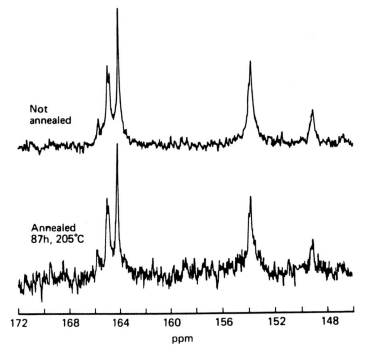

Fig. 16. ^{13}C NMR spectra of PHBA/PHNA (73/27) in PFP at 80 °C

this polymer has been prepared by a rather complex sequence of steps where a mixture of the three monomers of AcBA, TA and AcBP are heated to ~ 300 °C in Therminol 66 solvent. Since the TA is insoluble until one approaches 300 °C, one tends to form low molecular weight blocks of the other two monomers. Only after one exceeds temperatures of 275 °C does the TA begin to react. At this point, the melting point of the reaction mixture rises too rapidly to permit continued stirring. Hence the material is usually removed, ground and polymerization continued as a slurry, to yield a polymer with a melting point of 408 °C. ^{13}C NMR analysis of the soluble oligomer (after heating to 290 °C) indicates a blocky structure [38]. Hence one might expect that the polymer during solid state polymerization should retain the blocky structure. On the other hand, Blackwell has reported that melt spun fibers show a random sequence based again on simulations of the X-ray diffraction pattern [39]. We have recently examined this issue and concluded that the as-prepared copolymer is indeed blocky. However, heating at elevated temperatures of ~ 450 °C to permit fiberization could lead to randomization of the blocky structure units. In fact, as shown in Fig. 17, we have found that heating the as-received PHBA/BPT (2/1) to 456 °C results in a material with a much lower T_{cn} of 370 °C. Presumably, this lower melting material is more random than the sequence distribution of the as-prepared copolymer.

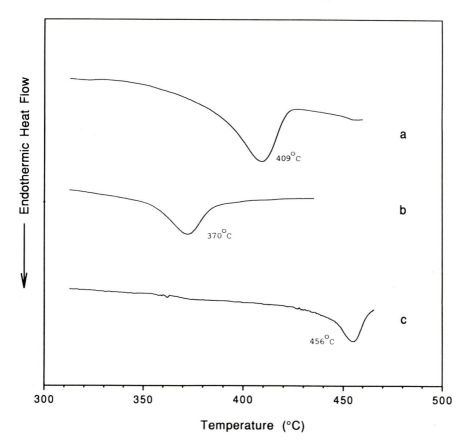

Fig. 17a–c. DSC traces of LPC 2:1 Xydar 300: **a** original; **b** randomized at 440 °C for 15 min and annealed at 270 °C for 5 h; **c** ordering of the randomized sample from (b) at 346 °C for 40 h

The potential for rapid randomizing processes in the copolyesters at elevated temperatures has been demonstrated conclusively by heating a mixture of the two homopolymers of PHBA and PHNA at 450 °C at a pressure of around several hundred psi [40]. Within a few seconds a viscous melt was observed to extrude from the cracks in the mold. Analysis of this material showed a structure consistent with the random 50/50 copolymer of HBA/HNA (see Figs. 18 and 19). We estimate that at this very high temperature the rate of interchain transesterification reactions corresponds to 1000 ester interchanges/chain/10 s.

On the other hand, as opposed to the randomizing reactions which occur in the nematic melt if one anneals these copolyesters near their crystal nematic transition a completely different process appears to be operative. Thus several workers [11, 14], have reported that heating the HBA/HNA system near its melting point results in a dramatic increase in T_{cn} by approximately 50 °C. As

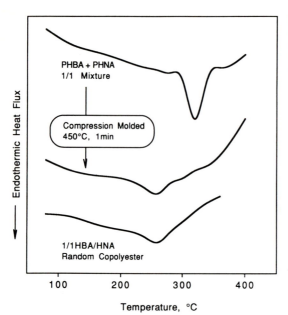

Fig. 18. Randomizing processes in the aromatic copolyesters: transesterification of PHBA/PHNA mixture during compression molding at 450 °C

shown in Fig. 20, this kind of process occurs throughout the entire compositional range of the HBA/HNA system. Furthermore, on heating above the new T_{cn}, the melting point reverts back to the original value. Some workers have chosen to interpret this process as a physical ordering involving melting of the crystallites at T_{cn} followed by nucleation and crystallization of longer ordered sequences which are already present within the random copolyester [11]. We feel that this ordering process may occur by ester interchange reactions within the existing crystallites at T_{cn} to produce more ordered sequences. The driving forces for ordering include improved packing of the chains in the crystallites and a corresponding higher density and increased dipolar interactions between chains. This kind of process can only occur in the existing crystallites since in the coexisting nematic phase (noncrystalline phase) randomizing reactions are occurring. We believe that the increased melting point of $\sim 50\,°C$ in the HBA/HNA copolymer system arises from some improved ordering of the sequences but not necessarily to the point where complete ordering occurs. One would not expect to maintain the mobility within the crystallites at the original T_{cn} and the potential for such reactions would drop significantly as the T_{cn} value increases by $40–50\,°C$. This mobility within the crystallite is essential to permit the interchain transesterification reactions to proceed. It follows that one should be able to induce further ordering by increasing the annealing temperature so that it is close to the new T_{cn}. Unfortunately, in our study of the HBA/HNA system, we have observed an unexpected degradation reaction occurring in such

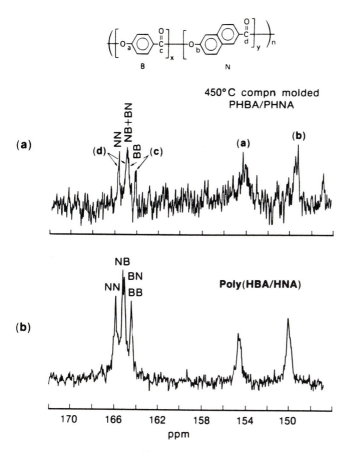

Fig. 19a,b. Comparison of ^{13}C NMR spectra of a compression molded mixture of the PHBA + PHNA with that of random HBA/HNA (50/50) copolymer

experiments to the point that we can no longer observe a T_{cn} from DSC nor do we see any melting even at 450 °C under a polarizing microscope [43]. On the other hand, with the HBA/BPT copolyesters we have indeed been able to observe this kind of process. Thus, heating the random copolymer of HBA/BPT (4/1) near its T_{cn} of 320 °C, the T_{cn} increases to a value of 401 °C. Further heating at 360 °C results in an even higher T_{cn} of 422 °C (see Fig. 21). Hence a self consistent picture can be proposed based on these experiments [43]. Clearly, a more definitive analysis is still required to prove conclusively that the ordering process is chemical rather than physical and such experiments are underway. Based on our present knowledge, one can summarize the respective roles of chemical and physical processes at elevated temperatures in the LCPs as shown in Table 5.

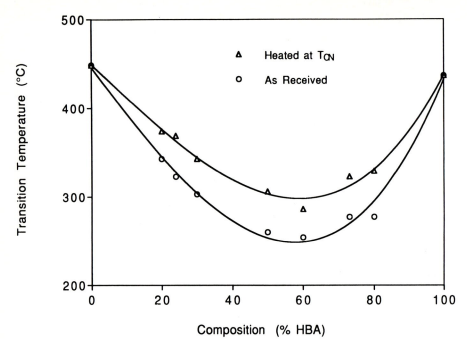

Fig. 20. Transition temperatures of ordered and randomized HBA/HNA copolyesters as a function of the composition

Table 5. Chemical vs physical process in the LC polyesters

Temperature	Dominant Process	Changes in Microstructure or Morphology
T_{cn}	Chemical	Randomization
T_{cn}	Physical	Crystallization
Annealing (near T_{cn})	Chemical	Crystal Ordering
Annealing (below T_{cn})	Physical	Further Crystallization
ca, T_{ng}	Physical	Nematic Glass

5. Processing of LCP

In this section, the mechanical properties resulting from melt processing of LCPs are not covered since they have been discussed in detail in earlier publications. Rather, an effort is made to focus on the problems associated with melt processing and to explore possible solutions. As noted earlier, one of the

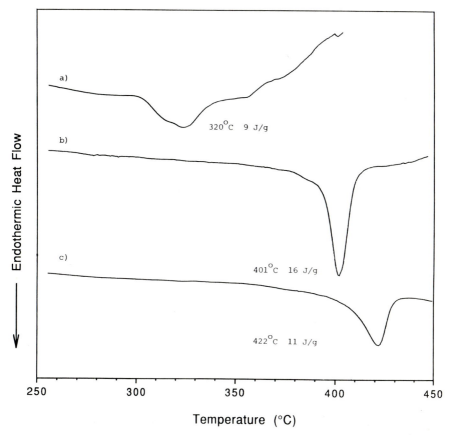

Fig. 21a–c DSC traces of random LCP 4:1 PHBA/BPT, MW 10k: **a** product from melt polymerization; **b** annealed at 300 °C for 48 hrs; **c** annealed at 360 °C for 24 h

most serious concerns with the LC polyesters are the microfibrillar and skin core morphologies that arise from melt processing these systems under shear. These morphologies most likely have no relation to the nematic domains present in the melt. Rather, they arise from local stresses that develop during cooling of the highly oriented morphologies. For example, in the chain direction, one would observe little tendency for contraction on cooling since the coefficient of thermal expansion is very low or even negative. In the perpendicular direction, the correspondingly large contractions which should occur are impeded by cooling of the surface. As the surface freezes the more mobile interior can accommodate to the fixed volume by forming these fibrillar morphologies. This kind of morphology has been effectively described by Sawyer [44] as shown in Fig. 22. In fact, there appears to be a hierarchy of microfibrillar morphologies extending down to diameters of 50 Å. This kind of hierarchy is often looked upon as being

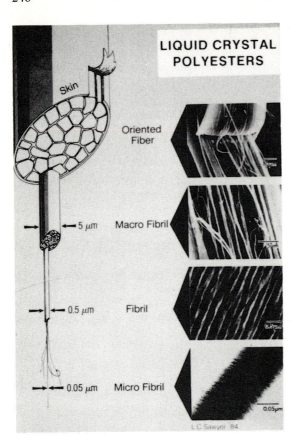

Fig. 22. Microfibrillar
morphology of LC Polyesters
[45]

beneficial in developing very strong structures. On the other hand, in the case of fibers one can almost draw a direct correlation with the poor crushing resistance of the fiber with the incidence of these morphologies. One can further interpret the differences in compressive properties between polyethylene (Spectra), aromatic copolyester, and Aramids with polyethylene being the worst and Aramid the best by comparing the kinds of lateral forces present in these systems. Thus, with Spectra one would anticipate only van der Waal forces between chains while in the polyesters one would also observe strong dipole interactions. With the Aramids one would also have the potential for strong hydrogen bonding. This kind of correlation would appear useful in trying to design anisotropic fibers which could reduce problems associated with poor compressive properties.

If, indeed, these kinds of morphologies are detrimental to the mechanical properties of highly oriented LCPs, one might ask whether use of solid state forming might obviate at least some of these problems. There was one very early report in the literature where high energy rate forging (HERF) was used to fabricate the PHBA homopolymer [1]. In this case the sample was cold sintered into the desired shape and then heated to 300 °C followed by forging (HERF) at ~ 20 000 ft/1b. These samples showed a high degree of orientation as shown from X-ray diffraction (see Fig. 23) and the elastic modulus in the plane was 2.3 times the isotropic values. There was no indication of any unusual morphologies and examination of break surfaces showed that fusion had occurred. The cost of the HERF equipment precluded commercial development of such a route. On the other hand, these results suggested a possible alternative to melt processing using solid state forming.

Recently, some work has been initiated to examine use of solid state forming for fabricating LCPs. Attempts to process an LCP such as 24/76 HBA/HNA in a capillary rheometer at temperatures well below the crystalline melting point showed that, indeed, the LCPs could be made to flow at 40 °C below the T_{cn} of 330 °C. As shown in Fig. 24, the change in viscosity with shear below T_{cn} showed a similar reduction in viscosity as compared to that of the melt state [45]. Furthermore, examination of the morphology of such specimens displayed a high degree of microfibrillation similar to what would be observed in melt processing (see Fig. 25). Hence one can conclude that the shear forces in a capillary rheometer are sufficient to melt the local crystalline regions that acts as crosslinks. This is especially true with polymers where the degree of crystallinity is relatively low.

Attempts to process more crystalline LCPs by solid state forming have only met with modest success. For example, attempts to compact PHBA/BPT at

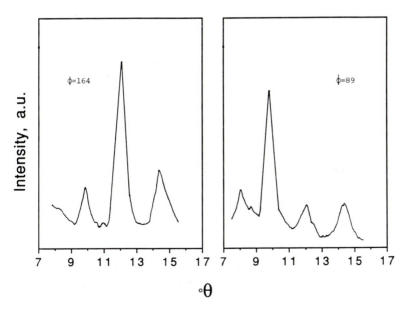

Fig. 23. X-ray of PHBA homopolymer – orientation during HERF

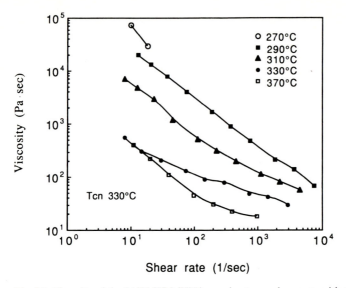

Fig. 24. Viscosity of the 24/76 HBA/HNA copolyester vs shear rate with temperature

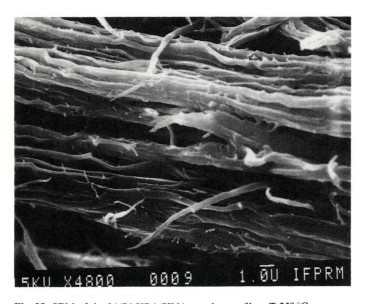

Fig. 25. SEM of the 24/76 HBA/HNA copolyester fiber, T 250 °C

30 °C below the T_{cn} at 290 MPa (42 KPS) and up to 1.0 Hz only led to sintering at the grain boundaries [46]. Clearly higher pressures and rates of forming are required if one is to approach the kind of results observed over 25 years ago using HERF.

6 New Directions

During the past decade much of the effort in industry on LCPs has been directed at preparing variations on the systems that were commercialized in the 1970s and early 1980s. In two very recent papers [17, 18] a completely new direction for use of LCPs has been disclosed. In these papers it was shown that LCPs such as the HBA/HNA system could be coated onto metal substrates such as aluminium or steel to form a tightly adherent coating. Furthermore, when two such coatings were brought in contact with each other at elevated temperatures and at modest pressures one could form a reasonable adhesive joint with a lap shear strength of $\sim$ 1500 psi. Failure always occurred cohesively and never at the polymer metal interface. These adhesive bonds are very resistant to moisture and are inert to boiling water even after immersion for 100 h. The adhesive strength remains high well above the T_g at $\sim$ 110–125 °C. For example, the 73/27 copolymer retains practically all of its mechanical properties up to 150 °C and then drops off quickly by 180 °C (see Fig. 26). In the case of the 24/76 HBA/HNA copolymer, a system that melts $\sim$ 40–45 °C above that of the 73/27 copolyester, one observes a corresponding increase in adhesive use temperature to $\sim$ 200 °C.

The above results are certainly unexpected in light of an earlier report on adhesion of polyimide films [47]. Here it was shown that adhesion between two polyimide films could only be achieved when polyimide chains from the one surface extended approximately 300 Å into the other surface. Because of the extended nature of the polyimide chains the potential for significant entanglements near the respective interfaces is low, thus requiring much deeper penetration by the PI to provide the necessary frictional forces for good adhesion. In the

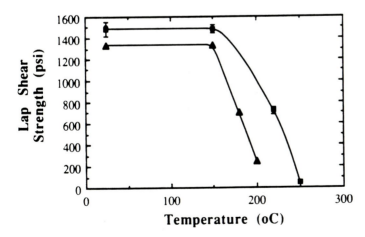

Fig. 26. Change in lap shear strength with temperature: ▲ 24/76 HBA/HNA copolyester; ■ 73/27 HBA/HNA

case of the rod-like LCP the likelihood of entanglements is even lower. Since it is not reasonable to expect rapid diffusion and penetration of rigid chains across the interface to depths of several hundred Angstroms, one, must invoke a different mechanism. Obviously one should consider the potential of rapid ester interchanges between chains at the interface as a possibility. It has already been noted in the earlier section on "Microstructure" that compression molding of a physical mixture of PHBA + PHNA at 450 °C yielded the random copolyester in well under a minute (possibly 10 s). Thus at temperatures normally used to obtain these adhesive bonds, it seems reasonable to anticipate that interchain transesterification reactions at the interface could lead to a homogeneous bond at the interface. Additional evidence in support of this thesis is the fact that the adhesive bond can be formed at temperatures below T_{cn}. In fact, as will be discussed shortly, one can prepare coatings of crosslinked aromatic copolyesters which, when brought in contact with each other, can form strong adhesive bonds at the interface. In this latter case the potential for rapid diffusion of chains is almost non-existent. Finally, in some related work we have observed that briefly heating the 24/76 HBA/HNA at 360 °C or 35 °C above its T_{cn} and then quenching results in complete retention of the crystalline order (see Fig. 27)

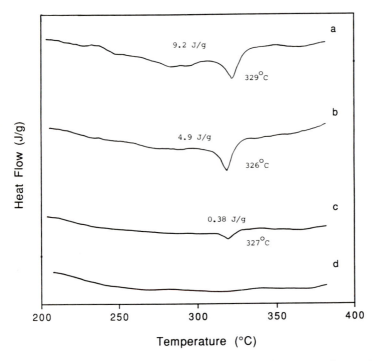

Fig. 27a–d. Memory effect after heating of PHBA/PHNA 24/76 copolyester above the T_{CN} for short time: **a** original; **b** sample heated at 360 °C for 3 min; **c** sample heated at 360 °C for 15 min; **d** sample heated at 360 °C for 30 min

[48]. Only after heating for 15 min does the crystallinity disappear after quenching (see Fig. 27c). This experiment suggests that diffusion processes in the melt are very slow and that there is a strong memory which persists well into the nematic state. In fact it is not clear that the primary mechanism acting to eliminate the memory is diffusion or interchain transesterification or possibly both. At this point it is safe to argue that both diffusion and interchain transesterification reactions are operative over the 15 min time span at 35 °C above T_{cn}.

The above concept of forming adhesive bonds in the solid state has been used to demonstrate the possibility of the parallel processing of a multi-chip module substrate, consisting of a multilayered polymer substrate with circuitry embedded on each polymer layer via lithographic processing [43]. In this case, it is essential that the polymeric layer retains its dimensional stability so that registration and interconnections between the layers can be achieved using a Pb-Sn solder (see Fig. 28). A copolyester which appears to be ideally suited for this purpose is the 4/1 PHBA/BPT which melts at 320 °C in the randomized form

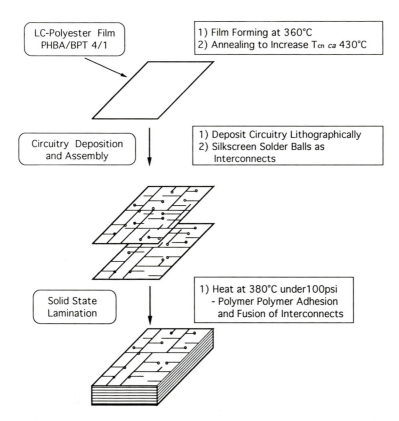

Fig. 28. Schematic for LC polyester thin film solid state processing of multichip electronic modules

and at 435 °C in the ordered version (see Fig. 21). This system can be easily processed into a film at 360 °C and then annealed at 350 °C and 390 °C to raise the T_{cn} to 435 °C. Lithographic deposition of copper circuitry followed by silk screening of Pb-Sn solder balls for use as the interlayer interconnects should permit solid state consolidation of the structure at ~ 380 °C, including fusing of the interconnects. So far the possibility of forming the adhesive bond at 380 °C has been demonstrated.

The tendency for ordering on heating near T_{CN} has, up to now, been only demonstrated in the rigid rod-like LCPs. The question that arises is whether such processes could occur in different kinds of morphologies and whether they would have any practical utility. To test this possibility a copolymer of sebacic acid with biphenol and hydroquinone (40/60) was prepared. As can be seen in Fig. 29, such a copolymer could be annealed near 200 °C to produce a more ordered system with a melting point of 240 °C [43]. Heating at 260 °C appeared

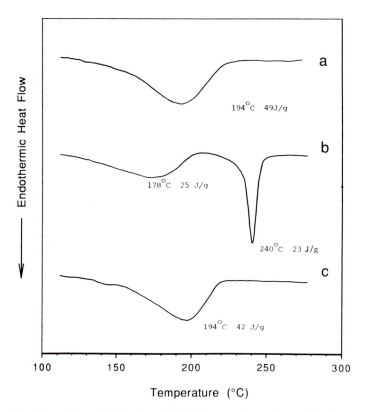

Fig. 29a–c. Thermal behavior of aliphatic/aromatic copolyesters based on HBA: **a** annealed at 220 °C for 4 h; **b** sample from (a) annealed at 200 °C for 62 h; **c** sample from (b) randomized at 260 °C

to randomize the ordered sequence yielding the melting point of the originally prepared polymer. It is noteworthy that the transition enthalpy of the random copolymer is at least 4–8 times the values observed with the rigid-rod like aromatic polyesters. Since this kind of ordering process appears to work in a broad range of polyesters, we have examined such a possibility even with thermoplastic elastomers based on polyester units. As shown in Fig. 30, one can significantly increase the melting point of the indicated system by 20 °C by heating near the melting point [49].

Coatings of the LCPs on metal substrates such as steel have been shown to provide excellent resistance to corrosion and abrasion suggesting the possibility

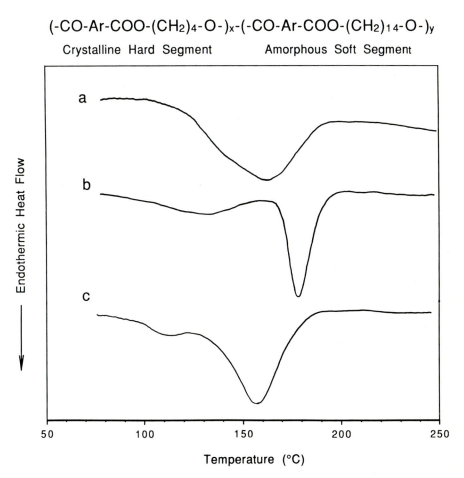

Fig. 30a–c. Ordering process in thermoplastic elastomers based on polyester units after annealing near T_m: **a** original; **b** annealed at 150 °C; **c** heated at 220 °C

of an alternative to galvanization of steel [50]. With relatively thin LCP coatings of 5–10 microns it is possible to cold draw the steel ~ 50% without damaging the coatings (see Fig. 31).

To facilitate processing of these copolyesters it would seem reasonable to redesign them in the form of low viscosity, branched, liquid crystalline oligomers which could be crosslinked (see Fig. 32). In fact, it was found that low melting

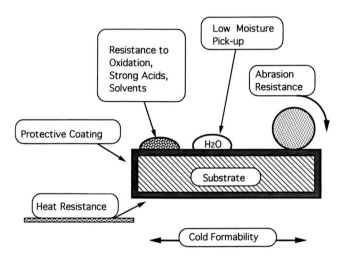

Fig. 31. Protective LCP coating on metal substrate indicating possible advantages

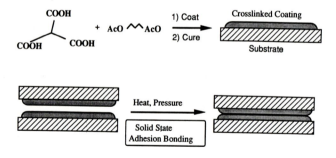

Processing Condition

Oligomer Mixture	Curing Temp.(°C)	Laminating Temp.(°C)	Lap Shear Strength (psi)
All Aliphatic	200	220	685
Aliphatic/Aromatic	240	220	2480
All Aromatic	330	400	1600

Fig. 32. Methodology for synthesis and processing of crosslinked polyesters as adhesives

oligomers could be synthesized that were easily processible as coatings at much lower temperatures than are used for processing of the corresponding thermoplastic LCPs. These coatings could then be advanced by heat to yield cross-linked structures. This kind of process has already been demonstrated using all aromatic LCP oligomers to form strong adhesive bonds by heating two such coatings at elevated temperatures under pressure. Equally as important, it has now been demonstrated that these new oligomers can be used as matrices for graphite fiber reinforced composites. Preliminary testing of these composites indicates excellent translation of fiber mechanical properties to the composites from room temperature to approximately 200 °C, with 50% retention in properties at 300 °C [51].

In a somewhat related process one can also generate structural foams which are very tough and stable to 400 °C under nitrogen. Heating two foams in contact with each other under 100 psi at 400 °C for 10 s led to formation of a good adhesive bond at the interface without collapse of the foam [51].

7 References

1. Economy J, Novak BE, Cottis SG (1970) SAMPE J 6: 6; Economy J, Cottis SG, Novak BE (1972); 3,962,314 (1972) 3,772,250 (1973) US Pat. 3,637,595
2. Jackson WJ Jr, Kuhfuss HF (1976) J Polym Sci Polym Chem 14: 2043
3. Calundann GW (1979) US Pat 4,161,407, 4,184,996 (1976)
4. Obtained from American Performance Product Personnel
5. Roviello A, Sirigu A (1975) J Polym Sci Polym Lett 13: 455
6. Noel C, Shibaev V (1991) Definitions of terms relating to low molar mass and polymer liquid crystal. IUPAC Recommendation 199
7. De Meuse, M. T., Jaffe M (1988) Mol Crystal Liq Cryst Nonlin Opt 157: 535
8. Lenz RW, Jin, J-II, Feichtinger, KA (1983) Polymer 24: 327
9. Quack L, Hornbogen E, Volksen W, Economy J (1989) J Polym Sci Part A Polym Chem 27: 775
10. Joseph E, Wilkes G, Baird D. (1985) Polymer 26: 689
11. Lin YG, Winter HH (1988) Macromolecules 21: 2439
12. Butzbach GD, Wendorf JH, Zimmerman HJ (1986) Polymer 27: 337
13. Mühlebach A, Economy J, Johnson RD, Karis RD, Lyerla J (1990) Macromolecules 23: 1803
14. White JL (1985) J Appl Polym Sci Sympos 41: 241
15. Muir MC, Porter RS (1989) Mol Cryst Liq Cryst 169: 83
16. Ciferri A, Krigbaum WR, Meyer RB (1982) "Polymer Liquid Crystals", Academic Press, N.Y.
17. Economy J, Gogeva T, Habbu V (1992) J Adhesion 37: 215
18. Economy J, Andreopoulos A (1993) J Adhesion 40: 115
19. Sirigu A. (1991) "Polymer Liquid Crystals", Ciferri, A. Ed., VCH Publishers, Inc., N.Y. p 261
20. Gilkey R, Caldwell JR (1959) J Appl Polym Sci 2: 198
21. Klepl (1883) J Prakt Chem 28: 194
22. Fischer E (1909) Berichte 42: 215
23. Economy J, Storm RS, Matkovich V, Cottis SG, Novak BE (1976) J Polym Sci Polym Chem Ed 14: 2207
24. Eareckson WM (1959) J Polym Sci 40: 399
25. Kwolek SL (1971) U.S. Patent 3,600,356
26. Cottis SG, Economy J, Wohrer LC (1976) U.S. Patent 3,975,487
27. Lieser G (1983) J Polym Sci Polym Phys Ed 21: 1611
28. Yoon DY, Masciocchi N, Depero LE, Viney C, Parrish W (1990) Macromolecules 23: 1793

29. Economy J, Volksen W, Viney C, Geiss R, Siemens R, Karis T (1988) Macromolecules 21: 2777
30. Lyerla JR, Economy J, Maresch GG, Mühlebzch A, Yannoni CS, Fyfe CA (1990) ACS Sympos Series 435: 359
31. Davies GR, Ward IM, in "High Modulus Polymers", Zachariades, AE, Porter RS, Eds., Marcel Dekker, Inc., N.Y., 1988, Chapter 2.
32. Unpublished data, Johnson RD, Lyerla J, Mühlebach A, Economy, J.
33. Fahie BJ, Fyfe CA, Facey GA, Mühlebach A, Niessner, Economy J, Lyerla JR (1991) Mol Cryst Liq Cryst 203: 127
34. Mühlebach A, Johnson RD, Lyerla J, Economy J (1988) Macromolecules 21: 3115
35. Economy J, Johnson RD, Lyerla JR, Mühlebach A (1990) ACS Symposium Series No 435: 129
36. Blackwell J, Biswas A (1985) Macromolecules 18: 2126
37. Hanna S, Windle AH (1988) Polymer 29: 207
38. Johnson RD, Economy J, Lyerla J, Mühlebach A (1989) Amer Phys Soc Polym Section, St. Louis, MO
39. Blackwell J, Cheng HM, Biswas A (1988) Macromolecules 21: 39
40. Mühlebach A, Economy J, Johnson RD, Karis T, Lyerla J (1990) Macromolecules 23: 1803
41. Cheng SZD (1988) Macromolecules 21: 2475
42. Kachidza J, Serpe G, Economy J (1992) Makromol Chem Macromol Sympos 53: 65
43. Potter C, Lim JC, Serpe G, Economy J Macromol. Chem. Macromol Sympos. (in press)
44. Sawyer L, Jaffe M (1986) J Mater Sci 21: 1897
45. Unpublished data, Economy J, Serpe G
46. Unpublished data, Economy J, Goranov C
47. Brown HR, Yang ACM, Rusell TP, Volksen W, Kramer EJ (1987) Polymer 20: 1116
48. Unpublished data, Economy J, Serpe G
49. Economy J, Fisher C (1994) Polymer for Adv. Tech. 5: 295
50. Frich D, Economy J (1993) ACS, PMSE, Chicago, 69: 438
51. Economy J, Frich D, Goranov K, Lim JC (1994) ACS, PMSE, San Diego, 70: 398

Rigid-Rod Polymers and Molecular Composites

Fred E. Arnold, Jr.[1] and Fred. E. Arnold[2]
[1] Institute of Polymer Science, University of Akron, Akron, Ohio 44325, USA
[2] Air Force Wright Laboratory, Materials Directorate, Wright-Patterson AFB, Ohio 45433, USA

The last two decades of research on high performance rigid-rod polymers and molecular composites have produced a wide variety of material systems for evaluation. The purpose here is not to present a comprehensive review but rather to describe these relatively recent research developments with regard to the synthetic methods and approaches that have been utilized. In doing this some properties of the polymers are discussed as well as some of the problems encountered in conjunction with their synthesis and evaluation. In an effort to limit the scope of the review, we will address only aromatic heterocyclic rigid-rod polymers, extended chain polymers and derived molecular composites.

List of Symbols and Abbreviations

BBB   – polybenzimidazobenzophenanthroline
BBL   – polybenzimidazoisoquinoline
PPA   – polyphosphoric acid
PBZT   – polybenzobisthiazole
PBZO   – polybenzobisoxazole
PBZI   – polybenzobisimidazole
PBZX   – polybenzobisazole class of materials
$[\eta]$ – intrinsic viscosity
TGA   – thermal gravimetric analysis
TG-MS   – thermal garavimetric-mass spectrometry
ABPBT   – polybenzthiazole
ABPBO   – polybenzoxazole
ABPBI   – polybenzimidazole
PMDA   – pyromellitic anhydride
BPDA   – biphenylenetetracarboxylic dianhydride
C_{cr} – critical concentration
PPQ   – polyphenylquinoxaline
PEEK   – polyetheretherketone
PEK   – polyetherketone
PPMA   – solvent composed of methanesulfonic acid/P_2O_5 (10:1w/w)

1 Introduction

Since the late 1970s the Air Force Wright Laboratory and Air Force Office of Scientific Research has had a very energetic program in lyotropic liquid crystalline rigid-rod polymers. The program encompassed theoretical aspects, synthesis, solution properties, processing, mechanics and morphology in an effort to exploit the potential of using the ordered polymers approach to structural materials for aerospace applications. The overall objective of this program was to provide the technology for a monolithic plastic material having the mechanical properties of high performance metals with the economics and ductility of a nonmetallic.

In the early 1970s research at Wright Laboratory, Polymer Branch was directed toward the synthesis of highly fused aromatic heterocyclic polymer systems for high temperature applications. One noteworthy example containing eight consecutive fused rings was poly (6,9-dihydro-6,9-dioxobenzimidazo [2,1-b:1°,2'-j] benzo [1mn] [3,8]phenanthroline-3,12-diyl), referred to as BBB [1]. Weight losses of less than 10% at 600 °C in air and less than 5% at 700 °C in nitrogen were observed for the material. At that time it was predicted [2] that a completely fused ring structure or double strand ladder polymer would be even more resistant to degradation than an analogously structured non-ladder system. The homolog of BBB is poly[(7-oxo-7,1OH-benz[de]imidazo[4',5':5,6]benzimidazo[2,1-a]isoquinoline-3,4:10,11]-tetrayl)-10-carbonyl], (BBL), which was prepared in high molecular weight with a high degree of structural perfection [3]. Both the BBB and BBL polymers were prepared by the polycondensations of either 3,3',4,4' -tetraaminobenzidine or 1,2,4,5-tetraaminobenzene with 1,4,5,8-naphtalenetetracarboxylic acid in polyphosphoric acid (PPA).

BBB   BBL

Although a high increase in thermal properties was not realized, the research uncovered a unique film forming phenomenon. Dilute methane sulfonic acid solutions (0.05 to 0.5%) of BBL were found to precipitate in the form of two dimensional microscopic sheets. Upon drying the precipitated solids coalesced to form a tough, durable film having a gold metallic luster. The sizes of the sheet-like particles found were dependent upon the initial concentration of polymer solution used in the precipitation and the method of mixing a nonsolvent with the solution. When the acid solution was dilute (e.g. about 0.1%) and the method of stirring was a magnetic stirring bar, the resulting sheet-like forms measured 50–150 microns across. It was also found that nonsolvent dispersions of BBL could be spray coated on a variety of substrates. By utilizing this unusual property displayed by the polymer, films and coatings could be obtained with relative ease and without residual acid impurities. This coalescence of discreet particles of solid matter without going through a melt phase was the key factor in the origin of the Air Force's ordered polymer research program.

Other aromatic heterocyclic polymers were examined to permit comparisons from a structural point of view. BBB was an ideal polymer to examine since one could compare the behavior of ladder structure to the very similar nonladder structure. When dilute methanesulfonic acid solutions (0.05% to 0.3%) of BBB were precipitated in methanol, tiny globular particles were formed, and upon collection and drying, no film was obtained. By considering model forms of the BBL chain structures, one expects that the ladder polymer closely resembles a rigid-rod with no opportunity for bending or twisting. On the other hand, the nonladder BBB backbone incorporates a single bond link in the chain between each unit structure which can produce kinks in the chain up to about 75° depending upon the specific unit structure isomers being linked and the way units are rotated with respect to each other. Although highly unlikely, it would be possible in the extreme for a segment of BBB chain to loop back upon itself within a distance of about five unit structures (Fig. 1).

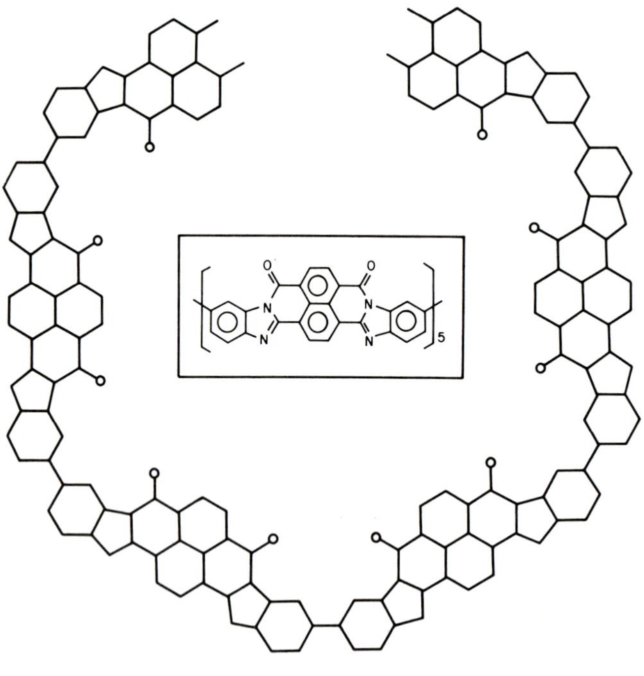

Fig. 1. Diagrammatic model of BBB loop

Attention was then turned to aromatic heterocyclic ladder polymers other than BBL. Due to the complexities in their synthesis, ladder polymers were not reported extensively in the literature at that time. A sample of polyfluoflavine (**I**) having an inherent viscosity in methanesulfonic acid of 2.5 was obtained from professor C. S. Marvel at the University of Arizona. The ladder polymer was prepared [5] from the A-B polycondensation of 2,3-dihydroxy-6,7-diamino-quinoxaline hydrochloride in PPA. Transparent blue sheets were observed on precipitation of this polymer from methanesulfonic acid, and when collected and dried, it formed gold films much like the color of the BBL films. It was felt at that time that all ladder polymers of sufficient molecular weight would form precipitated films.

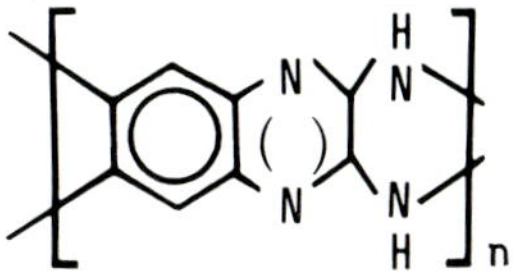

I

In an effort to vary the molecular geometry of the BBL ladder structure, a series [6, 8] of imidazo-isoquinoline ladder polymers was prepared and evaluated with respect to their film forming properties. The polymers were prepared

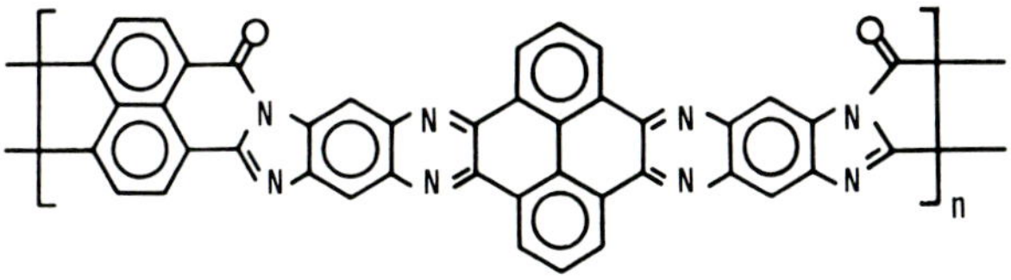

II

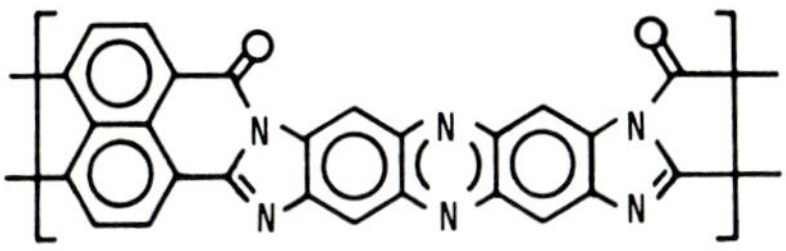

III

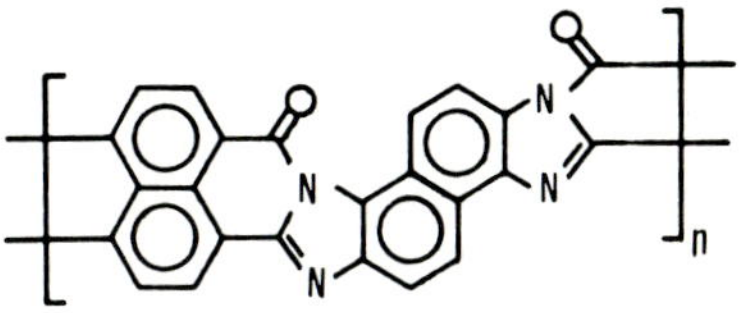

IV

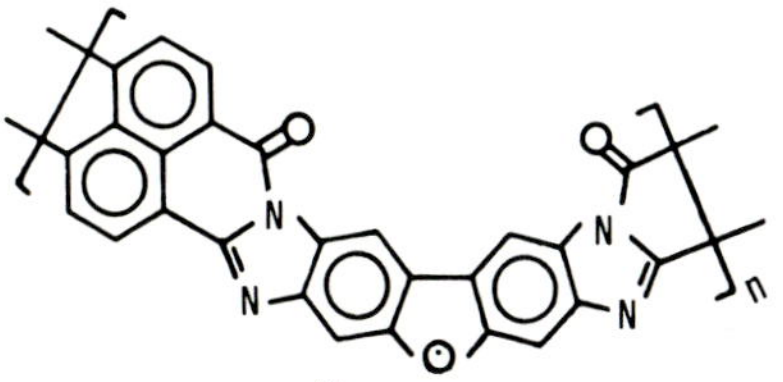

V

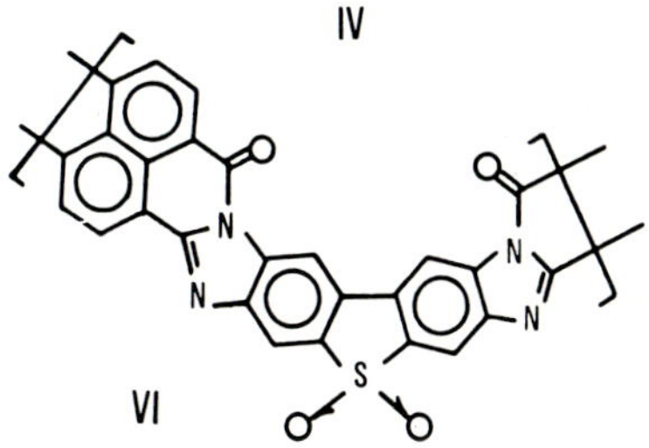

VI

from various aromatic and aromatic heterocyclic fused tetraamines by polycondensation with 1,4,5,8-naphtalenetetracarboxylic acid in PPA. It was found
that polymers **II**, **III**, and **IV** all precipitated from acid solution in the form of
transparent sheet-like material. When the precipitates were collected and dried
on glass frits, films with a silver metallic luster were formed from polymers **II**, **III**,
and a film with a gold-metallic luster was formed from polymer **IV**. Polymers **V**
and **VI** did not exhibit the unusual aggregation film forming properties found
for BBL. The fused aromatic heterocyclic monomers, 2,3,7,8-tetraaminodiben-
zofuran and 2,3,7,8-tetraaminodibenzothiophene-5,5-dioxide, due to their molecular geometry, produce kinks along the polymer backbone which makes the
chains more difficult to order in solution and pack in the solid phase. In fact,
these polymers in the extreme could also form a loop within a distance of about
seven unit structures (Fig. 2), much like the BBB polymer. In contrast (Fig. 3),
the four geometric isomers of the BBL polymer describe an essentially straight
line which affords the ability for optimum packing of chains.

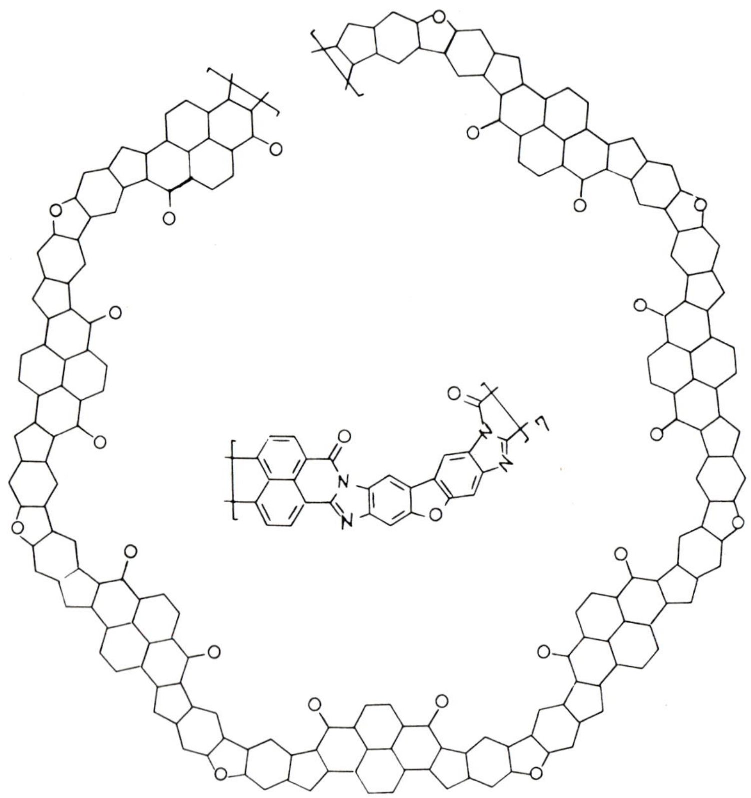

Fig. 2. Diagrammatic model of ladder loop

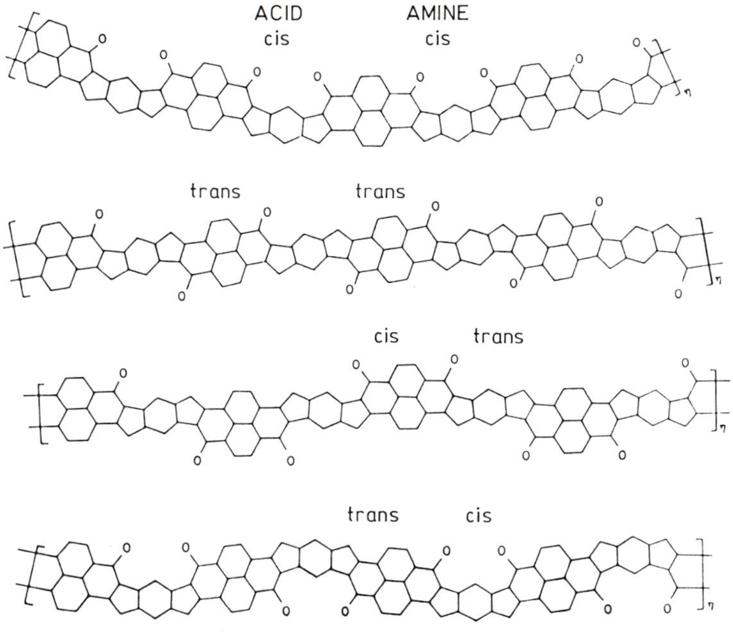

Fig. 3. Geometric BBL isomers

It becomes quite clear that the critical factor in determining the film forming tendencies was the molecular geometry of the polymer backbone and not the ladder structure. It was felt that the sheet-like precipitates were the result of the rod-like materials ordering in solution and that order being maintained on precipitation. These sheets have been found to possess sufficient internal molecular order to exhibit birefringence properties. The high strength of these unique films stemmed from their "composite" character. They were considered "composite" since their formation was the result of aggregation and coalescence of individual microscopic sheets of precipitated polymer. In conventional composite technology, the individual microscopic sheets would be considered as plies, and the collection in a random fashion would produce an isotropic molecular composite.

In the mid 1970s, researchers at DuPont reported [9] that extended chain para-ordered polyamides gave high strength, high modulus fibers when processed from liquid crystalline solutions. The fact that one could maximize the order in the liquid crystalline state offered an excellent opportunity to design and process para-ordered polymer systems to reach the objectives of the program. The synthetic effort at Wright Laboratory began to focus on aromatic heterocyclic systems, other than ladder polymers, with a para-ordered geometry. One class of heterocyclic polymer systems that would meet all the harsh requirements for advanced aircraft and aerospace applications and could be obtained with the appropriate geometry was the benzobisazole materials.

2 Polybenzobisazole Systems

2.1 Synthetic Methods

The rigid-rod benzobisazole polymers that have received the greatest attention in the literature are the poly([benzo(1,2-*d*:4, 5*d'*)bisthiazole-2,6-diyl]-1,4-phenylene) (PBZT), poly([benzo(1,2-*d*:5,4-*d'*)bisoxazole-2,6-diyl]-1,4-phenylene) (PBZO) and poly([1,7-dihydrobenzo(1,2-*d*:4,5-*d'*) bisimidazole-2,6-diyl]-1,4-phenylene) (PBZI) systems. All three polymers have been prepared [10-12] by the polycondensation of terephthalic acid with 1,4-dimercapto-2,5-diaminobenzene, dihydrochloride, 1,3-dimercapto-4,6-diaminobenene dihydrochloride, or 1,3-diamino-4,6-(ditoluenesulfamido)benzene in PPA. PBZT and PBZO were both found [11, 13] to form neumatic liquid crystalline solution in methane sulfonic acid at relatively low concentrations (5–7%). If polymerizations in PPA are carried out at polymer concentrations high enough to promote anisotropic solutions, these reaction mixtures can be used directly for processing into high modulus fibers [14–15] and films [16].

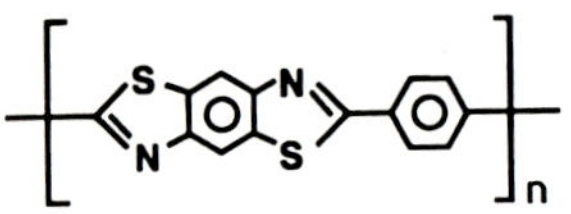

PBZT 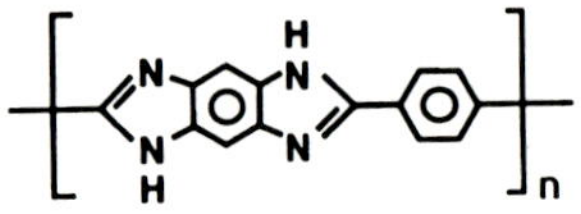   **PBZI**

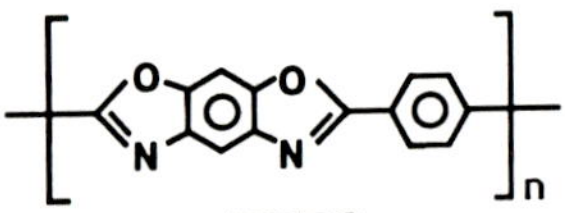

PBZO

Air sensitivity of the amino-monomers prevents their use in the polycondensation reaction; therefore, hydrochloride salts or other protective groups must be used and always maintained under a nitrogen atmosphere. The process for preparing the rigid-rod benzobisazole polymers involves two stages. In the first stage the dehydrochlorination of the amino monomer is carried out at 60–80 °C under a reduced nitrogen pressure to facilitate the removal of the hydrogen chloride. In the second stage a stoichiometric amount of terephthalic acid is added, the mixture slowly heated to 190–200 °C and maintained at that temperature for 8–10 h.

Wolfe and Sybert, while working on the scale-up of PBZT, found a more versatile synthetic method in PPA that allowed polymerizations to be carried out at higher concentrations and resulted in a significant increase in both polymerization rate [17] and the achievable molecular weight. The new synthesis method, P_2O_5 Adjustment Method [18], maintains the effectiveness of the PPA while its composition changes on hydrolysis by the condensation byproduct. PPA is a mixture of condensed phosphoric acid oligomers depending on the ratio of water to P_2O_5 content. During the initial dehydrochlorination phase, the P_2O_5 content is kept low, approximately 77%. This provides a PPA with a substantially reduced viscosity, allowing the hydrogen chloride gas to escape the reaction vessel. After the dehydrochlorination is complete, the P_2O_5 content is adjusted to 82–84% which favors the dehydrating nature of the PPA and drives the polycondensation reaction to completion. A much higher P_2O_5 content than 84% increases the bulk viscosity to a point that limits the processing of fiber or film.

PBZI is the only polymer of the three that cannot be obtained as a lyotropic liquid crystalline solution. The polymer does not display sufficient solubility in acidic solvents to exhibit mesogenic behavior. Polymerizations utilizing 1,2,4,5-tetraaminobenzene tetrahydrochloride or 1,3-diamino-4,6- (p-toluenesulfamido) benzene at low concentrations (1–2%) provide low molecular weight polymers with intrinsic viscosities between 2–4 dL/g. Polymerizations above the theoretical critical concentration result in the polymer crystallizing out of solution and being completely insoluble in all acidic solvents. PBZI is a better dehydrating agent than PPA, and somehow the water of condensation acts as a crystallizing agent during the polycondensation process. A variety of high molecular weight benzobisimidazole polymers containing various pendant groups have been obtained [19–21] as anisotropic liquid crystalline reaction mixtures. It is felt that the pendent groups retard the crystallizing process by preventing the close packing of chains.

2.2 Processing and Properties

The benzobisazole family of rigid-rod polymers is soluble in acidic solvents such as PPA, methanesulfonic acid, chlorosulfonic acid, 100% sulfuric acid and Lewis acid salts such as antimony trichloride and bismuth trichloride. More recently, PBZT has been reported [22] to form liquid crystalline solutions in nitromethane containing aluminum trichloride or gallium trichloride. Since the glass transition temperature of these materials is above their decomposition temperature, they must be processed from solution.

Processing rigid-rod polymers from the liquid crystalline state produces very high modulus, high strength materials in the form of both fiber and film. The materials are processed by dry-jet wet spinning techniques. Anisotropic polymer dopes are extruded through a die into a coagulation bath, normally water, where the extrudate is elongated by taking up the fiber or film at a faster rate than that of the dope leaving the jet. The ratio of these rates is called the spin-draw ratio, and the geometry and configuration of the die determine monofilament, multifilament yard or film. The as-spun fibers or films are then heat treated under tension by passing through a tube oven at temperatures between 500 and 650 °C under an inert atmosphere [23, 24]. Residence time that the fiber or film is in the oven ranges from 10 to 30s depending on the temperature.

Exceptional mechanical properties in fiber and film have been obtained for
PBZT and PBZO materials. Modulus values of 300 GPa with tenacity values of
3.0 GPa have been reported [14] on wet-spun fibers after heat treatment. Films
with 2 GPa strength and 270 GPa modulus have been produced [25] with
controllable planar orientation. Studies of the morphology [26, 27] which
develops during coagulation of the oriented PBZT solution have shown the
formation of a network of microfibrils, the width of which is less than 10 nm.
This fibrillar structure and the lack of transverse coupling between the fibrils
contribute to a microbuckling type of instability resulting in low compressive
properties. Poor axial compressive strength is the limiting factor for using these
materials in many aerospace applications. The reported compressive strength
for PBZT fiber is 345 MPa [28] and 200 MPa for PBZO [29].

Dilute and concentrated solution properties of PBZT and PBZO have been
studied [30] in detail by Berry et al. Their research has provided important
relationships for the solutions of these polymers such as the response of the
viscosity to temperature, shear, polymer concentration, polymer molecular
weight and solvent composition. The relationship of intrinsic viscosity to
molecular weight has been determined for both PBZT and PBZO [31].

$$PBZT : [\eta] = 1.65 \times 10^{-7} MW^{1.8}$$

$$PBZO : [\eta] = 2.77 \times 10^{-7} MW^{1.8}$$

Of all the high temperature, high performance polymers, the benzobisazole
rigid-rod polymers are one of the most thermally stable systems known. A
variety of thermal techniques have been used to investigate their stability such as
thermal gravimetric analysis (TGA), thermal gravimetric-mass spectrometry
(TG-MS) and isothermal aging studies. A TGA plot shown in Fig. 4 is indicative

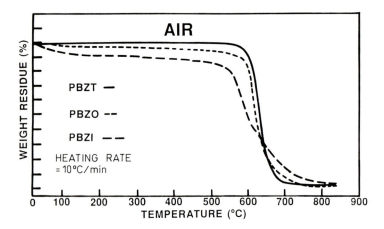

Fig. 4. TGA of PBZX

of the relative thermal stabilities of these materials. The early weight loss exhibited by these materials from ambient temperature to approximately 200 °C is the result of water loss. The relatively high water as seen for PBZI is typical of that polymer. The onset of weight loss in these materials is seen in the range of 600 °C, with PBZT and PBZO exhibiting the greatest weight retention with an extrapolated onset of degradation at 620 °C. A more detailed evaluation [12] employs TG-MS analysis. The ion intensity as a function of temperature for PBZT is shown in Fig. 5. The analysis was performed in vacuo at a heating rate of 3 °C/min. The evolution of hydrogen sulfide, followed closely with hydrogen cyanide, indicates that the thermal decomposition of one of the heterocyclic rings begins near 600 °C with a maximum rate near 700 °C. The total weight loss of 28% at 1000 °C is consistent with the loss of 1 mole of hydrogen sulfide, 1 mole of hydrogen cyanide, and 0.25 mole of carbon disulfide per mole of repeat units. TG-MS analysis is also routinely used to define maximum temperature parameters when processing and heat treating film and fiber. Isothermal aging studies in air have been carried out [32] at a number of different temperatures and morphologies, such as bulk polymer, as spun fiber and heat treated fiber.

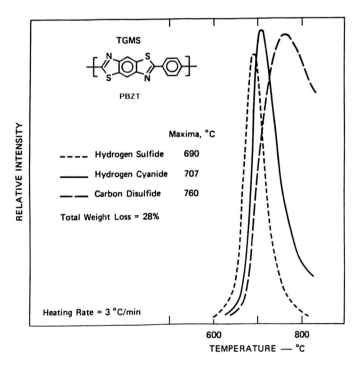

Fig. 5. TG-MS of PBZT

3 Structural Variations

3.1 Pendent Groups

Synthetic research efforts on the benzobisazole rigid-rod polymers more recently have been devoted to structural modifications of the rod-like polymers to broaden and uncover new material options for these materials. A number of different structural modifications have been carried out at Wright Laboratory, Materials Directorate. In contrast to the extended-rod polymers that have both translational and rotational degrees of freedom, the rigid-rod benzobisazole systems have only a rotational degree of freedom around the polymer axis. This difference translates to a more facile liquid crystalline transition and superior mechanical properties. Structural modifications for the most part have been made by maintaining the para-ordered geometry while modifying the structure with pendent groups.

As noted earlier, the primary limitation of these materials has been their poor compressive properties. Structural tailoring with pendents has allowed various approaches to be investigated to improve the compressive properties of high modulus PBZX films and fibers. In this context it must be assumed the intrafibrillar interaction within the microfibril domains has the dominant role in the compressive performance of these fibers. Four material approaches have been taken for improvement of the compressive properties which include increasing lateral order, disrupting packing order, chemical crosslinking, and providing systems with a ribbon-like morphology.

In any materials approach a model polymer system is visualized for synthesis that would adequately test the viability of the approach. The new composition is then synthesized and processed, and the critical property is tested. This is extremely difficult in the area of improved compressive properties since the critical test requires not only high quality high molecular weight polymer for spinning, but very high quality fiber must be processed for testing. As with any new composition, only small quantities are available for processing. The test method for compression must also utilize small quantities of monofilament fiber and be consistent with composite data. Of all the testing techniques that have been developed in recent years, it has been shown [33] that the tensile recoil test [34] is more comparable to actual testing of the fiber in a composite.

In an effort to increase the lateral order between polymer chains, heterocyclic pendent groups were attached to the rigid-rod PBZX systems. It was anticipated that with the heterocyclic's high degree of dipole/dipole interactions, a three dimensional order would be obtained. Novel polybenzobisazoles containing pendent benzothiazole groups were prepared [19] by the reaction of 2-benzothiazole or 2,5-benzothiazole terephthalic acids with the appropriate aminomercapto, aminohydroxy or tetraamino monomers in PPA. Intrinsic viscosities in the range of 4–19 dL/g were measured in MSA.

In all cases the anisotropic polymerization mixtures (10% by weight) could be used directly in the formation of dry-jet wet-spun fibers. Monofilament fibers were obtained by coagulation in water, tension dried at 150 °C and heat treated at 500–600 °C with a 30s residence time. The best fibers were obtained from the high molecular weight PBZT polymer (**VII**) which exhibited modulus values that ranged between 172 GPa and 207 GPa and tenacity values up to 2.4 GPa. Unfortunately, the compressive property as measured by the tensile recoil test was only 380 MPa, showing only a slight improvement over PBZT.

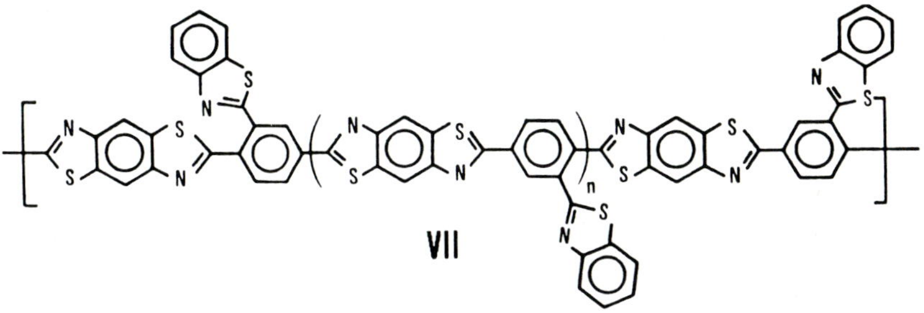

VII

Bulky pendent groups derived from a phenylbenzthiazole- substituted *p*-terephenylene dicarboxylic acid (**VIII**) [35] were utilized to disrupt the packing order and potentially increase the compressive properties. Monomer **VIII** when polymerized [36] with 1,4-dimercapto-2,5-diamino-1,4-benzene dihydrochloride, provided anisotropic solution in PPA. Concentrations as high as 10% by weight could be employed in the polymerizations providing polymer with intrinsic viscosities between 20 and 30 dL/g. More interesting, the attainment of these molecular weights was achieved at temperatures below 100 °C. At higher temperatures (180–190 °C) normally used in the polymerizations, gel was observed upon isolation of the material. Although anisotropic reaction mixtures were obtained, high bulk viscosities and the formation of small amounts of gel at high spinning temperatures prevented the polymer from being processed into fiber.

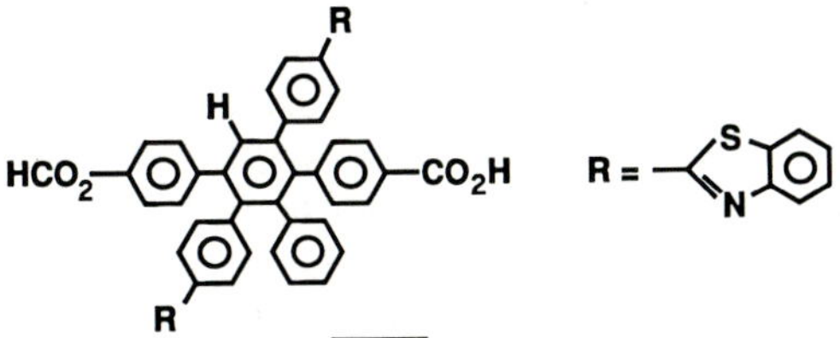

VIII

Random copolymers were prepared [37] in PPA by replacing a small part (1–2 mol %) of the terephthalic acid with monomer **VIII**. Polymerizations were carried out at 10% by weight concentrations to give copolymers (IX) with

IX

R = [structure]   X = 1-2 mol %

intrinsic viscosities between 20–28 dL/g. Anisotropic reaction mixtures were spun at two or more different spin draw ratios to characterize the effect of molecular orientation. A heat treatment temperature of 550 °C was used for all the copolymers which is comparable to the typical heat treatment temperature of 600 °C for PBZT. Modulus values for the copolymers were between 193 and 283 GPa with tensile strengths as high as 2.5 GPa. The fiber compressive strengths measured for the copolymers, 344–482 MPa, were only marginally better than those reported for PBZT. It remains to be seen from current morphological studies that the proposed mechanism for improved compressive strength was tested.

Crosslinking of rigid-rod polymers has been investigated as a method of improving their shear modulus for improved compressive properties. Significant two and three dimensional backbone structure is seen in graphite and silicon carbide fiber that possess very good compressive strength. Monomers containing latent crosslinking functionality are polymerized to form systems with reactive pendent groups. These groups are activated during or after heat treatment of the fiber to generate a network structure.

PBZX containing pendent liable methyl groups were prepared [20] from 2-methyl and 2,5-dimethyl terephthalic acids by condensation with the appropriate amino monomers. Thermal properties were investigated using TG-MS analysis. Figure 6 shows a typical TG-MS of the monomethyl pendent PBZT in vacuo at a heating rate of 3 °C/min. The initial product of thermal degradation was methane, starting at 420 °C and maximizing at 580 °C. Other products derived from the thermal degradation of the polymer backbone were hydrogen sulfide (690 °C), hydrogen cyanide (700 °C), and carbon disulfide (750 °C). To simulate thermal treatment of as-spun fiber, bulk polymer samples were heated for various times and temperatures. It was found that after exposure of samples at 500–550 °C (nitrogen atmosphere), between 30 and 40s, they were found to be completely insoluble in all acidic solvents. Using the same time frame but under an air atmosphere, a temperature of only 300–350 °C was required to render the samples insoluble.

Benzobisthiazole random copolymers (**X**) [38] were also prepared using 2-methyl terephthalic acid and terephthalic acid with 1,4-dimercapto-2,5-diamino-1,4-benzene dihydrochloride. Various copolymers provided materials with a wide range of crosslink densities. High quality fibers with spin draw ratios as high as 50 to 1 were obtained by dry-jet wet spinning of anisotropic polymerization mixtures. Heat treatment of the fibers was carried out in a tube oven

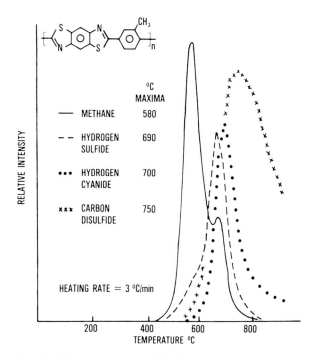

Fig. 6. TG-MS of monomethyl PBZT

between 350–500 °C in a nitrogen atmosphere. In the inert atmosphere the proposed crosslinking mechanism involved the loss of the pendent methyl group with a subsequent biradical coupling.

Y = 0 TO 50 mol %

X

Sweeny at DuPont recently employed the analogous crosslinking mechanism using halo-substituted PBZTs [39]. The method was based on the thermolysis of active aryl halides contained in the polymer repeat unit and coupling of the free radicals formed. High molecular weight PBZTs were made from 1,4-dimercapto-2,5-diaminobenzene dihydrochloride and chloro, bromo, and iodoterephthaloyl chlorides in PPA (85% P_2O_5). Copolymers were also employed to vary crosslink densities. Fiber was air-gap spun into water from the PPA reaction mixtures at 140–180 °C. The fibers were heat treated under argon

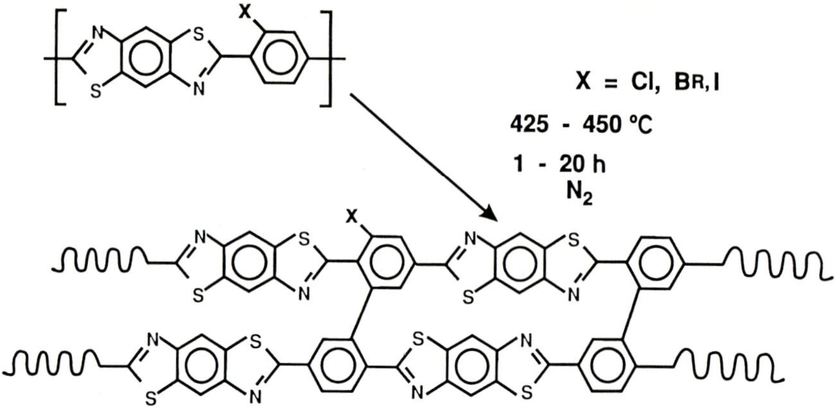

$$X = Cl, Br, I$$
$$425 - 450 \,°C$$
$$1 - 20 \, h$$
$$N_2$$

by passing through a 30 centimeter tube heated to 400–500 °C under the highest tension without breakage.

Highlights of both studies are summarized and shown in Table 1. The primary difference between the two studies was the time at which the polymers were thermally treated. Both systems were heat treated at 500 °C in an inert atmosphere. Samples of the halo pendent system were further treated by heating between 425–450 °C for 1–20 h. As-spun monomethyl PBZT exhibited the highest measured compressive properties; unfortunately, after heat treatment, the polymer lost 60% of its compressive strength. Within the series of fibers, there was no systematic trend of correlation between methyl pendent content and compressive strength. Gamma radiation of the as-spun fiber showed no

Table 1. Compressive properties from crosslinking studies

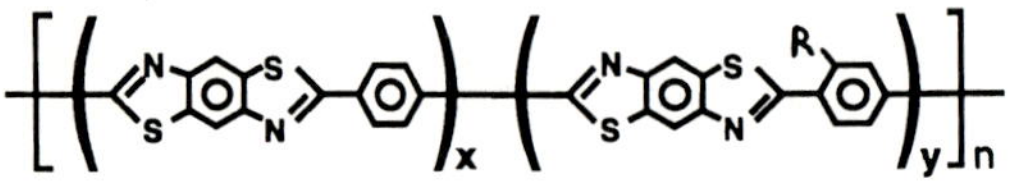

| Mole % | | R | HT[b] (°C) | | Mod (GP$_A$) | TS (GP$_A$) | CS[a] (MP$_A$) |
X	Y						
0	100	CH$_3$-	As-Spun		103	1.4	689
0	100	CH$_3$-	Irradiated Igrad		124	1.3	793
0	100	CH$_3$-	500		220	2.4	206
90	10	CH$_3$-	500		193	1.6	517
50	50	CH$_3$-	500		282	3.2	206
0	100	BR-	450/2 h[c]		103	2.6	452
0	100	BR-	450/2 h		108	3.7	689
0	100	⊢	425/2 h		73	1.3	393
50	50	⊢	425/1 h		126	2.9	496

[a] Compressive strength measured by recoil test.
[b] Heat treatment.
[c] Aging temp/time after heat treatment 500 °C.

appreciable increase in compressive strength. Of the two pendent polymer systems, the halo pendent polymers showed the greatest increase in compressive properties after thermal treatment. It is not clear as yet if this increase is due to the difference in pendent leaving groups or the extended thermal aging utilized on the halo pendent materials.

Ladder polymers were also briefly investigated in an effort to simulate the ribbon-like morphology found for carbon fiber. Other than the imidazoisoqui-noline type ladder polymers, very few systems exist that exhibit the solubility required to make liquid crystalline solutions. Recently, some very interesting systems were discovered [40] when making other structural variations to the PBZX backbone, namely, hydroxy pendent groups. The hydroxy pendents were introduced via 2,5-dihydroxyterephthalic acid. It was found that the intra-molecular hydrogen bonding between the hydroxyl group and the adjacent nitrogen atom of the benzoazole unit confers the characteristics of aromatic heterocyclic ladder polymers to the resultant dihydroxy-PBZXs. The structures are referred to as pseudo-ladder structures since the hydrogen bonds are much weaker than their covalent counterparts and therefore easier to be broken thermally with the polymeric structures reverting to those of single strands.

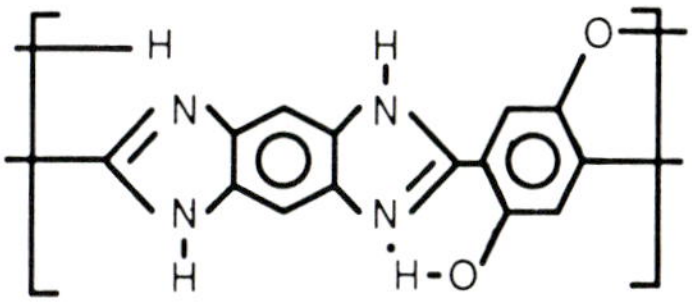

DIOH-PBI

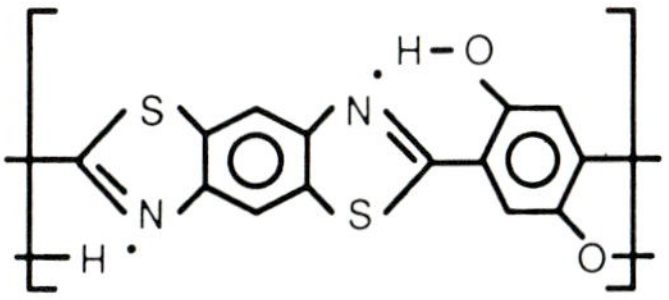

DIOH-PBZT 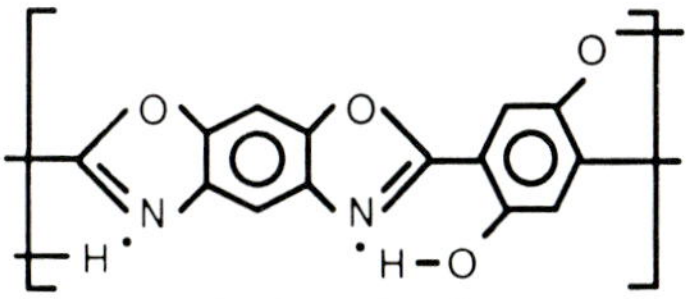   DIOH-PBO

BBL, the material that originated the order polymer technology program at Wright Laboratory, was recently polymerized in the liquid crystalline state. Previous polymerizations at 1–2% by weight concentrations resulted in poly-mer with intrinsic viscosities of 2–3 dL/g; however, polymerizations at concen-tration between 10–15% by weight provided anisotropic reaction mixtures and material that exhibited intrinsic viscosities of 17–30 dL/g. This is just another example that shows polymerization in the ordered state provides higher molecu-lar weight polymer.

Both anisotropic dopes of BBL [41] and the pseudo ladder DIOH-PBZT [42] were dry-jet wet spun into high modulus fibers and their compressive

Table 2. Ladder polymer mechanicals

Structure	Heat treated °C	Modulus (GP$_A$)	Tenacity (GP$_A$)	Compressive strength (MP$_A$)[a]
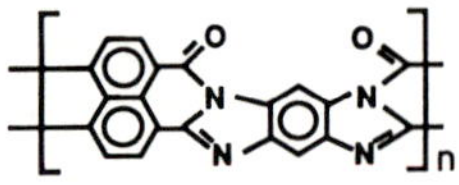	300	117	0.83	406
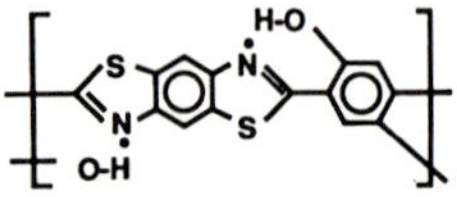	435	241	2.0	207

[a] Compressive strength measured by recoil.

strength measured by the fiber recoil test (Table 2). The ladder and pseudo-ladder systems showed no improvement in compressive properties. Although the compressive strength of DIOH-PBZT was poor compared to that of PBZT, this was the first indication that the presence of the pendant had actually imposed a negative effect on the compressive strength. The system may provide clues as to the mechanism of compressive failure in rigid-rod ordered polymers. On the other hand, its ribbon-like ladder structure with a uni-directional character is expected to lend itself to other areas of applications, namely, electronic and photonic transports.

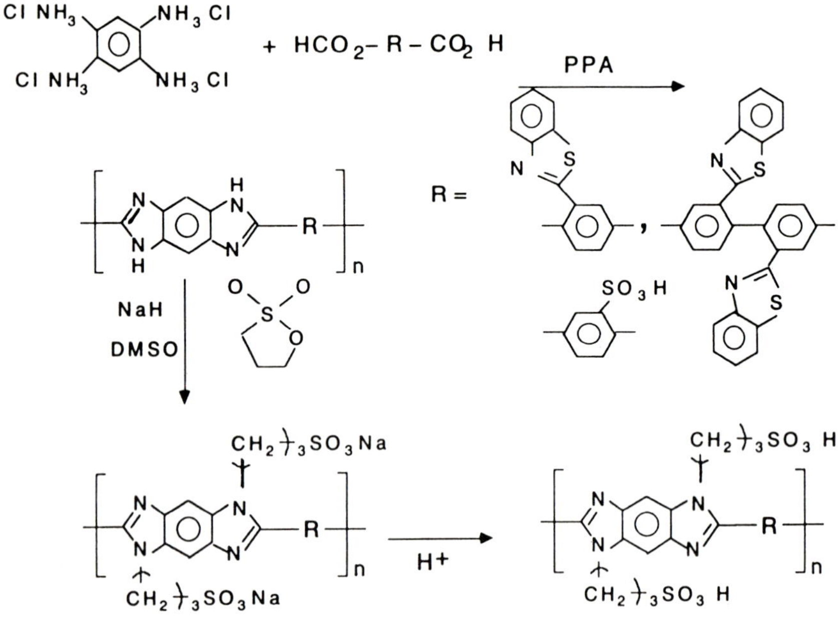

Pendent groups have also been used to promote solubility in solvents other than corrosive, acidic solvents. Dang and Arnold have demonstrated [43] the most dramatic change in solubility properties of the rigid-rod benzobisazole polymers. With post polycondensation reactions utilizing polyanions of various PBZIs, they were able to render these rigid-rod materials soluble in water at concentrations greater than 50% by weight. PBZIs containing pendent be-nzthiazole or sulfo groups were prepared by the polycondensation of 4,4'-dicarboxy- 2,2' bisbenzthiazolylbiphenyl, 2-benzthiazoleterephthalic acid or 2-sulfoterephthalic acid with 1,2,4,5-tetraaminobenzene tetrahydrochloride in PPA. These materials were then derivatized by the formation of the polyanions with sodium hydride in dimethylsulfoxide (DMSO) followed by reaction with 2-propanesulfone to form propane sulfonate pendent water soluble systems. The pendent benzthiazole and sulfo groups were required to maintain solubility in PPA for propagation as well as partial solubility in DMSO for the formation of the polyanion. The propane sulfonate polymers could be converted to the sulfonic acids by simple treatment with hydrochloric acid. Lyotropic behavior of these water based systems was found to occur at concentrations greater than 35% by weight.

Dealing with the same type of rigid-rod ionomers, Wallow and Novak recently prepared [44, 45] a series of carboxy pendent polyphenylenes (**XI**). Using a Pd(o) water soluble catalyst, p-ordered polyphenylenes were prepared

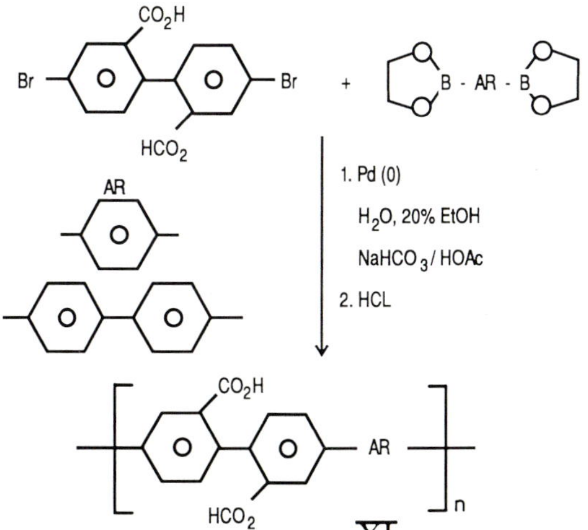

from dibromobiphenic acid and ethylene glycol diesters of 1,4-phenylene and 4,4' -biphenyl bis-boronic acids. The derived water soluble sodium salts could be isolated as the free acids by addition of dilute hydrochloric acid. Concentrated solutions (> 5% by weight) could be prepared in basic aqueous dimethyl-formamide (DMF) (75:25 water:DMF); however, the solutions tended to form extremely viscous gels. To date, no mesogenic behavior of these systems has been reported. Thermal analysis of the materials showed no glass transitions; however, they do exhibit dehydration at 300–330 °C to form the polyanhydrides which was also confirmed by IR spectroscopy.

Polypyridinium salts are another example of ionically charged rigid-rod polymers obtained by Harris and Chuang [46]. The desired polymers involved the reaction of aromatic diamines with bis-pyrylium salts. During the polymerization, the oxygen containing pyrylium ring essentially undergoes a transmutation to the nitrogen containing pyridine ring. The aromatic monomers that were used in the polymerization included the 1,4-phenylene, 4,4'-biphenylene and 4,4''-terphenylene diamines, all leading to a complete p-ordered geometry. The DSC thermograms of the polymers contained strong melting endotherms in the range of 380 to 412 °C. Although the polymers were not soluble in water, they did exhibit solubility in a number of aprotic solvents. However, the extent of their solubility was insufficient to form liquid crystalline solutions.

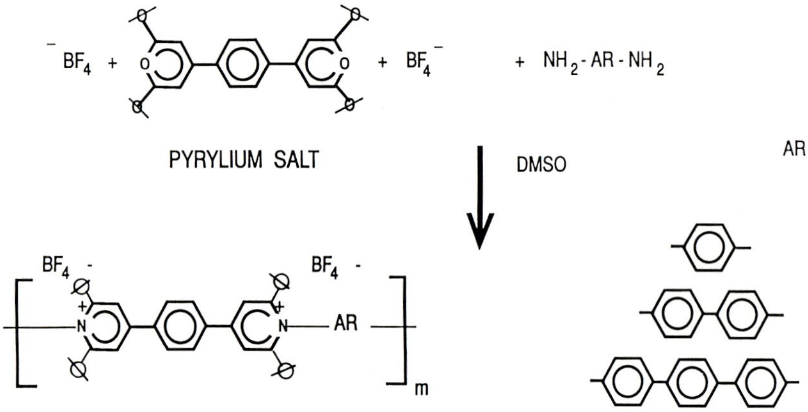

3.2 Backbone Deviations

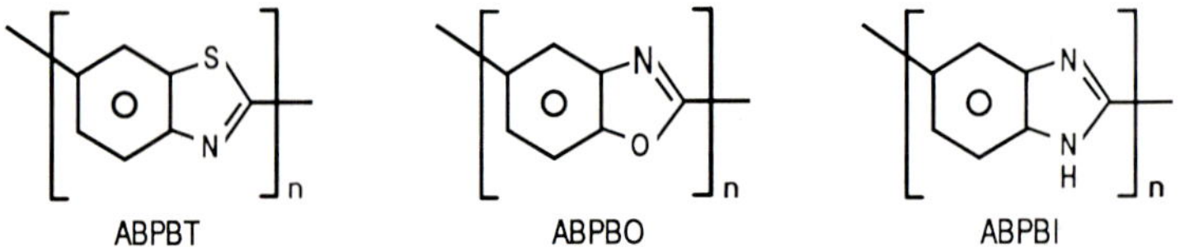

Other structural variations on the rigid-rod PBZXs have encompassed a variety of changes that affect the backbone geometry. Deviation from 180° para-catenation has been investigated by a number of researchers for improved processability. Solution properties are of particular interest in an effort to determine concentration effects on the ability to form liquid crystalline solutions. Most notable backbone deviations have been the ABPBT, ABPBO and ABPBI systems which are characterized by catenation angles of 162°, 150°, and 150° respectively. They are classified as extended chain systems because of the unrestricted rotation between the repeat units. The polymer backbone can assume either a coil-like conformation (*cis*) or an extended chain conformation (*trans*). Dilute solutions favor a random distribution of *cis* and *trans* conformations. As the concentration of polymer is increased, the trans extended chain conformation is believed to dominate and generate the liquid crystalline phase.

The ABPBX systems are obtained by the polycondensation of the AB-monomers, 3-mercapto-4-aminobenzoic acid hyrochloride, 3-amino- 4-hydroxybenzoic acid hydrochloride and 3,4-diaminobenzoic acid. The early work [47] of Imai, Taoka, Uno and Iwakura first demonstrated the polymerization of the AB-monomers in PPA. Initial attempts were conducted at (1–2% by weight) monomer resulting in polymer with viscosities in the range of 1–2 dL/g. A more comprehensive study by Wolfe et al. [48] of ABPBT and ABPBO utilizing the P_2O_5 adjustment method at high concentrations up to 21% by weight provided polymer with intrinsic viscosities as high as 24 dL/g. The minimum concentrations for the formation of the neumatic phase was determined to be 13% by weight for ABPBT and 14.5% by weight for ABPBO. The Mark–Houwink–Sakurada constants for both systems were determined by light-scattering measurements, and the relationship to intrinsic viscosities is as follows:

$$ABPBT \ [\eta] = 1.26 \times 10^{-4} \ Mw^{1.00}$$

$$ABPBO \ [\eta] = 1.09 \times 10^{-4} \ Mw^{1.02}$$

It is interesting to note that the AB-monomer, 3,4-diaminobenzoic acid is an exception, and higher molecular weights are obtained from dilute polycondensations in PPA.

Substitution of the 1,4-phenylene unit with 2,5-thiophene has also been investigated as a backbone deviation in PBZT [49] and PBZO [50] systems. A catenation angle of 148° was introduced in both systems resulting in anisotropic reaction mixtures at concentration above 10% by weight. Evers et al. polymerized the mono, di and tri-thiophene moieties as diacid chlorides with 1,4-dimercapto-2,5-diamino-1,4-benzene dihydrochloride in PPA to give polymers with intrinsic viscosities in the range of 5–8 dL/g. All these materials

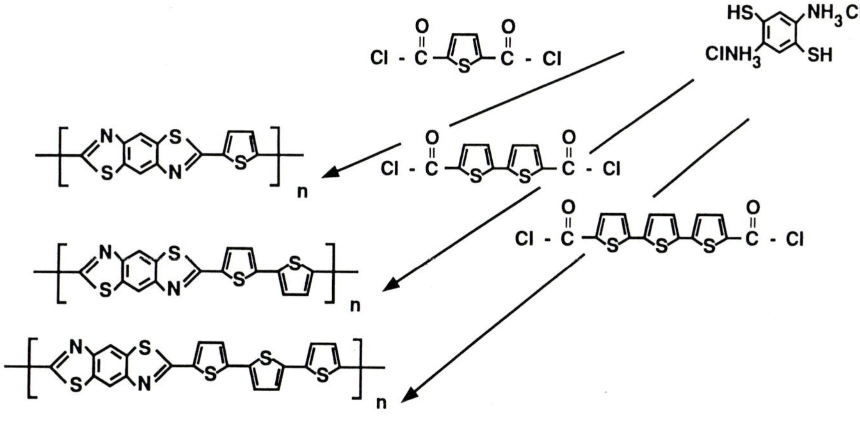

exhibited liquid crystalline behavior except the one obtained from the tri-
thiophene diacid chloride.

Structural tailoring by backbone changes for such properties as color and
transparency has been investigated by Wright Laboratory [51, 52]. An obvious
approach to render the PBZXs colorless is to synthetically modify their
molecular structure so as to disrupt their conjugation system. Cage-like hydrocarbon molecules such as diamantane and bicyclo[2.2.2]octane have been
incorporated into PBZT via the diacid monomers. The hydrocarbon molecules
are non-chromophoric due to lack of π-electrons and satisfy the all para
molecular geometry requirements. High molecular weight polymers (**XII**) and
(**XIII**) exhibiting intrinsic viscosities of 10-30 dL/g were obtained from the PPA
anisotropic polymerizations of diamantane-4,9-dicarboxylic acid and bicyclo-
[2.2.2]octane-1,4-dicarboxylic acid chloride with the PBZT amino monomer.

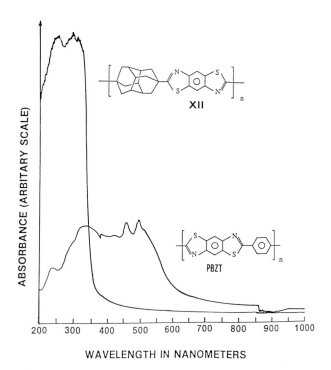

Fig. 7. Absorption spectra of XII compared to PBZT

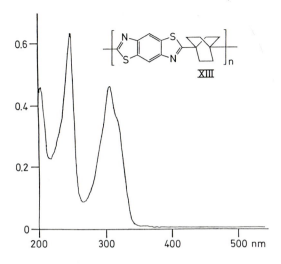

Fig. 8. Absorption spectra of XIII

The absorption spectra for the two polymers are depicted in Figures 7 and 8 showing a complete transparency in the visible and near-infrared regions. Both polymers showed exceptional thermal and thermaoxidative stabilities due to their cage-like structures. No weight loss was detected by TGA before reaching 432 °C in air and 670 °C in helium. Thermal analysis of the polymers gave no indication of any glass transitions.

4 Polyimide Systems

4.1 Rigid-Rod

Aromatic poly(pyromellitimides) have drawn widespread industrial interest from their excellent mechanical and electrical properties, outstanding thermal stability [53, 54] and relative attractive economics as high performance materials. Typically these polymers are composed of diimide biradicals derived from pyromellitic anhydride which are chain linked through aromatic biradicals via an aromatic diamine. With the appropriate p-catenated aromatic diamine, they meet all the requirements of a rigid-rod polymer system. They display very high softening temperatures due to the consecutive three fused rings, analogous to the benzobisazole structures. Most importantly, the rigid-rod poly(pyromellitimides) do not enjoy the solubility properties that are exhibited by the PBZX materials and are either insoluble or exhibit limited solubility in both aprotic and protic solvents. Over the past thirty years of research on these polymer systems, no liquid crystalline solutions have been reported.

Synthesis of rigid-rod poly(pyromellitimides) is represented by a two step process including addition (propagation) and cyclodehydration (imidization).

Conditions for low temperature solution polymerizations of pyromellitic dianhydride (PMDA) have been developed for a wide variety of aromatic 1,4-phenylene [54, 55] and 4,4′-biphenylene [56–58] diamine monomers in a number of aprotic solvents to give high molecular weight prepolymers referred to as polyamic acids. Since the imidized structures are insoluble, they must be processed in the form of their polyamic acids which are subsequently imidized thermally or by chemical dehydrating agents. Although this procedure is acceptable for thin film or fibers, the fabrication of thick parts is complicated by the water of imidization.

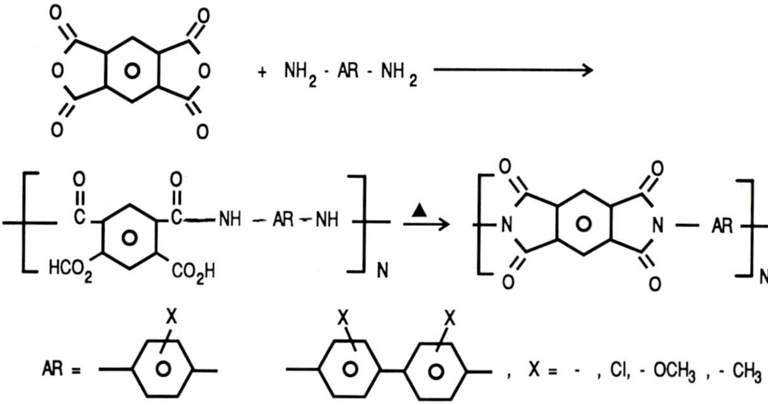

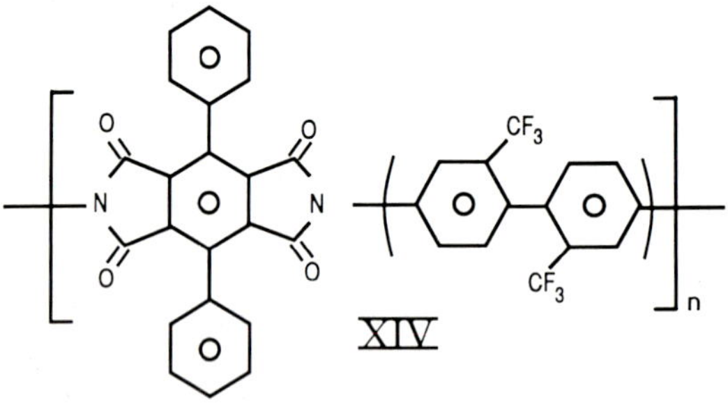

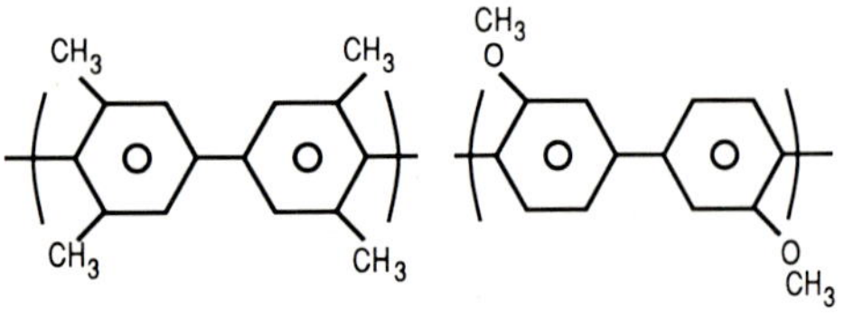

Harris and Hsu modified the structure of the rigid-rod poly(pyromellitimides) so they would display solubility in common organic solvents [59]. The approach involved the synthesis of 3,6-diphenylpyromellitic dianhydride and its polymerization with various pendent 4,4′-biphenylene diamines. The polymers, represented by structure **XIV**, were prepared in refluxing m-cresol containing isoquinoline as a catalyst for the cyclodehydration. The water of imidization was allowed to distill from the reaction mixture so that the polyimides were generated in a single step synthetic process. Intrinsic viscosities of the polymers ranged between 0.91 and 2.6 dL/g. Solubility in *m*-cresol at ambient temperature was approximately 3% by weight. The polymer containing per-fluoromethyl groups was soluble in *m*-cresol above 85 °C at concentrations as high as 12% by weight. Although the ability of the polymers to form lyotropic solutions has not been reported, a microscopic examination of the gels formed on cooling from 85 °C through cross-polarizers revealed the presence of liquid crystalline order [60].

4.2 Extended Chain

Kaneda et al. synthesized [61] a series of high molecular weight extended chain copolyimides (**XV**) by the reaction of PMDA and 3,3',4,4'-biphenyltetra-carboxylic dianhydride (PPDA) with 3,3'-dimethyl-4,4'-diaminobiphenyl. Solvents used for the one-step synthesis to the fully cyclized imide structure were phenol, *p*-chlorophenol, *m*-cresol, *p*-cresol and 2,4-dicholorophenol. The polycondensations were performed at 180 °C for 2 h with a monomer concentration of 6% by weight and *p*-hydroxybenzoic acid used as a catalytic accelerator. A maximum of 50 mol % of PMDA could be used before the copolymer precipitated from solution. Reconstituted copolymers as isotropic dopes (8–10% by weight) in *p*-chlorophenol were dry-jet wet spun between 80 and 100 °C [62].

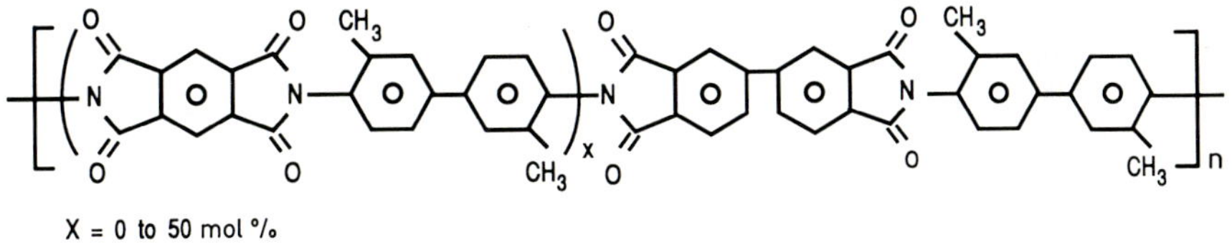

X = 0 to 50 mol %

<u>XV</u>

Table 3. Mechanical properties of copolyimide XV

Dianhydride		As-spun	Modulus (GPa)	Heat treated[a]	
mol % BPDA	PMDA	Strength (GPa)		Strength (GPa)	Modulus (GPa)
100	0	0.68	19	1.7	73
90	10	0.72	27	2.1	91
80	20	0.74	26	2.8	125
70	30	0.62	25	2.6	151
60	40	0.43	13	2.2	108
50	50	0.55	15	1.7	118

[a] Heat treated at 450 °C (nitrogen atmosphere).

Table 3 shows the mechanical properties obtained on the as-spun and heat treated (450 °C) fibers as a function of dianhydride composition. The results summarized in the table reveal an amazing effect of heat treatment on the mechanical properties of the copolyimide fibers.

A more recent study by Cheng and Harris on an extended chain polymer system, (XVI), involved BPDA and 2,2'-bis(trifluoromethyl)-4,4'-diaminobiphenyl [63]. Here again, a one-step polymerization was utilized and gave the completely cyclized imide structure. The polymerization was carried out in refluxing m-cresol containing isoquinoline with a polymer concentration of 10% by weight [64]. Polymer with an intrinsic viscosity of 4.9 dL/g was obtained in m-cresol at 30 °C. TGA analysis of powdered XVI samples at a heating rate of 10 °C/min showed a 5% weight loss under nitrogen and air at 600 °C. TGA-mass spec (Fig. 9) shows the degradation products formed as a function of temperature, at a heating rate of 3 °C/min. The initial breakdown of the polymer occurs

XVI

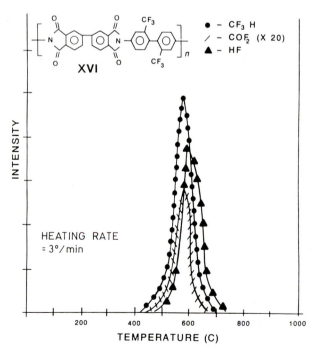

Fig. 9. TG-MS of XVI

from the perfluoromethyl pendant which is remarkably stable. Other products not shown were carbon monoxide, hydrogen cyanide and ammonia, all typical of the main chain heterocyclic degradation.

Isotropic solutions (12–15% by weight) of XVI in *m*-cresol were dry-jet wet spun into a coagulation bath of water/methanol. The as-spun fibers were drawn at temperatures above 380 °C to give fibers having a tensile strength of about 3.2 GPa and a modulus of 130 GPa. The annealed fibers displayed distinct wide-angle X-ray patterns from which a monoclinic unit cell was determined.

5 Molecular Composites

5.1 Blends

As described earlier in the chapter, the objective of the Air Force ordered polymer research program was to provide a monolithic plastic material having the properties of high performance metals. For their weight the rigid-rod polymeric materials fabricated from the liquid crystalline state far exceed the strength and modulus properties of structural metals used currently in advanced aircraft and aerospace systems. Although fiber and film could be readily obtained, lamination to provide large bulk specimens proved difficult since the glass transition temperatures of these materials are above their decomposition temperatures. The inability to laminate such materials initiated a new concept in material technology, the molecular composite concept [65].

A molecular composite is defined as rigid-rod molecules molecularly dispersed in a matrix of flexible coil polymer such that the rods act as the reinforcing elements. The flexible coil matrix, having a glass transition temperature, can be used as a vehicle to consolidate various monoliths into bulk structures. Other interest in this area is based on the improvements in fracture and impact toughness, as well as dimensional stability over those of conventional reinforced composites. The concept of reinforcement on the molecular level using intrinsic rigid-rod polymers represents one of the most exciting, emerging technologies in the field of structural materials.

The proof of concept and the fabrication of the first true rigid-rod molecular composite were initially demonstrated [66] using PBZT as the rigid-rod reinforcement and ABPBI as the continuous flexible coil matrix. To insure the molecular dispersity of the rigid-rod polymer in the blend, one can only process from solutions at lower than its corresponding critical concentrations (C_{cr}) [67, 68] to prevent the segregation of rigid-rod reinforcement. Above C_{cr} the solutions become biphasic where the rigid-rod segregates into liquid crystalline domains which are dispersed in an entangled flexible coil matrix.

The mechanical properties (Table 4) of the resulting molecular composite fiber films demonstrated a virtually complete translation of rigid-rod polymer modulus on a volume fraction basis to the host matrix. This implied that the

Table 4. Mechanical properties of PBZT/ABPBI molecular composites

Composition	Form	Modulus (GPa)	Tensile Strength (MPa)	Elongation %	Ref.
PBZT	(Fiber)	300	3000	1.1	[70]
ABPBI (AS)	–	16.8	675	29	[66]
ABPBI (HT)	–	36.0	1110	5.2	[70]
30% PBZT/70% ABPBI	–				
As-spun (AS)	–	16.8	1161	7.2	[66]
Heat treated (HT)					
427 °C	–	71.7	1215	2.5	–
525 °C	–	109.7	1283	1.8	–
30% PBZT/70% ABPBI	(Film)				
As processed	–	30.4	690	5.5	[69]
Heat treated 540 °C	–	88.2	918	2.4	–
$C > Cc_r^a$	–	1.1	35	5.6	[70]

[a] Fabricated at a concentration above critical concentration.

Table 5. Mechanical properties of PBZT/thermoplastic molecular composites

Composite system	Composition (wt/wt)	Form	Modulus (GPa)	Tensile strength (MPa)	Ref.
PBZT/PPQ	30/70	Fiber	17.5	355	[73]
PBZT/Nylon	30/70	–	36.0	345	–
PBZT/PEEK	50/50	Film	11.0	248	[74]
PBZT/PEEK	50/50	Bulk	15.8	105	–
PBZT/Nylon-6,6	50/50	Bulk	20.8	183	[72]

composite rule of mixtures had been obeyed, and the concept of forming a composite on the molecular level analogous to microscopic chopped fiber composite was valid. Morphology studies [66, 69, 70] employing small-angle and wide-angle X-ray diffraction, optical microscopy and scanning electron microscopy on these blends showed no phase separation to the resolution limit of the techniques employed.

The model PBZT/ABPBI molecular composite system is limited since the rod and the matrix do not possess glass transition temperatures for subsequent post form consolidation. In an effort to improve the processability for molecular composites, thermoplastics were used as the host matrix. Processing from acidic solvents requires the thermoplastic host to be soluble and stable in methanesulfonic acid. Thermoplastic matrices were investigated including both amorphous and semicrystalline nylons [71, 72], polyphenylquinoxaline (PPQ) [73] and polyetheretherketone (PEEK) [74]. Table 5 shows the mechanical properties obtained for various processed PBZT thermoplastic molecular composite systems. As an example, the PBZT/Nylon systems showed 50–300% improvement over uniaxially aligned chopped fiber composite of comparable compositions. However, the thermally-induced phase separation during consolidation was evident. Distinct phase separated PBZT-rich domains 2–7 micrometers in size were observed after thermal treatment. This drive toward phase separation was demanded by the unfavorable thermodynamic forces when sufficient mobility was thermally imparted to the system. Acid stable thermoset systems were also briefly investigated [75] utilizing the glass transition temperature of thermoset for consolidation. Problems associated with phase separation during the coagulation process prevented their success.

An approach [76] that allows the use of a wide variety of commercial thermosets and thermoplastic resins takes advantage of the in situ generation of a rigid-rod polymer. Conceptually, the process involves the blending of two compatible coil-like polymers in which one would thermally isomerize to a rigid-rod after fabrication, thus providing the reinforcing element of the composite. In this approach the key element is the thermally induced isomerization of a coil-like precursor to the respective rigid-rod within the matrix. Presumably, since the isomerization process is taking place after the coil-like precursor and the matrix material have been uniformly mixed, there is a high probability that the aggregation of the rigid-rods may be prevented. The success of this approach is primarily dependent on the efficiency of the isomerization process which must be high enough to generate the necessary aspect ratio for reinforcement.

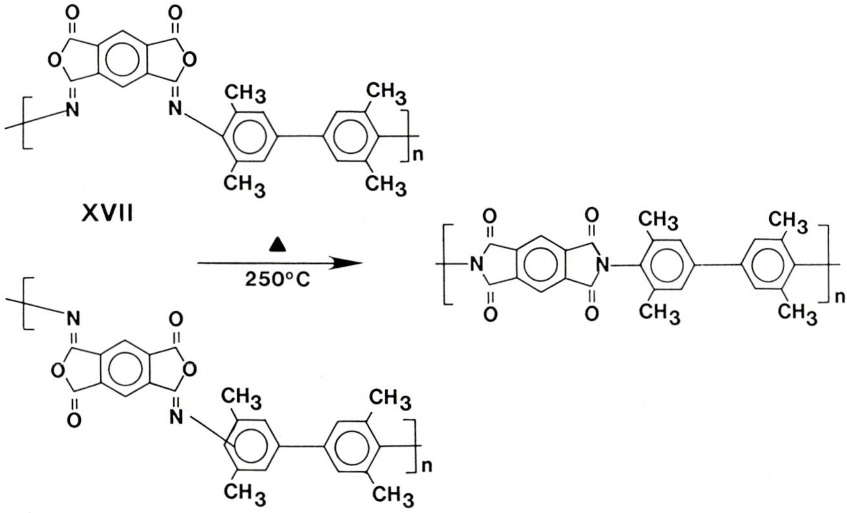

The well known thermally induced isomerization of an isoimide to an imide was the chemistry selected to test the concept. A series of high molecular weight polyisoimides was prepared based on PMDA and pendent aromatic diamines that on thermal treatment would exhibit the required geometry for reinforcement. Polymerizations of the diamines with PMDA were carried out in DMAC (10% by weight) at room temperature in a dry nitrogen atmosphere. Subsequent cyclodehydration of the polyamic acid to the corresponding polyisoimide was effected by the addition of *N,N'*- dicyclohexyldicarbodiimide. Extreme care was taken to keep the materials anhydrous since polyisoimides are sensitive to water and revert back to the polyamic acids. The highest molecular weight poly-isoimide (**XVII**) ([η] = 1.89 dL/g) was derived from 4,4'-diamino-3,3',5,5'-tetramethylbiphenyl.

Films of the polyisoimides were cast from DMAC at 55 °C under reduced pressure (0.1 mm). A study of the isomerization reaction was conducted by FTIR and showed that the isomerization began at approximately 100 °C and was complete after 3 h at 250 °C. In all cases the thermally treated films were insoluble in all solvents tested. Composite films were produced with XVII and three commercial matrix systems: a polyarylsulfone (Radel), a polysulfone (Udel), and an acetylene terminated isoimide thermosetting resin (IP-600). Films of the matrix and XVII were cast from DMAC. Slightly cloudy films, indicating some phase separation, resulted with both the Radel and Udel systems. Composite films cast with IP-600, however, were completely clear and showed no signs of phase separation. The structural similarity of the IP-600 resin and XVII may account for the greater homogeneity of the system. Property assessment of these films before and after thermal treatment is currently underway.

5.2 Block and Graft Copolymers

One-part molecular composite systems based on block copolymers represent one of the most attractive approaches in resolving problems associated with phase separation. In the block copolymer the rigid-rod constituent is chemically bonded to the flexible coil matrix, while in the physically blended molecular composite, the reinforcing rigid-rod polymer is only physically entangled in the matrix. Hence, the translation of stress or strain from the reinforcing rigid-rod molecule in the matrix would be more efficient in the block copolymer than in the physical blend. This would not only provide higher tensile strength but also fracture toughness and dimensional stability.

As a model system Tsai, Arnold, and Hwang synthesized a series of novel ABA block copolymers [77] containing a rigid-rod (B) block for reinforcement and a flexible coil (A) block as the matrix. PBZT was the rigid-rod (B) block utilized in the study and was polymerized in such a way as to provide carboxylic acid end-groups. The carboxy-terminated PBZTs ([η] = 10 to 24 dL/g) were copolymerized with the AB monomer, 3,4-diaminobenzoic acid, which generates a benzimidazole (A) block as well as grafts the blocks together. Composition of the blocks was varied by the weight of the AB monomer used in the copolymerization.

Young's modulus of the block copolymer fibers compared favorably with that of the physical blends as shown in Table 6, and in general they follow the linear rule of mixtures. The modulus data suggested that one does not need very large PBZT molecules to have the reinforcing efficiency. From the tensile data, one clear trend is that the tensile strength of the block copolymer system is much

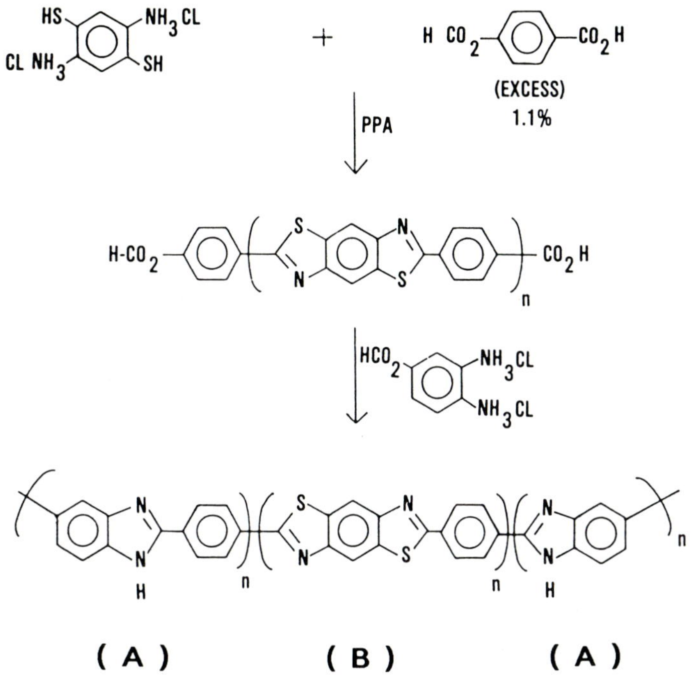

Table 6. Comparative tensile properties of block copolymer blended molecular composite fibers

Composition PBZT/ABPBI	[η] (dL/g) PBZT-Rod	[η] (dL/g) copolymer	Modulus (GPa)	Tensile (MPa) strength	Elongation %
30/70	10.7	8.5	102.7	1696	2.3
25/75	12.9	10.7	94.5	1566	2.5
30/70	17.7	7.3	115.8	1600	1.4
30/70	Physical Blend[a]		116.5	1268	1.4

[a] Physical blend of 30% by weight PBZT ([η] = 31 dL/g) and 70% by weight ABPBI ([η] = 16.7 dL/g).

better (up to 33%) than that of the physical blend. This enhancement in tensile strength is most illuminating when one considers the fact that the molecular weight of both the flexible coil block (A) and the central rigid-rod block (B) is much lower than that in the physical blend. Morphology studies [78] on the copolymer fibers showed no phase separation down to 3 nm with oriented crystallites of PBZT and ABPBI no larger than 3 nm. From the morphology studies and enhancement in tensile properties, the ABA block copolymer systems clearly show the advantages in preventing the segregation of PBZT molecules and in the efficiency of translation of strength. Although the tri-block copolymers are attractive from their properties, they exhibited no glass transition temperatures for consolidation.

More recently [79], a carboxy-terminated PBZT ($[\eta] = 4.8$ dL/g) was reacted with *m*-phenoxybenzoic acid via a Friedel Craft procedure in a methanesulfonic acid/P_2O_5 mixture. This provided an ABA block copolymer in which the outer blocks (A) are composed of flexible coil polyetherketone (PEK) and a center block (B) which contains the rigid-rod PBZT. Thermomechanical analysis showed that 20 PBZT/80 PEK and 10 PBZT/90 PEK compositions exhibited glass transition temperatures of 157 °C and 135 °C respectively. Consolidation studies have not been investigated to date.

The grafting of flexible coil side-chains onto rigid-rod polymers represents another approach in obtaining a one-part molecular composite system. One of the early works [80, 81] on grafting flexible coils involved the polyanion of the benzthiazole pendent PBZI. The polyanion generated from sodium methylulfinylmethide in DMSO reacted with propylene oxide to afford polypropylene oxide side chains. In addition to increasing the C_{cr} for solution processing, the propylene oxide grafts were expected to improve the softening behavior for bulk pendant consolidation. While the parent PBZI readily exhibited mesogenic behavior in acidic solvents, lyotropic solutions of the graft copolymer could not be obtained up to 8% by weight, concentration above which the polymer was not completely soluble. Although tensile bars could be consolidated at 190–232 °C 5 MPa pressure), they proved to be fairly brittle. The relatively poor thermooxidative stability of the aliphatic chains limited the upper temperature that could be used for consolidation. The processing and characterization of the graft copolymers have not been sufficiently pursued to the point of establishing the viability of the system.

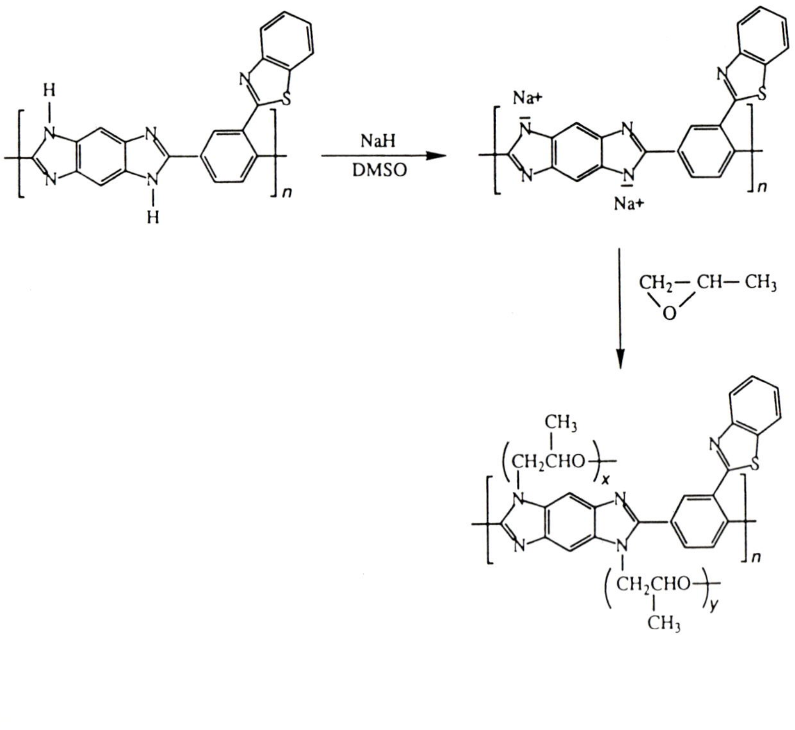

A more thermally stable flexible coil side chain was sought that could be grafted to the rigid-rod backbone. The side chains would have to withstand the high temperatures of the consolidation as well as depress the softening behavior of the rod. Since the rigid-rods exhibit solublity only in acidic solvents, the number of potential grafting reactions is very limited. Evers et al. investigated [82, 83] the Friedel Craft AB-polymerization of *m*-phenoxybenzoic acid as a method of obtaining thermally stable polyetherketone side chains. Such polymerizations were carried out in PPMA, a solvent composed of methanesulfonic acid/P_2O_5 (10:1 w/w). Two PBZT rigid-rod copolymer systems were explored by making structural backbone changes for grafting sites.

In one case articulated PBZT copolymers (**XVIII**) were prepared [84] containing randomly distributed 4,4'-(*o*-diphenoxybenzene) units between the rigid-rod PBZT segments. The diphenoxybenzene moieties served as the graft sites for the mPEK pendents reacting para to the diphenylether group. The copolymers were prepared through the copolymerization of 1,4-dimercapto-2,5-diamino-1,4-benzene dihydrochloride with terephthaloyl chloride and 4,4'-(*o*-phenylenedioxy) dibenzoyl chloride in PPA. Graft site concentrations were varied by the stoichiometry of the reaction mixture. By varying the mole ratio of diphenoxybenzene, different lengths of rod segments between the points of articulation were obtained for various reinforcement aspect ratios. The articulated copolymers were reacted in PPMA with *m*-phenoxybenzoic acid to form the polyetherketone and at the same time graft to the copolymer backbone. Unreacted mPEK homopolymer was removed by extraction with tetrahydrofuran.

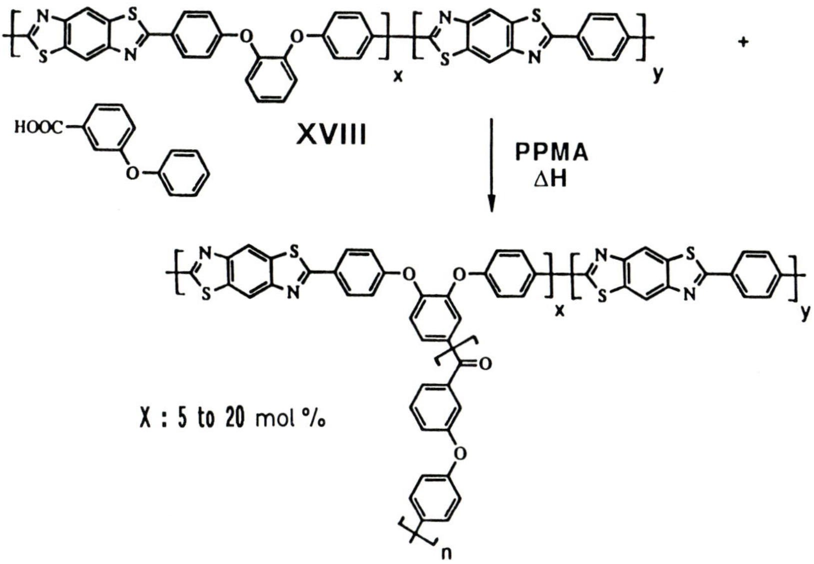

Compression molding at elevated temperature and pressure was used to achieve bulk specimens with three dimensionally isotropic properties. Successful consolidations were achieved at 285 °C with a pressure of 6.9 MPa. X-ray scattering and electron microscopy were employed to elucidate the consolidation process. Copolymers with low rod content showed tensile modulus as predicted by the Halpin-Tsai equation [66] with minimal rod aggregation. Copolymers with high rod content exhibited phase separation resulting in low reinforcement efficiency for the PBZTsegments within the bulk ridgi-rod molecular composites. A shortcoming of this approach was the limited flexibility in graft sites or points of articulation in the polymer backbone.

An alternate approach was the synthesis [85, 86] of PBZT copolymers (**XIX**) containing pendent 2,6-dimethylphenoxy graft sites. Such copolymers do not lead to any breaks within the rigid-rod backbone; thereby, they have no adverse effects upon rod reinforcement efficiency. Using the conventional PBZT polymerization procedure, 2-(2,6-dimethylphenoxy)terephthaloyl chloride was substituted for terephthaloyl chloride up to 30 mol %. The pendent dimethylphenoxy copolymers were then reacted with m-phenoxybenzoic acid in PPMA.

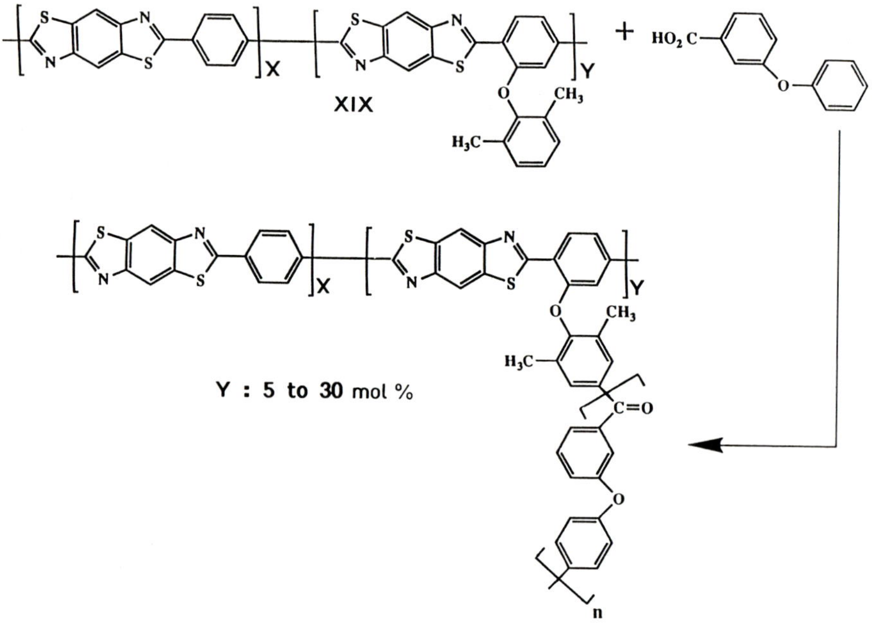

XIX

Y : 5 to 30 mol %

Dynamic mechanical analysis [87] of the graft copolymers indicated softening temperatures in the range of 180 to 265 °C. Consolidated molded specimens were obtained through the application of pressure up to 40 MPa and temperatures 40 to 50 °C above their softening temperature. Wide angle X-ray diffraction studies [88] confirmed that the grafting approach was effective in preventing rod aggregation during the thermal processing to bulk forms. Tensile moduli and strengths of 9.7 GPa and 40 MPa respectively were obtained for a representative sample composed of 46% by weight PBZT and 54% by weight

PEK. The poor tensile strength of these graft copolymers might be related to the relatively short *m*-PEK side chains, insufficient to provide effective chain entanglement required for tenacity. Recent molecular dynamic simulations [89] were performed using CHARMm as implemented by the QUANTA molecular modeling software package. The lengths of the PBZT rod and *m*-PEK side chains were 101 Å and 50 Å respectively. Animation of the dynamic trajectories revealed that the *m*-PEK side chains are more inclined to associate with the PBZT rod segment than they are to coil about each other.

While the processing and characterization of the graft copolymers have not been sufficiently pursued to this point to establish the viability of the concept, these research efforts have demonstrated that rigid-rod polymer fusibility could be substantially modified through the introduction of flexible coil side chains. Unfortunately, in spite of the careful processing and the attainment of excellent consolidation as well as minimal phase separation, the tensile properties are less than expected.

Finally, Tan and Arnold introduced [90–93] a novel concept which is a rather radical departure from all the precedents utilized to tackle the phase separation problem related to physically blended molecular composites. In this approach a coil-like or extended precursor polymer undergoes a thermally induced elimination reaction. The expelled molecular fragments containing suitable reactive groups can then polymerize in situ via an addition process to form a thermoset or thermoplastic matrix. In other versions the expelled fragments can be a preformed polymeric component (thermoplastic) or oligomeric component (plasticizer) containing appropriate, monofunctionalized end-groups. Systems based on the concept are referred to as In Situ Molecular Composites.

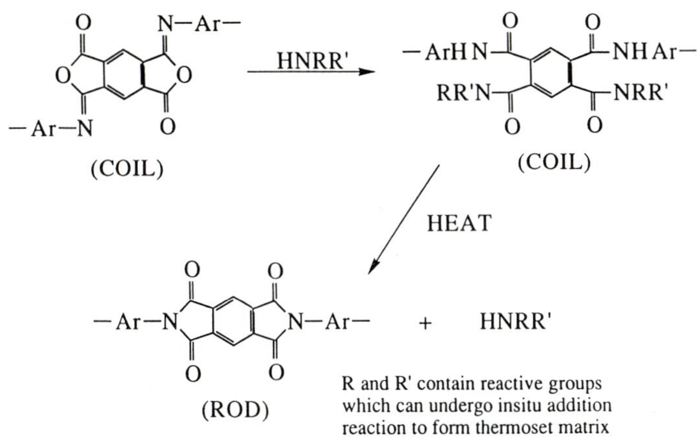

The selection of the chemistry to demonstrate the concept is all important in providing a material that meets all the required processing criteria. Initial attempts to provide a system which would demonstrate the concept involved the synthesis of high molecular weight polyamicdialkylamides. The polymer system is conveniently prepared [94] by the reaction of a dialkylamine with an aromatic polyisoimide. In this system the dialkylamine would contain the appropriate reactive oligomer structurally designed to thermally polymerize, preferably by an addition mechanism to form a thermoset matrix. The poly-isoimide (**XVII**) obtained from PMDA and 4,4′-diamino-3,3′,5,5′-tetramethyl-biphenyl was utilized in the study since it exhibited the required geometry for reinforcement after the cyclodeamination/imidization process.

Two classes of reactive secondary amines (**XX** and **XXI**) were synthesized [90] for the thermoset version of the In Situ Molecular Composite concept. The thermoset chemistry of the bisnadimide (**XX**) involved that of PMR systems (polymerization of monomeric reactants) [95]. The secondary amine (**XXI**) embodied the more recent ringopening polymerization of benzocyclobutene (BCB) [96]. Polymerization exotherms for both thermoset amines occurred with an onset at 225 °C, maximizing at approximately 260 °C as evidenced by DSC at a heating rate of 10 °C/m.

XX

XXI

Model polyamicdialkylamides were prepared by the reaction of the second-ary amines (**XX**) and (**XXI**) with the polyisoimide (**XVII**) in DMAC at room temperature. Unfortunately, the resultant polyamicdialkylamides only exhibited limited softening behavior for processing into bulk specimens. The inability to provide molded specimens prevented the validation of the In Situ Molecular Composite concept. Research is currently being carried out at a number of academic and industrial institutions on new thermoset chemistry, polyamices-ters, and extended chain polyimides.

6 Conclusions

We have thus far discussed in general terms rigid-rod polymers and molecular composites as technical opportunities for advanced aircraft and aerospace systems. Rigid-rod polymer systems show the greatest promise where low weight, high strength and stiffness, high temperature capability, environmental resistance, and radar transparency are demanded. The radar transparency of PBZX fibers and films is a significant advantage over graphite fibers in many applications to reduce radar and infrared signatures. Spacecraft and space structures are applications where ordered polymers can play a significant role in manned space stations, satellites, and probes. PBZX fibers show better elongation at break than graphite and exhibit far superior thermal and thermo-oxidative properties than Kevlar fibers, which make them a more attractive choice where damage tolerance is required.

Commercial applications have been identified primarily in the electronics industry where requirements for dimensional stability, mechanical properties, and high temperature resistance make these systems attractive in advanced circuit board technology. Other commercial applications include high temperature membranes and filters where these materials offer performance improvements over glass, Kevlar, and graphite composites. Industrial development of these types of materials will most likely be dependent on monomer cost and advances in various product properties requirements.

Although several high modulus materials have been identified in this chapter as well as current commercial materials, a more wide spread acceptance of these materials will occur if their compressive properties are vastly improved. We have discussed several synthetic approaches to improve the compressive properties by structural changes of PBZX systems with small increments of improvement. From what has currently been done, crosslinking seems to have shown the greatest improvement. An in-depth study on a rigid-rod polymer system that forms three dimensional crosslinks by an addition mechanism without the evolution of volatile byproducts is required. For the most part, innovative approaches, both chemical and physical, are required to attack and solve this deficiency.

Historically, polyimides have always been the leading heterocyclic material utilized for high temperature applications. The favorable economics of dianhydrides and diamines has been the primary driving force for these materials. Although many rigid-rod polyimides have been prepared, they lack the solubility required to promote liquid crystalline solutions. Great strides have recently been made in strength properties of extended chain polyimide systems by high temperature drawing. Modulus values for these systems are somewhat low which inhibit them for many applications. It is anticipated that in the near future, rigid-rod polyimides will be synthesized that exhibit the necessary solubility characteristics for the attainment of liquid crystalline solutions. At that time modulus values will be vastly increased.

The concept of reinforcement on the molecular level using intrinsic rigid-rod polymers represents one of the most exciting, emerging technologies in the field of structural materials. If successful, molecular composites will offer a wide variety of new material forms. The technology is in its infancy, and a number of problems exist which will have to be overcome. As previously discussed, the critical problem is the inability to consolidate without phase separation. We have discussed the latest thinking and literature approaches to attack this problem; however, it is too early at this time to predict success or failure. Many academic and industrial organizations throughout the world are actively engaged in research to provide new material systems based on the concept. It remains to be seen if such activities will be met with success.

7 References

1. Van Deusen RL, Goins OK, Sicree AJ (1968) J Polym Sci A6: 1777
2. Tessler MM (1966) J Polym Sci Al: 252
3. Arnold FE, Van Deusen RL (1969) Macromolecules 2: 497
4. Arnold FE, Van Deusen RL (1971) J Appl Polym Sci 15: 2035
5. DeSchryver F, Marvel CS (1967) J Polym Sci 5: 545
6. Arnold FE (1970) J Polym Sci 8: 2079
7. Arnold FE, Sicree AJ, VanDeusen RL (1974) J Polym Sci 12: 265
8. Kovar RF, Arnold FE (1974) J Polym Sci 12: 401
9. Kwolek SL, Morgan PW, Schaefgan JR, Gulrich JR (1977) Marcromolecules 10: 1390
10 Kovar RF, Arnold FE (1976) J Polym Sci Polym Chem Ed 14: 2807
11. Wolfe JF, Arnold FE (1981) Macromolecules 14: 909
12. Wolfe JF, Arnold FE (1981) Marcromolecules 14: 915
13. Choe EW, Kim SN (1981) Macromolecules 14: 920
14. Allen SR, Filippov AG, Farris RJ, Thomas El, Wong CP, Berry GC, Chenevey EC (1981) Macromolecules 14: 1135
15. Chenevey EC, Helminiak TE (1986) U.S. Pat 4,606,875
16. Feldman L, Farris RJ, Thomas EL (1985) J Mater Sci 20: 2719
17. Chow AW, Sandell JF, Wolfe JF (1987) Polymer 29: 1307
18. Wolfe JF, Sybert PD, Sybert JR (1985) U.S. Pats 4,533,692, 4,533,693, 4,533,724
19. Tsai TT, Arnold FE (1989) High Perf Polym 3: 179
20. Tsai TT, Arnold FE (1988) Polym Prepr Am Chem Soc Polym Div 29: 324
21. Dang TD, Tan LS, Arnold FE (1990) Am Chem Soc Polym Mat Sci Eng Proc 62: 86
22. Jenekhe SA, Johnson PO (1990) Macromolecules 23: 4419
23. Allen SR, Farris RJ, Thomas EL (1985) J Mater Sci 20: 2727
24. Chenevey EC, Timmons WD (1989) Matrl Res Soc Symp Proc 134: 245
25. Lusignea RW (1989) Matrl Res Soc Symp Proc 134: 265
26. Cohen Y, Thomas EL (1985) Polym Eng Sci 25: 1093
27. Cohen Y, Thomas EL (1988) Macromolecules 21: 433
28. Allen S (1987) J Mater Sci 22: 853
29. Kumar S, Adams WW, Helminiak TE (1988) J Reinforced Plast Comp 7: 108
30. Berry GC (1985) Mat Sci Eng 52:82 and references therein
31. Berry GC, Metzger PC, Venkatraman S, Cotts DB (1979) Polym Prepr Am Chem Soc Polym Div 1, 20: 82
32. Denny LR, Goldfarb IJ, Soloski EJ (1989) Matrl Res Soc Symp Proc 134: 395
33. Kumar S, Helminiak TE (1989) Matrl Res Soc Symp Proc 134: 363
34. Allen SR (1987) J Mater Sci 22: 324
35. Burkett J, Arnold FE (1987) Polym Prepr Am Chem Soc Polym Div 1, 28: 2
36. Arnold FE (1989) Matrl Res Soc Symp Proc 134: 117
37. Wang CS, Burkett J, Bhattacharya S, Chuah HH, Arnold FE (1989) Proc Am Chem Soc Div Polym Matrl Sci Eng 60: 767
38. Chuah HH, Tsai TT, Wei KH, Wang CS, Arnold FE (1989) Proc Am Chem Soc Div Polym Matrl Sci Eng 60: 175
39. Sweeny FW (1992) J Polym Sci Polym Chem Ed 30: 1111
40. Dang TD, Tan LS, Arnold FE (1990) Proc Am Chem Soc Div Polym Matrl Sci Eng 62: 86
41. Wang CS, Lee CY-C, Arnold FE (1992) Matrl Res Soc Symp Proc 247: 747
42. Dang TD, Tan LS, Wei KH, Chuah HH, Arnold FE (1989) Proc Am Chem Soc Div Polym Matrl Sci Eng 60: 424
43. Dang TD, Arnold FE (1992) Polym Prepr Am Chem Soc Polym Div 1, 33: 912
44. Wallow TI, Novak BM (1991) Polym Prepr Am Chem Soc Polym Div 1, 32: 191
45. Wallow TI, Novak BM (1992) Polym Prepr Am Chem Soc Polym Div 1, 33: 908
46. Harris FW, Chuang CK (1989) Polym Prepr Am Chem Soc Polym Div 2, 30: 433
47. Imai Y, Taoka I, Uno K, Iwakura Y (1965) Makromol Chem 83: 167
48. Chow AW, Bitler SP, Penwell PE, Osborne DJ, Wolfe JF (1989) Macromolecules 22: 3514
49. Dotrong M, Tomlinson RC, Sinsky M, Evers RC (1991) Polym Prepr Am Chem Soc Polym Div 32: 85
50. Promislow JH, Samulski ET, Preston J (1992) Polym Prepr Am Chem Soc Polym Div 1, 33: 211

51. Dang TD, Archibald TG, Malik AA, Bonsu FO, Baum K, Tan LS, Arnold FE (1991) Polym Prepr Am Chem Soc Polym Div 2, 32: 199
52. Dotrong M, Dotrong H, Moore GJ, Evers RC (1991) Polym Prepr Am Chem Soc Polym Div 1, 32: 201
53. Adroua NA, Bessenov MI, Louis LA, Rudakoo AP (1970) Polyimdes, Technomic Stanford, CT
54. Mittal KL (1984) Polyimide, Plenum, NY
55. Makino H, Kusuki Y, Harada T, Shimazaki H (1993) U.S. Pat 4,370,290
56. Ohta T, Yanaga Y, Hino S (1984) U.S. Pat 4,438,256
57. Khune GD (1980) J Macromol Sci Chem 14: 687
58. Jinda T, Matsuda R, Sakamoto M (1984) Sen Gakkaishi 40: 42
59. Harris FW, Hsu SLC (1989) High Perf Polym 1: 3
60. Cheng SZD, Lee SK, Barley JS, Hsu SLC, Harris FW (1991) Macromolecules 24: 1883
61. Kaneda T, Katsura T, Nakagawa K, Horio M (1986) J Appl Polym Sci 32: 3131
62. Kaneda T, Katsura T, Nakagawa K, Horio M (1986) J Appl Polym Sci 32: 3151
63. Cheng SZD, Wu Z, Eashoo M, Hsu SLC, Harris FW (1991) Polymer 32: 1803
64. Harris FW, Hsu SLC, Tso CC (1990) Polym Prepr Am Chem Soc Polym Div 2, 31: 342
65. Helminiak TE, Benner CL, Husman GE, Arnold FE (1980) U.S. Pat 4,207,407
66. Hwang WF, Wiff DR, Benner CL, Helminiak TE, (1983) J Macromol Sci Phys 22: 231
67. Hwang WF, Wiff DR, Benner CL, Helminiak TE, Adams WW (1983) Polym Eng Sci 23: 784
68. Hwang WF, Wiff DR, Verschoore C (1983) Polym Eng Sci 23: 790
69. Wellman M, Husman G, Kulshreshtha AK, Helminiak TE, Wiff DR, Benner CL, Hwang WF (1980) Am Chem Soc Div Org Coat Plast Chem 43: 783
70. Krause SJ, Haddock T, Price GE, Lenhert GP, O'Brien JF, Helminiak TE, Adams WW (1986) J Polym Sci Poym Phys 24: 1991
71. Kyu T, Helminiak TE (1987) Polymer 28: 2130
72. Wang CS, Goldfarb IJ, Helminiak TE (1988) Polymer 29: 825
73. Hwang WF, Wiff DR, Helminiak TE, Adams WW (1983) Org Coats Appl Polym Sci Proc Am Chem Soc 48: 919
74. Gabriel CA, Farris RJ, Malone MF (1986) Nonwovens conference 1: 255
75. Chah HS, Tan LS, Arnold FE (1989) Polym Eng Sci 29: 107
76. Wallace SJ, Tan LS, Arnold FE (1990) Polymer 31: 2411
77. Tsai TT, Hwang WF, Arnold FE (1989) J Polym Sci Polym Chem Ed 27: 2839
78. Krause SJ, Haddock TB, Price GE, Adams WW (1988) Polymer 29: 195
79. Coopper KL, Arnold FE (1992) Polym Prepr Am Chem Soc Polym Div 1, 33: 1006
80. Dang TD, Evers RC (1988) Polym Prepr Am Chem Soc Polym Div 2, 29: 244
81. Dang TD, Evers RC, Moore DR (1991) J Polym Sci 29: 121
82. Bai SJ, Dotrong M, Soloski EJ, Evers RC (1991) J Polym Sci Polym Phys Ed 29: 121
83. Dotrong M, Dotrong MH, Bai SJ, Evers RC (1992) Sci Adv Matrls Proc Eng Tech Symp Ser 37: 1004
84. Dotrong M, Dotrong MH, Bai SJ, Evers RC (1992) SAMPE Ser 37: 1004
85. Dotrong M, Dotrong MH, Evers RC (1991) Am Chem Soc Div Polym Matrl Sci Eng Prepr 65: 38
86. Dotrong M, Dotrong MH, Evers RC (1992) Polym Prepr Am Chem Soc Polym Div 1, 33: 477
87. Vakil UM, Wang CS, Lee CY-C, Dotrong MH, Dotrong M, Evers RC (1992) Polym Prepr Am Chem Soc Polym Div 1, 33: 479
88. Song HH, Price GE, Vakil UM, Dotrong MH (1992) Polym Prepr AM Chem Soc Polym Div 1, 33: 319
89. Trohalaki S, Dudis DS (1992) Polym Prepr Am Chem Soc Polym Div 1, 33: 704
90. Tan LS, Arnod FE (1991) Polym Prepr Am Chem Soc Polym Div 1, 32: 636
91. Lee CY-C, Chen YF, Tam LS, Arnold FE (1991) Polym Prepr Am Chem Soc Polym Div 1, 32: 55
92. Tan LS, Arnold FE (1991) Polym Prepr Am Chem Soc Polym Div 1, 32: 51
93. Tan LS, Jones EG, Soloski EJ, Benner CL, Lee CY-C, Arnold FE (1991) Polym Prepr Am Chem Soc Polym Div 1, 32: 53
94. Tan LS, Arnold FE (1987) Polym Prepr Am Chem Soc Polym Div 2, 28: 316
95. Serafini TT, Delvigs P, Lightsey GR (1972) J Appl Polym Sci 16: 905
96. Tan LS, Arnold FE (1988) J Polym Sci Polym Chem 26: 1819

Received March 1993

High Performance Polymer Blends

Michael Jaffe, Paul Chen, Eui-Won Choe, Tai-Shung Chung
and Subhash Makhija
Hoechst Celanese Corporation, 86 Morris Avenue, Summit, New Jersey 07901,
USA

This chapter is an overview of the science and technology of high temperature polymer blends, an area which has become a major topic of scientific investigation in the last two decades. The morphology and, hence, the property spectrum of polymer blends is controlled by the miscibility and the phase behavior of the mixture. Approaches to address the interactions between the morphology and the miscibility in these blends are discussed. The chapter specifically reviews the examples of PBI blends, LCP blends, semi-interpenetrating networks and molecular composites. This review is not meant to be totally comprehensive, but is intended to give the reader an overview of the important technological trends in this emerging field.

Advances in Polymer Science, Vol. 117
© Springer-Verlag Berlin Heidelberg 1994

1 Introduction

Over the past several decades, many synthetic polymers exhibiting high levels of performance in areas such as use temperature, chemical stability and tensile properties have been described [1–3]. The commercial acceptance of these materials, however, has been disappointing. For example, the use of high temperature stable polymers as matrices for structural composites for use at elevated temperatures (T > 200 °C) is highly restricted because of the unresolved paradox between part high use-temperature and part fabrication, i.e., polymers which exhibit acceptable thermal stability tend to be difficult to fabricate, whereas easily processible polymers tend to fall short of performance requirements. The issues inherent in successful material development range from the scientific understanding of the limits of behavior through synthesis, scale-up, characterization and part fabrication to part certification. Exacerbating the problem is the observation that the solutions of different aspects of the problem are often contradictory, e.g., polymers which exhibit long-term thermo-oxidative stability (thousands of hours) are often difficult to fabricate into useful large parts. Finally, the cost commitment necessary for the scale-up of attractive candidates to quantities sufficient for materials screening, detailed characterization, part fabrication and part evaluation is very large, limiting the number of polymers brought forward for consideration by the user community. Polymer blending, especially the blending of polymers available in scale, represents an attractive route to the minimization of these problems and the identification of high performance polymers.

The attractiveness of the blending route lies in the lower costs and faster times associated with blend development when compared to new chemistry. Blending also lends itself to property tailoring and allows for almost infinite possibilities. For the purposes of this discussion, blend and alloy mean the same (a material containing more than one polymer), miscible means single phase, compatible means polyphasic with interaction at the phase boundaries, inter-penetrating network means co-continuous phases (one of which is usually crosslinked) and a "molecular composite" is a miscible blend containing "rod-like" and flexible polymers. The focus of this chapter will be either on blends where both components of the blends are "high performance" or systems where the introduction of a high performance polymer lends high performance character to the blend. The "toughening" of high performance polymers through the introduction of a discreet rubbery phase, analogous to the toughening of engineering resins will not be treated. Specifically, polymer blends comprised of one or more aromatic heterocyclic polymers (polyimides, etc.), and blends containing at least one liquid crystalline polymer component will be discussed.

The examples described in this chapter were chosen to give the reader an overview of the science and technology of polymer blending as well as insights into specific polymer blend systems chosen to correct or tailor specific processing or performance deficiencies.

2 Background

The science and technology of polymer blends has been extensively studied and reviewed [4–7]. The thermodynamics of polymer mixing of polymer 1 and 2 can be described by the Flory-Huggins [8–9] equation,

$$\frac{\Delta Gm}{RT} = \frac{\Phi 1}{M1} \ln \Phi 1 + \frac{\Phi 2}{M2} \ln \Phi 2 + \Phi_1 \Phi_2 \chi$$

where Φ is volume fraction of the polymer, M is weight average molecular weight and χ is an "interaction parameter". Suffixes 1 and 2 refer to two polymers. In most cases, where both components are high molecular weight polymers, the entropy of mixing is negligible (sum of the first two terms), and therefore miscibility is essentially a function of the χ parameter alone and polymers are miscible only in those rare instances when the χ parameter is negative. The above equation assumes that the polymers consist of flexible chains and there are no specific interactions (such as hydrogen bonding) between the polymers. Both these assumptions can be removed by modifying the above equation appropriately. In the case of blends containing copolymers (or more complex chain architectures), Paul, Karasz and others showed that miscibility can be engineered through manipulation of the interchain and intrachain χ_s [10].

It was the insight that miscibility could be engineered through specific interactions such as hydrogen bonding and through use of copolymers that led to the activity of the 1980s to tailor the performance of high performance polymers through the formation of miscible blends. The advantage of miscible systems is that they are characterized by single glass transition temperatures (thermodynamically single homogeneous phases), allowing the modification of processing temperatures. Other properties tend to follow the rule of mixtures type behavior, although both positive and negative synergies are noted. It must be pointed out that true miscible blends (one phase systems) are not always required to achieve desirable properties. Certainly, in some systems what is desired is not molecular miscibility but compatibility. Compatible blends are two phase materials with properties controlled by the properties and geometry of each phase and the nature of the connectivity between phases (compatibilizers modify/improve this interface). However, miscible systems do offer the advantage of tailoring ultimate properties to desired needs. In principle, every miscible blend is bounded by upper (Lower Critical Solution Temperature, LCST) and lower (Upper Critical Solution Temperature, UCST) temperature limits. However, in reality the LCST may be higher than the degradation temperature of the blend and UCST may be lower than the glass transition temperature and hence can not be determined. In principle, miscible blends may be processed as single phase systems and then phase separated in a controlled manner to achieve a desirable morphology. Polymers which can crystallize and/or form mesophases possess additional routes to phase separation, complicating process control, but

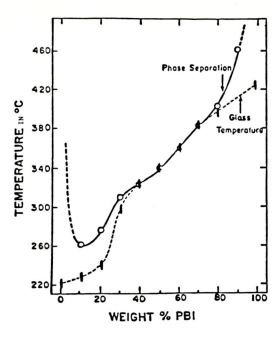

Fig. 1. Phase diagram of PBI/UL-TEM

broadening the extent of morphological richness, hence, property range, achievable with a given blend system. A typical phase separation diagram for a miscible blend containing two amorphous polymers is shown in Fig. 1.

This paper covers the following specific examples:
PBI blends (with polyimides, polyarylates, poly (4-vinyl pyridine));
Semi-interpenetrating Polyimide Network;
LCP blends;
Molecular Composites;

3 Blends of PBI with Polyimides

Poly[2,2'-(m-phenylene-5,5'-benzimidazole)] (PBI) is a very high glass transition temperature (Tg ~ 430 °C), commercially available material. It possesses excellent mechanical properties, but is difficult to process into large parts and has high moisture regain and poor thermo-oxidative stability at temperatures above approximately 260 °C. Polyimides, especially the thermoplastic polyimides, offer attractive thermo-oxidative stability and processibility, but often lack the thermal and mechanical characteristics necessary to perform in applications such as the matrix for high use-temperature (over 300 °C) structural composites (for example, carbon fiber reinforced) for aerospace use. The attempt to mitigate the weaknesses of PBI through blending with a variety of polyimides was investigated by a multisector team involving the University of Massachusetts, Virginia Polytechnic and State University, Lockheed Aeronautical Systems, General Electric Aircraft Engine Business Group and Hoechst Celanese. The breadth of this team allowed the investigation of chosen blend systems from studies of the fundamental phase nature of the blends to part evaluation.

It was found that PBI was miscible with a broad range of polyimides, including ether imides, fluoro-containing imides and others. Two blends were chosen for in-depth investigation and downstream evaluation: an 85/15 blend of PBI/Polyetherimide (PEI) and the 10/90 PBI blend with an experimental copolyimide containing 37.5 mole percentage of 4,4'-hexafluoroisopropylidene-diphthalic anhydride (6FDA). The PEI used in this study was the General Electric Company product known a ULTEM™ 1000. Thus, PEI and ULTEM are used interchangeably in this article. Glass transition temperatures and other key parameters of the blend components are shown in Table 1. The excellent compressive properties conveyed to the blends by PBI is evident and carries forward to the blends' performance as illustrated in Tables 2 and 3. It should also be noted that PEI and 6FCoPI are attacked and softened by common solvents such as methylene chloride and acetone, while PBI and the two blends described appear to be insoluble, even after two-year exposure as thin films.

Solution blends of 20–25% by weight were formed in DMAc, with conventional dry spinning and film casting techniques used to produce blend fiber and film, respectively. Blend powders were prepared by precipitating the dope with a non-solvent (water). All materials were extensively washed in methanol or water to reduce residual solvent to less than 1 wt %. Neat resin tensile bars and plaques were compression molded from both powder and fiber.

Figure 1 shows a typical PBI/polyimide solution blended "phase diagram" after solvent removal. It is clear that, in the absence of solvent, the single phase

Table 1. Structure and properties of candidate high temperature matrix polymers

Properties	PBI	PEI	ec6FCoPI#2
Tg (°C)	420	220	340
Tensile			
Strength (MPa)	100	108	97.2
Elongation (%)	1.8	33	4.4
Modulus (GPa)	5.68	3.18	3.36
Flexural			
Strength (MPa)	100	143	153
Modulus (GPa)	6.32	3.39	3.83
Compressive			
Strength (MPa)	397	–	183
Modulus (GPa)	6.46	–	3.71

Table 2. Mechanical properties of 85/15
PBI/PEI blend

Tensile	
Strength (MPa)	158
Elongation (%)	3.4
Modulus (GPa)	5.34
Flexural	
Strength (MPa)	248
Modulus (GPa)	5.73
Compressive	
Strength (MPa)	300
Modulus (GPa)	5.18

Table 3. Mechanical properties of 10/90
PBI/6FCo PI blend

Tensile	
Strength (MPa)	103
Elongation (%)	4.8
Modulus (GPa)	3.36
Flexural	
Strength (MPa)	156
Modulus (GPa)	3.90
Compressive	
Strength (MPa)	187
Modulus (GPa)	3.81

system is metastable over much of the composition range, although this material shows a single Tg over the entire composition range on a first heating, as shown by dynamic mechanical analysis in Fig. 2.

At both ends of the composition range (i.e., 85/15 PBI/PEI), there are windows where the single phase nature of the material can be utilized in processing and the resulting blends are thermally useful to above 250 °C. It can be shown that the origin of the miscible behavior in this system is a strong hydrogen bond interaction between the imidazole hydrogen and the carbonyl of the polyimide. Figure 3 shows that during the long annealing times above Tg often associated with the molding cycles for these materials, the carbonyl IR absorbance shift induced by the hydrogen bonding disappears. This indicates phase separation has taken place in the blend, confirming the metastable nature of the observed miscibility. Figure 4 shows the weight loss exhibited by the 85/15 PBI/PEI blend during exposure in air at 315 °C.

It is noted that the rate of degradation is faster than that of either component taken separately. The basic degradation products are consistent with published models of PBI degradation [11] with the PBI probably acting as a base to

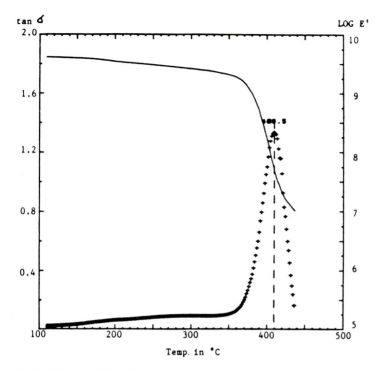

Fig. 2. Glass transition of PBI/ULTEM blend

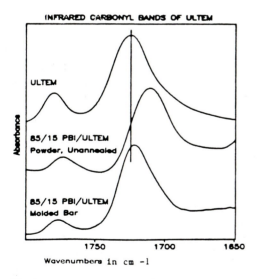

Fig. 3. FTIR evidence for miscibility

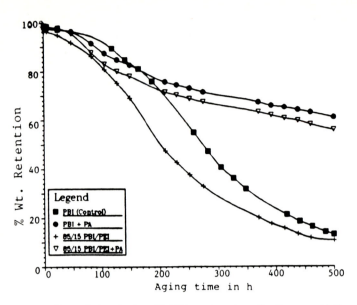

Fig. 4. Effects of phosphoric acid (PA) treatment on the thermo-oxidative stability of PBI and PBI/PEI blend films aged in air at 315 °C

accelerate the degradation of the PEI. It is clear that the overall rate of degradation is exacerbated by the strong intimate interaction present within the blend. Treatment shown to improve the thermo-oxidative stability of PBI, i.e., protonation of the imidazole ring by, e.g., phosphoric acid, will also improve the thermo-oxidative performance of the blend (see Fig. 4). Table 2 summarizes the mechanical properties of the PBI/PEI blend, showing excellent tensile and compressive performance. The total evaluation of the PBI/PEI system suggests it to be a viable candidate for matrix use up to about 260 °C.

Blend systems with significantly improved thermo-oxidative performance can be achieved through incorporation of carefully designed polyimide molecules. As shown in Table 1, a copolyimide containing the sulfone and 6F moieties which exhibits a Tg above 300 °C (see Fig. 5), as well as extraordinary short-term thermo-oxidative stability can be synthesized.

The approximately 100 °C increase in Tg in comparison to the PEI, allows the blend composition to be shifted to 90% polyimide, keeping a blend Tg in the range of 350 °C, but now showing a greatly reduced thermal-oxidative weight loss after a 1000 h exposure to air at 315 °C, as illustrated in Fig. 6. The excellent mechanical properties listed in Table 3 indicate that the 10/90 PBI/6FCoPI is a matrix candidate for use at temperatures up to about 315 °C.

It has been shown that miscible polymer blends and copolymerization offer complementary routes to polymer systems of tailored properties. The recognition that miscibility (at least in a transient sense) is much more common with aromatic heterocyclic polymers than is observed with low temperature flexible

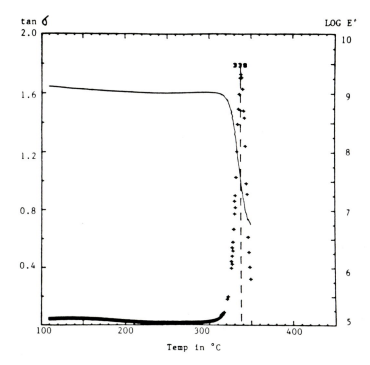

Fig. 5. Glass transition of 6FCoPI#2

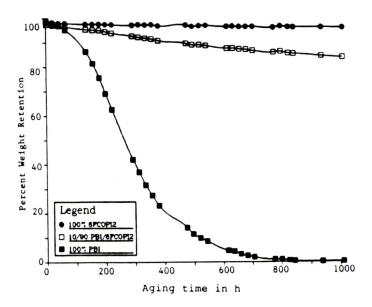

Fig. 6. Thermo-oxidative stability of end-capped PBI and 6FCoPI#2 blend films aged in air at 315 °C

polymers opens a broad class of materials to investigation and eventual commercial development. For example, in addition to the PBI/PI blends described here, similar miscibility has been reported in PI/PI blends [12] PES (polyethersulfone)/PI blends. In most cases, the heightened polymer/polymer interactions can be attributed to strong, specific interactions between the polymers being blended, with the reduced backbone flexibility associated with these polymers also playing a contributory role. Copolymerization allows for heightened miscibility and the control of polymer cost-performance. Identification of the optimum solution to a given problem, i.e., blending, copolymerization or a combination, can only be determined after analysis of the required processibility/performance profile. It is also evident that the small scale evaluation described here can only identify materials of promise, and that the ultimate evaluation to show true end-use utility and cost-effectiveness can only be achieved with much larger quantities of material, tested in specimen or final part configurations in a manner meaningful to both the materials user and the materials supplier.

4 Polybenzimidazole/Polyarylate (PBI/PAr) Blends

Polyarylates (PAr) are wholly aromatic polyesters derived from aromatic dicarboxylic acids and diphenols or their derivatives. They are amorphous in nature with good injection moldability. Figure 7 shows the typical formula structure of PAr.

PAr is soluble in similar polar organic solvents (e.g., NMP, DMAc, DMSO, etc.) which dissolve PBI. It was observed that miscible solution blends of PBI and PAr could be formed. For example, NMP dopes containing 10 wt % PBI and PAr are visually homogeneous and contain no insolubles as formed. After being kept at room temperature for a period of time (e.g., several days), a PBI-rich phase starts to form precipitate, but this polyphasic material can be easily redissolved into a single phase with a mild heating (i.e., 100 °C for 20 min). Based on the haze level, the stability of the PBI/PAr/NMP solutions appeared to increase with the increase of the relative PAr concentrations.

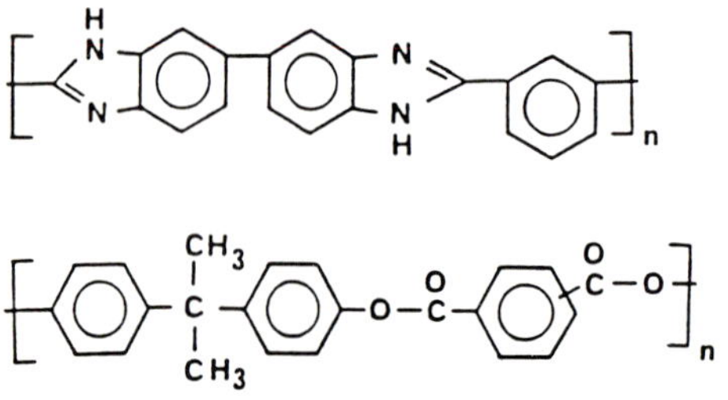

Fig. 7. Chemical structures of both PBI and polyacrylate

Figure 8 illustrates the relationship between inherent viscosity (IV) and concentration for PBI/PAr/NMP solutions. It is interesting to note that the IV of all solution blends exhibited normal polymer solution characteristics. At a fixed concentration (0.5%), it was noted that the IV of the solution blends exceeded the rule of mixtures (see Fig. 9) suggesting that PBI and PAr exhibit specific interactions in a dilute solution, such that the resulting hydrodynamic sizes of the blends were greater than that of the calculated averages based on each component.

Corroborating evidence of a PBI-PAr interaction was observed by FT-IR. Based on the carbonyl stretching of a pure PAr film and an 80/20 PBI/PAr film blend, it was found that the signal of the blend showed a dramatic downfield shift (i.e., from 1741 to 1730 cm^{-1}). This shift (see Fig. 10) indicated the existence of intermolecular H-bonding between PBI and PAr in the film blend.

It is known that PAr is soluble in common organic solvents, such as methylene chloride and tetrahydrofuran, whereas, PBI is insoluble in these solvents. After blending PAr with PBI, it was observed that the solvent resistance of PAr is significantly improved. For example, a PAr/PBI film blend having PBI as the minor (20%) component kept its physical integrity after soaking in methylene chloride for 30 min, whereas a pure PAr film would completely dissolve within 10 s. These experiments reconfirm that intermolecular interaction between PBI and PAr changes the nature of PAr to some degree, thereby improving its chemical resistance.

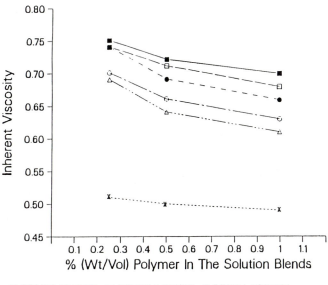

■ PBI/PA 1100/01; □ PBI/PA 180/201; ● PBI/PA 160/401;
○ PBI/PA 140/601; △ PBI/PA 120/801; × PBI/PA 10/1001;

Fig. 8. Correlations between the IV and the concentration of PBI/PA/NMP solution

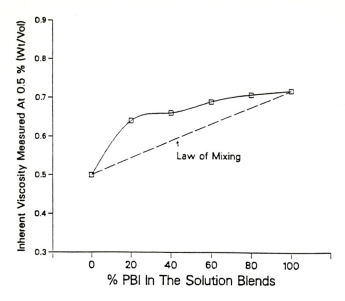

Fig. 9. Relationship between the blend IV and composition

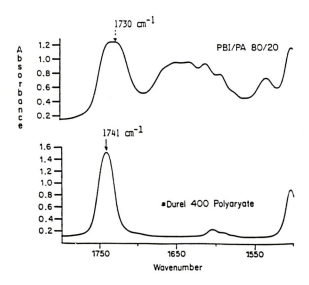

Fig. 10. FT-IR confirmation of the existing intermolecular hydrogen-bonding between PBI and PA

Mechanical properties of the blends were investigated in fiber form. The as-spun fiber tensile properties are summarized as shown in Table 4. These fibers were further drawn at elevated temperatures at various draw ratios. These drawn fibers were stronger than the as-spun fibers. Table 5 illustrates the effects of temperature and draw-ratio on the fiber properties. Drawing at 400 °C at a ratio

Table 4. Tensile properties of as-spun 80/20 PBI/PAr fibers

Sample	Denier (DPF)	Initial modulus (G/D)	Tenacity (G/D)	Elongation (%)
PBI/PA	5.271	42.6	1.53	87.3

Table 5. Properties of the (80/20) PBI/PAr fibers

Denier (DPF)	Draw Ratio	Temperature (°C)	Modulus (G/D)	Tenacity (G/D)	Elongation (%)
1.919	2.0	400	77.69	3.118	29.50
1.915	2.5	400	91.86	3.313	17.92
1.501	3.0	400	141.15	4.611	6.75
1.668	2.0	420	66.50	2.922	33.75
1.453	3.0	420	80.74	3.378	19.75
0.951	4.0	420	113.86	4.014	11.61

Table 6. Tensile properties of hot-drawn 80/20 PBI/PAr, 75/25 PBI/Ultem, and PBI fibers

Material	Denier (DPF)	Draw Ratio	Temperature (°C)	Modulus (G/D)	Tenacity (G/D)	Elongation (%)
80/20 PBI/PA	1.501	3.0	400	141.2	4.611	6.75
75/25	0.96	3.0	420	112.0	4.32	9.9
PBI/Ultem	1.700	2.0	440	91.0	4.07	31.9
PBI Control						

of 3:1 gave the best tensile modulus and strength. Table 6 provides a comparison among PBI/PAr and that of PBI/PEI and PBI fibers. This comparison is based on the highest modulus and tenacity of PBI/PAr and PBI/PEI obtained from Table 6 and from the literature [12]. Dry-spun PBI/PAr has very impressive tensile modulus and strength. The properties are at least comparable, or superior, to those of PBI/PEI fibers. Compared to standard PBI fibers, the PBI/PAr blend fiber has higher tensile moduli and strengths than those of PBI fibers, whereas, the elongations at break of the former are inferior to those of the latter. These properties suggest that the PBI/PAr fibers may be suitable for engineering and aerospace applications.

5 Polybenzimidazole and Fluoro-Containing Polyamideimide Blends

Polyamideimides (PAI) are generally prepared by the condensation polymerization of a trifunctional acid anhydride (e.g., trimellitic anhydride, TMA) with an aromatic diamine (e.g., 4,4'-methylene- or 4,4'-oxydianiline, MDA and

ODA). These polymers are characterized by excellent high temperature properties with Tgs typically above 270 °C and continuous service temperatures of
about 230 °C. The PAIs utilized here for blending studies were prepared by a
simple solution polymerization route, i.e., by reacting trimellitic anhydride acid
chloride and 6FDA and diamine monomer (ODA and MDA) in an appropriate
solvent (e.g., DMAc).

Using dissolution techniques, it was observed that the 10 and 25% 6F-PAI
(ODA-based) could be co-dissolved with PBI using DMAc. Solution blends with
PBI-PAI ratios ranging from 20/80 to 80/20, in 20% increments, were visually
homogeneous and contained no insoluble materials. At 15–20% solids concentration, these blends were processible and showed no sign of polymer precipitation for at least 24 h. Transparent, apparently miscible, blend films were
cast from the solution blends of PBI and 10% 6F-PAI.

Mechanical property studies of the 10% 6F-PAI and PBI blend films further
confirmed the existence of synergistic miscibility between the blend component
since the tensile modulus and strength of each corresponding file was either
equivalent to or exceeded the simple rule of mixtures (Fig. 11).

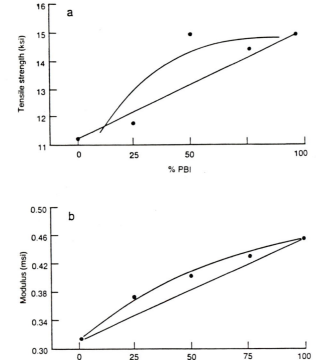

Fig. 11a. Tensile strength of 10% 6F-PAI/PBI blend film. **b** Modulus of 10% 6F-PAI/PBI blend
film

6 Fluoro- Containing Polyimide Blends

Because of their high inherent thermal stability and excellent electrical properties, fluoro-containing aromatic polyimides have been identified to have excellent potential for use in aerospace composites, aircraft wire and cable,
adhesives, and flexible wire boards. A variety of these polymers has been
synthesized and their composite properties have been reported by NASA
[13–15] and others [16–19]. Polymer blends of amides with imides or imides
with other conventional polymers have also received attention. For example,
MacKnight and Karasz and their coworkers [17–19] and Jaffe [20], have
reviewed and investigated polyimide blends as well as PBI/polyimide blends.
Some of the polyimides investigated contain fluoro-containing aromatic repeat
units. Yokota et al [21] reported their work on rigid rod polyimide blends. Yoon
et al. [22] have analyzed "molecular composite" films made from rigid and
flexible polyimides.

Recently, a series of hexafluoro-isopropylidene-containing polyimides have
become available, providing an opportunity to investigate the miscibility of pairs
of closely related polyimides. The miscibility criteria of these fluoro-polyimides
with one another and the use of the Flory-Huggins theory to predict the
miscibility in these systems will be summarized.

A series of hexafluoro-isopropylidene-containing polyimides were synthesized using solution polycondensation reactions and then chemically imidized
[23–25]. Table 7 shows their chemical compositions; Fig. 12 shows the monomer structures. The glass transition temperatures, Mw and Mn, of these
polyimides are summarized in Table 8. A Perkin-Elmer DSC was used to
determine these Tgs with a heating rate of 20 °C/min. All Tgs were measured at
the second heating.

Table 7. Compositions of 6F-Containing Polyimides Composition

Polymer	Dianhydride	Diamine	Molar Ratio[a]	Tg (°C)
6F1P	6FDA	4,4′ 6F Diamine	50/50	250.5
6F2P	BPDA	4,4′ 6F Diamine	50/50	267.0
6F3P	BTDA	4,4′ 6F Diamine	50/50	239.0
6F4P	ODPA	4,4′ 6F Diamine	50/50	224.5
6FC2P	PMDA + BPDA	4,4′ 6F Diamine	25/25/50	280.0
6FC3P	PMDA + BTDA	4,4′ 6F Diamine	25/25/50	261.0
6FC4P	PMDA + ODPA	4,4′ 6F Diamine	25/25/50	233.0
6F1M	6FDA	3,3′ 6F Diamine	50/50	318.5
6F2M	BPDA	3,3′ 6F Diamine	50/50	343.0
6F3M	BTDA	3,3′ 6F Diamine	50/50	304.0
6F4M	ODPA	3,3′ 6F Diamine	50/50	305.0
6FC2M	PMDA + BPDA	3,3′ 6F Diamine	25/25/50	375.0
6FC3M	PMDA + BTDA	3,3′ 6F Diamine	25/25/50	349.5
6FC4M	PMDA + ODPA	3,3′ 6F Diamine	25/25/50	347.0

[a] Dianhydrides/diamine ratio

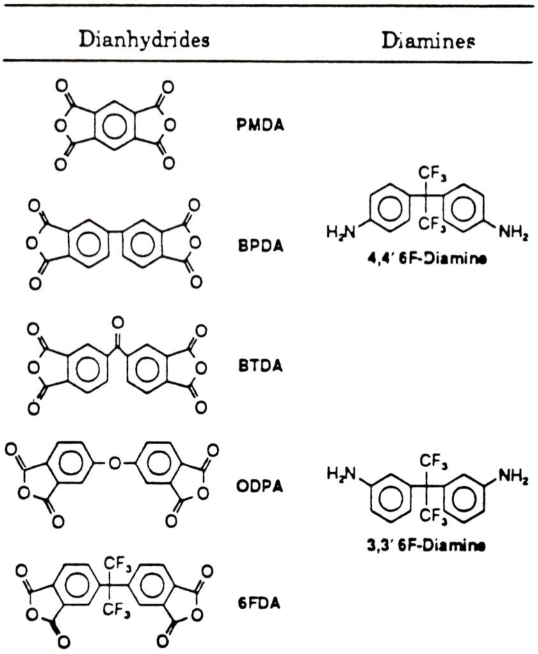

Dianhydrides		Diamines

PMDA

BPDA

4,4′ 6F-Diamine

BTDA

ODPA

3,3′ 6F-Diamine

6FDA

Fig. 12. Chemical structures of monomers for 6F-containing polyimides

Table 8. Mw, Mn, and Tg of 6F polyimides

Polymer	Mw	Mn	Mw/Mn	Tg(°C)
6F1M	105,300	58,500	1.8	250.5
6F2M	83,500	32,300	2.6	267.0
6F3M	63,000	37,700	1.7	239.0
6F4M	93,100	42,500	2.2	224.5
6FC2M	193,700	90,700	1.8	280.0
6FC3M	105,300	58,000	1.8	261.0
6FC4M	22,700	12,600	1.8	233.0
6F1P	194,900	69,300	2.8	318.5
6F2P	107,000	45,530	2.0	343.0
6F3P	55,000	32,700	1.7	304.0
6F4P	117,000	46,750	2.5	305.0
6FC2P	100,200	51,000	2.0	375.0
6FC3P	113,700	50,150	2.3	349.5
6FC4P	114,000	46,700	2.5	347.0

All polyimide alloys were solution blended at a 50/50 weight ratio by co-dissolving the polymer pairs in a common solvent, such as methylene chloride ($MeCl_2$) or a mixture of $MeCl_2$ and hexafluoroisopropanol (HFIP), co-precipitating in methanol and then drying overnight under vacuum at 100 °C.

7 Calculation of Parameters for the Flory-Huggins Theory

The Flory-Huggins theory of polymer solutions has been documented elsewhere [26, 27]. The basic parameters necessary to predict polymer miscibility are the solubility parameter δ, the interaction parameter χ, and the critical interaction parameter $(\chi)_{cr}$.

The solubility parameter is calculated from

$$\delta = \rho\Sigma F_i/M \tag{1}$$

where ρ is the density, M is the molecular weight, and Fi is the molar attraction constants of all the chemical groups in the polymer repeat unit. For simplicity of calculation, it was assumed that the densities for these polymers were the same as that of 6FIP (Table 7), which was 1.47 g/cc and Hoy's table was used to calculate and predict Fi for each chemical groups [25, 26, 28]. The unit for the molar attraction constant is $(cal/cc)^{0.5}$ mole.

The interaction parameter χ at 25 °C is defined as

$$\chi = (\delta_A - \delta_B)^2/6 \tag{2}$$

where δ_A and δ_B are solubility parameters of the polymers in each pair and their units are in $(cal/cc)^{0.5}$.

$(\chi)_{cr}$, the value for χ necessary to achieve system miscibility, is a function of the degree of polymerization, X_a, or the molecular weight, Mn, of a polymer. Since a precise measurement of X_a is not straightforward, Krause [26] proposed an approximate equation:

$$X_a = MN/100 \tag{3}$$

Based on X_a values, Krause [26] provided a table to predict χ_{cr} of two blending polymers having the same degree of polymerization.

Table 9 summarizes the calculations for χ from seven pairs of polyimide blends ranking them in order of increasing χ. From this order, some interesting observations could be made. For example, 6F1M/6F1P had the lowest calculated χ, while the polymers with rigid dianhydrides, such as PMDA or BPDA,

Table 9. Calculation of interaction parameter c

Polymer blend	δ difference	χ	Order[a]
6F1M/6F1P	0.1335	0.00297	1
6F2M/6F2P	0.1811	0.00546	7
6F3M/6F3P	0.1559	0.00405	2
6F4M/6F4P	0.1589	0.00421	3
6FC2M/6FC2P	0.1765	0.00518	6
6FC3M/6FC3P	0.1721	0.00494	4
6FC4M/6FC4P	0.1740	0.00504	5

[a] Based on χ values. The lower the χ, the lower the order

generally resulted in blends with a greater calculated χ. Furthermore, the addition of PMDA in polyimides, such as 6FC3M/6FC3P and 6FC4M/6FC4P yielded calculated χs between the more rigid pair (6F2M/6F2P) and the more flexible pair (6F1M/6F1P and 6FC2M/6FC2P). A low calculated χ implies that the effect on the solubility parameter of changing from the meta substituted 6F-diamine to the para substituted 6F-diamine is very small. A high calculated χ may indicate that when mixing two polymers which both have rigid and flexible monomers in their repeat units, homogeneity on the molecular level is relatively difficult to achieve.

According to gel permeation chromatographic (size exclusion) (GPC) data shown in Table 8, these polymers have relatively low Mn (about 30,000–70,000), with the degree of polymerization (DP) varying from about 50 to 200. This range of DPs is within the limits of the assumptions used by Krause defining the approximate levels of χ_{cr} necessary for blend miscibility [24]. Therefore, we can anticipate that these seven blends should be miscible when χ is smaller than χ_{cr} which is about 0.01 as predicted by Krause [26].

Table 10 summarizes the glass transition behavior of these polyimide blends and demonstrates that there is only one Tg for each blend. Similar results have been confirmed by Koros of the University of Texas [29]. This data confirms that as long as the dianhydride is the same in the composition, the change of 6F diamine from 3,3' to 4,4' does not alter solubility significantly, and the pairs are miscible. The relationship between the Tg and the Fox equation is discussed by MacKnight et al. work [17].

It has been shown that all seven of these pairs are miscible. However, based on trends of the calculated χ and Tg deviations defined above, it is implied that the strength of the molecular interaction resulting in miscibility among these pairs follows the order:

$$6F1M/6F1P \gg 6F3M/6F3P > \; = 6FC3M/6FC3P > \; = 6F4M/6F4P >$$

$$6F2M/6F2P > \; = 6FC4M/6FC4P \gg 6FC2M/6FC2P,$$

Table 10. Tgs of 6F/6F polyimide blends (50%/50%)

Polymer Blend	Tg1/Tg2[a] (°C)	Tg of 50/50[b] (°C)	Fox Eq.[c] (°C)	Temp. Diff.[d] (°C)	Order[e]
6F1M/6F1P	250.5/318.5	275.0	282.4	7.4	1
6F2P/6F2P	267.0/343.0	279.3	302.5	23.1	4
6F3P/6F3P	239.0/304.0	252.8	269.6	16.7	2
6F4P/6F4P	224.5/305.0	242.7	261.7	19.0	3
6FC2M/6FC2P	280.0/375.0	295.5	323.7	28.2	6
6FC3M/6FC3P	261.0/349.5	273.4	301.8	28.5	7
6FC4M/6FC4P	233.0/347	258.4	284.2	25.8	5

[a] Tg of individual polymer in a blend
[b] Measured Tg of a blend
[c] Calculated Tg from the Fox Equation
[d] Difference between the measured Tg and the calculated
[e] Based on temperature difference

An important insight into general rules for the blending of hexafluoroisopropylidene containing polyimides pairs is obtained, namely that these polyimides
are miscible with each other if, first, the dianhydride composition is the same in
each pair, and, second, the diamine is changed from the 3,3' to the 4,4' isomer.
This conclusion has been confirmed by DSC measurements showing a single Tg
for each blend pair. In addition, these results are consistent with calculation
based on the Flory-Huggins theory.

8 Blends of PBI and Poly(4-Vinyl Pyridine) (PVPy)

Poly [2,2'-(m-phenylene)-5,5'-bibenzimidazole] is also miscible with poly (4-
vinyl pyridine) (PVPy) over the entire composition range [30]. These blends
show single Tgs intermediate between the two pure polymers. Some low
molecular weight PBI blends with PVPy phase separates at temperatures above
their Tgs. Blends with high molecular weight PBI degrade at temperatures below
their phase separation temperature. The thermal stability of these blends is
similar to the stability of pure PVPy, and hence these blends cannot be
considered truly high performance blends. However, it should be pointed out
that in this study, using FTIR, it was shown that the -NH group of PBI forms a
hydrogen bond with the N of PVPy. Therefore, it opens up new avenues to form
miscible blends of polymers containing benzimidazole units with other high
temperature polymers containing pyridinium units or other N-containing
heterocylics in their backbone.

9 Semi-Interpenetrating Polyimide Network

Extensive research has been conducted and reviewed on interpenetrating
polymer networks during the past decade (31–33). IPNs can be categorized into
Semi-1-IPN and Semi-2-IPN. Semi-2-IPNs are those in which crosslinkable
oligomers or materials are polymerized into the linear matrix. Semi-interpenetrating networks could have combined polymers that have same or different
repeating units.
    Certain thermoplastic polyimides possess excellent resistances to high tem-
peratures and chemicals, with Tgs ranging from 217 to 371 °C. Certain polyimides also exhibit excellent toughness and dielectric properties. The melt
blending process of polyimides with other thermoplastic polymers is difficult due
to polyimides' high Tg, high melt viscosities, and incompatibility. A solution
process is used, therefore, to achieve a semi-interpenetrating polyimide network

(IPN) for developing tough and micro-cracking resistant high temperature matrix resins for the demanding applications in the aerospace, marine, and electrical/electronics industries. Solvents are removed by evaporation or freeze-drying. High performance semi-interpenetrating polyimide networks with easy processibility can be obtained by combining readily processable addition type brittle thermosetting polyimides with tough thermoplastic polyimides that are difficult to process. The addition type brittle thermosetting polyimides contain a reactive functionality such as acetylene [34–37], maleimide [36, 38–40] or epoxy [41].

St. Clair [36] synthesized and evaluated the properties of a semi-2-IPN comprising Thermid 600 (an acetylene terminated imide oligomer from National Starch and Chemical Company) and LaRC-TPI (a thermoplastic polyimide with a Tg of $\sim 257\,°C$). The composition having the ratio of 65:35 of thermosetting Thermid 600 to LaRC-TPI showed the best flexural strength at ambient temperature [36].

Semi-2-IPN has also been made using two polymers with identical polymer repeating units, polyimidesulfone ($PISO_2$) a linear thermoplastic polymer and acetylene-terminated polyimidesulfone ($ATPISO_2$) [42]. The advantages of using the identical functional units are mutual solubility and formation of homogenous phase. In fact, a single sharp damping peak was observed in the torsional analysis of the semi-2-IPN, eliminating phase separation problems usually encountered with IPNs. The semi-2-IPn showed improved strength, adhesion, thermo-oxidative stability, and processibility over $PISO_2$.

Pater [43] prepared a semi-2-IPN (LaRC-RP71) by imidizing and crosslinking a mixture having the ratio of 80:20 of acetylene terminated polyamic acid (Thermid LR-600) to linear LaRC-TPI polyamic acid [43]. The imidization and crosslinking of Thermid LR-600 took place in the presence of and during the imidization of LaCR-TPI polyamic acid. The curing profile for the neat resin was 1 h each at 250 and 316 °C at 2000 psi, and the neat resins were post-cured for 16 h at 316 °C in air. A dramatic increase (333%) in resin fracture toughness and some improvements in composite microcracking resistance compared to Thermid 600 neat resin and composite were obtained. Unidirectional composites from graphite fiber tows and these semi-2-IPNs or thermosetting resins, AS-4/LaRC-RP71 or AS4/Thermid-600, were prepared by compression molding using the following cure conditions: 135 °C/100 psi/1 h, 218 °C/100 psi/1 h, 316 °C/500 psi/1 h, and 350 °C/500 psi/1 h. The resulting composites were post-cured for 16 h at 316 °C in air. The composite from LaRC-TPI was prepared by applying 1000 psi pressure at 232 °C and then raising the cure temperature to 316 °C for 1 h and finally to 350 °C for 30 min. Mechanical properties of neat resins and unidirectional composites are shown in Tables 11 and 12 respectively.

Pater also prepared a semi-2-IPN, LaRC-RP41, from a mixture having the ratio of 80:20 of thermosetting PMR-15 to LaRC-TPI polyamic acid(44). LaRC-RP41 exhibited significantly improved toughness and micro-cracking resistance, but showed somewhat poorer mechanical properties at 316 °C, compared to those of PMR 15 near resin (see Fig. 13).

Table 11. Neat resin properties

Properties	LaRC-71	Thermid 600	LaRC-RP40	LaRC-RP41	PMR 15	LaRC-TPI
T_g, °C Dry	250,295	310	348,381	261,325	327	257
Wet	244,295	301	349,381	257,310	–	–
Aged[a]	261,312	–		335	–	–
Toughness, G_{1c}.J/m^2	368	85	368	476	87	1770
Moisture absorption, %	0.3	0.3	1.0	1.0	1.6	1.1
Temperature at 5% Wt. loss by TGA in Air °C	490	487	463			495
Isothermal wt. loss, %	14[c]	18[c]	6[b]	10.6[b]	8[b]	15[xxx]

[a] 1000 h at 316 °C in air
[b] at 316 °C for 1000 h in air
[c] 50 h at 371 °C in air by TGA

Table 12. Unidirectional Celion 6000 composite properties

Property	LaRC-71	Thermid 600	LaRC-RP40	LaRC-RP41	PMR 15	LaRC-TPI
T_g, °C	256,305	323	348,381	315	338	253
Fiber/Resin/Void, Volume %	59/37/4	58/35/7		59/38/3	60/39/1	58/33/9
Flexural Strength, Mpa						
25 °C	1303	1740	1840	1585	1846	972
316 °C	317	875	1199	218	1096	108
Flexural Modulus, Gpa						
25 °C	84	123	152.3	129	114	66
316 °C	25	99	139.2	54	91	19
Interlaminar shear strength, MPa						
25 °C	46	66	97.8	81	110	64
316 °C	22	31	47.5	50	55	15

Pater also prepared another semi-2-IPN, LaRC-RP40, from a mixture having the ratio of 80:20 of thermosetting PMR-15 to NR 150B2 polyimide precursor (a 6FDA-based material from DuPont) (Fig. 14) [45–46]. LaRC-RP40 blend molding powder was prepared at 80:20 weight percent of PMR-15 prepolymer to NR-150B2 polyamic acid [48]. A brown molding powder was obtained after concentrating the solution and then drying. The neat resin was cured as described above for LaRC-TPI. The LaRC-RP40-based IPN exhibited a significantly improved toughness and micro-cracking resistance compared to those of PMR 15 neat resin.

Pascal et al. [49–50] studied semi-2-IPN derived from linear thermoplastic polyimides and thermosetting bismaleimides from the same type of starting

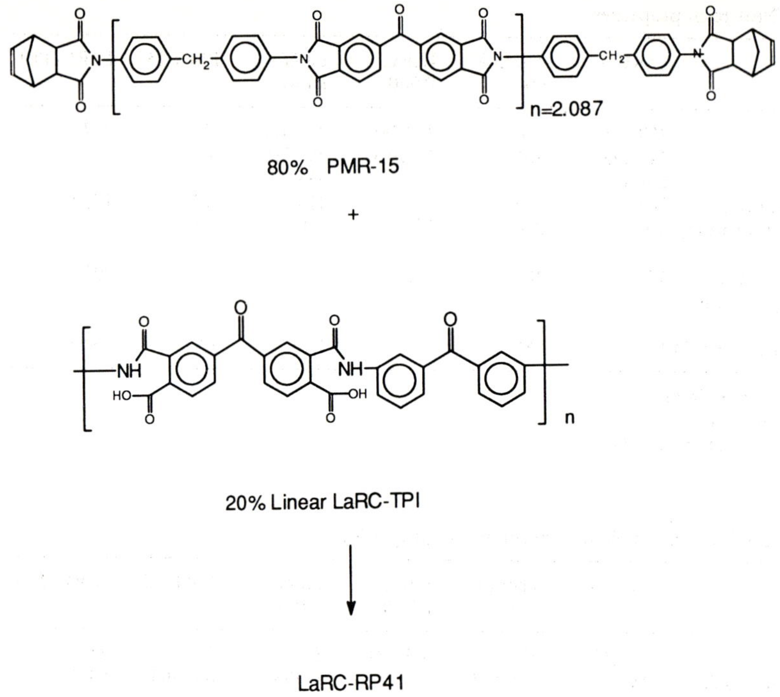

80% PMR-15

+

20% Linear LaRC-TPI

LaRC-RP41

Fig. 13. Synthesis of semi 2-IPN-LaRC-RP41

materials. Two diamines were 3,3′-[2,6-pyridinediylbis(oxy)]bibenzeneamine
(BAPPY) and 3,3′-[1,3-phenylene bis(oxy)]bisbenzeneamine (BAPB). The corresponding bismaleimides BAPPY-BMI and BAPB-BMI were synthesized and
blended with the corresponding polyimide in NMP. Two linear polyimides
BAPPY-BTDA and BAPB-BTDA were prepared by polycondensation between
these two amines and 3,3′,4,4′-benzophenonetetracarboxylic dianhydride,
BTDA [49–50]. TGA showed polyimide BAPPY-BTDA containing a pyridine
ring has lower thermal stability than the polyimide with only an aromatic ring.
The glass-transition temperatures were below 200 °C. Three types of mixtures at
varying ratios of bismaleimide thermosetting resin to linear polyimides, BAPPY-
BTDA/PPY-BMI, BAPPY-BTDA/MDA-BMI and BAPB-BTDA/PB-BMI,
were cured to obtain three IPNs. These IPNs showed a phase separation
between the linear polyimide and the crosslinked bismaleimide.

A freeze-drying method [51] for the preparation of high performance
interpenetrating polyimide networks has been developed recently. This method
provides controlled morphology of a high performance semi-interpenetrating
polymer network, and is a means of improving the physical and mechanical
properties of the neat resin and composite. A semi-IPN was prepared by
combining PMR polyimide system known as LaRC-RP46 with commercially
available NR-150B2, a linear thermoplastic polyimide [Fig. 15]. The physical
and mechanical properties and phase morphology of the neat resins and Celion

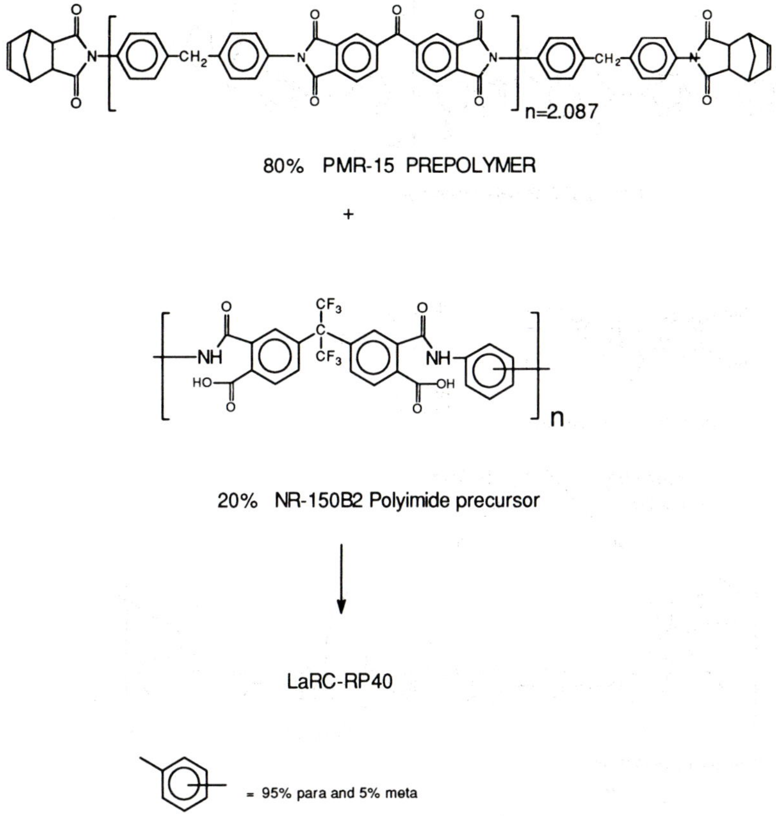

80% PMR-15 PREPOLYMER

+

20% NR-150B2 Polyimide precursor

LaRC-RP40

= 95% para and 5% meta

Fig. 14. Synthesis of semi 2-IPN-LaRC-RP40

6000 graphite fiber reinforced composites prepared by the freeze-drying conventional solution methods are compared in Table 13. The freeze-dry-processed neat resin and composites exhibited higher Tgs and a lesser extent of minor constituents phase separation than for the solution processed materials. The removal of solvent by polymer solution solidification and sublimation greatly retards constituents separation. There is no solvent present at higher temperature to plasticize the resin or produce voids. Thus, the Semi-2-IPNs prepared by the freeze-dry-process produces superior properties in both resin and composites to those made from the conventional solution process.

Other semi-2-IPNs [52] processed by the freeze-drying method included IPN from 4,4′-bismaleimido diphenylmethane (BMI) and linear BTDA/3,4′-ODA polyamic acid that were dissolved in 1,3,5-trioxane (Fig. 16). The resulting semi-2-IPNs exhibited higher Tgs and reduced phase separation, and contained no plasticizing solvent. A comparison of unidirectional properties of composites prepared by the freeze-dry process to those by traditional solvent evaporation process is presented in Table 13. The freeze-drying method for the preparation of IPNs appears to be superior to previous technology.

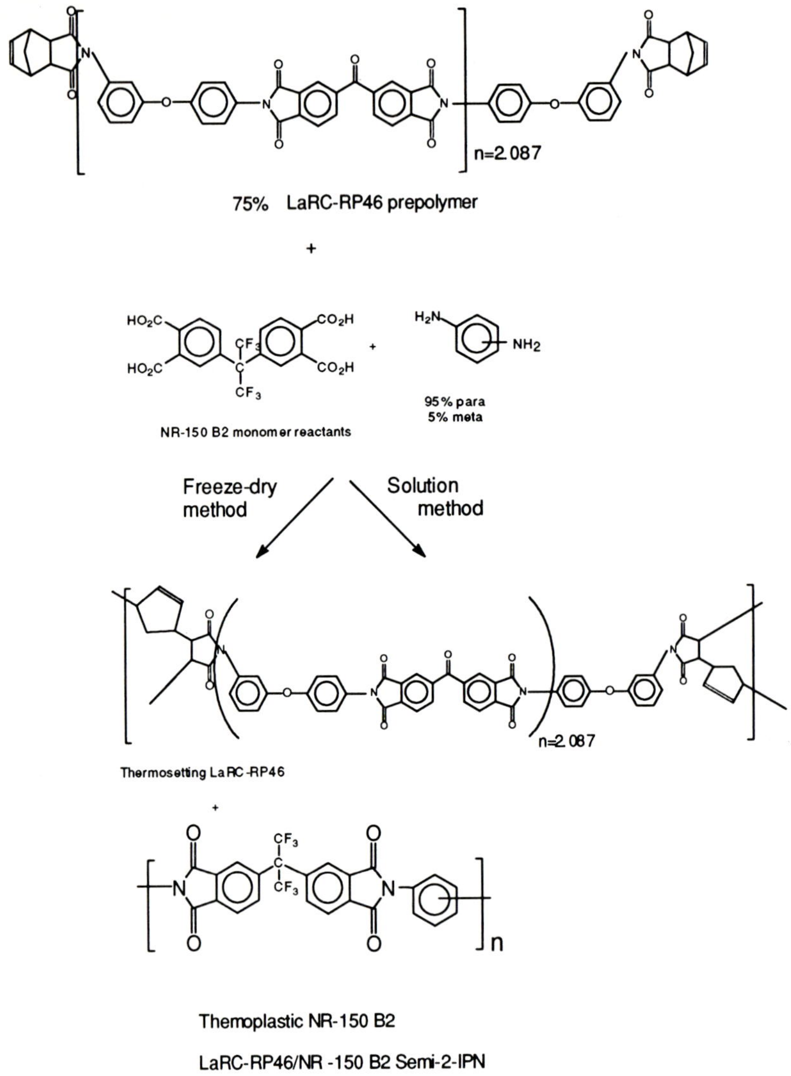
n=2.087
75% LaRC-RP46 prepolymer
+
HO2C
HO2C
CF3
CF3
CO2H
CO2H
+
H2N
NH2
95% para
5% meta
NR-150 B2 monomer reactants
Freeze-dry method
Solution method
n=2.087
Thermosetting LaRC-RP46
+
CF3
CF3
n
Themoplastic NR-150 B2
LaRC-RP46/NR-150 B2 Semi-2-IPN

Fig. 15. Synthesis of semi 2-IPN-LaRC-RP46

10 Multicomponent System containing Liquid Crystal Polymers (LCPs)

The motivations for blending LCPs with conventional polymers or with other LCPs are the same elements which make blending an attractive polymer modification option. These include cost-reduction, property tailoring, accelera-

Table 13. Unidirectional composite properties

Property	LaRC-RP46/NR-150B2		MDI BMI/BTDA/3,4'-ODA		(BTDA/3,4'-ODA) BMI/BTDA/3,4'ODA	
Method	Freeze-Dry	Solution	FD	Solution	FD	Solution
Tg, °C	301	293	293	285	292	255
Fiber/Resin/Void, Volume %	44/56/0.3	39/60/0.9				
C-scan	excellent	excellent				
Flexural Strength, Mpa						
25 °C	710	731	462	237	407	179
232 °C			358	145	283	166
316 °C	524	365				
Flexural Modulus, Gpa						
25 °C	51	46	62	51	24	19
232 °C			70	49	17	35
316 °C	35	14				
Interlaminar shear strength, MPa	–					
Room T			19	18	43	24
232 °C			20	11	33	26
316 °C	33					
Weight loss, %	2.2	2.0				

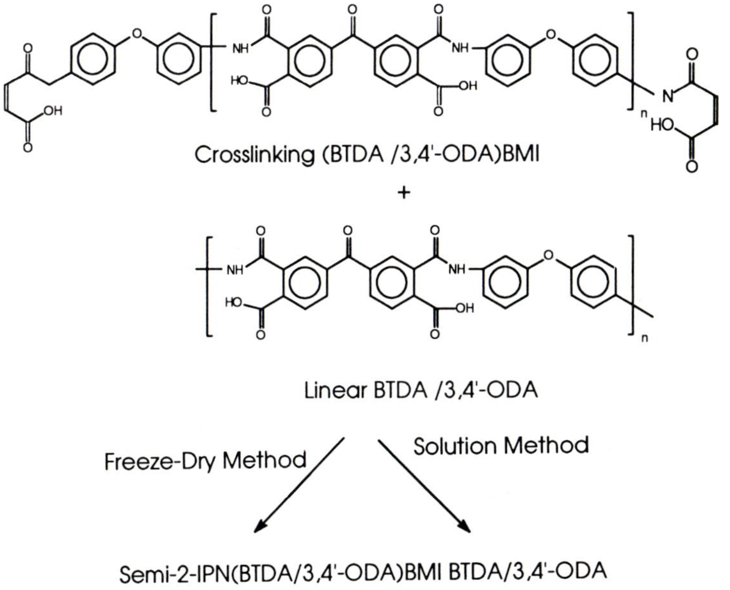

Crosslinking (BTDA /3,4'-ODA)BMI

+

Linear BTDA /3,4'-ODA

Freeze-Dry Method   Solution Method

Semi-2-IPN(BTDA/3,4'-ODA)BMI BTDA/3,4'-ODA

Fig. 16. Synthesis of semi-2-IPN(BTDA/3,4'-ODA)BMI BTDA/3,4'-ODA

ted new product development, and improved processibility. The cost-reduction objective is to provide an LCP-like property set at conventional property-like prices. Property tailoring is attractive from two points of view: First, with conventional polymers, LCPs can function as high modulus fibrous reinforcement and improved processibility and, second, with other LCPs, or at relatively low levels of conventional polymer addition, the objective is to mitigate LCP problems such as poor weld-line strength or high property anisotropy. Improved processibility focuses on utilizing the low viscosity of the LCP to improve the processibility of highly viscous conventional resins [53–56]. Finally, the blending of LCPs with other LCPs provides useful data for studying the nature of the structure, morphology, and chain/chain interaction of LCPs as new materials, while offering the opportunity of improved property sets.

11 LCP/Conventional Polymer Blends

Most of the work to date concerns the area with the greatest potential for commercial exploitation, the blending of LCPs with conventional polymers. While a few studies of solution blending with Kevlar do exist [57–61], most of the work has centered on melt blending thermotropic copolyesters (Vectra, Xydar) with engineering thermoplastics (PET, PC, PEI, etc.). For convenience, this work may be separated into three blend regions based on LCP content, namely:

Weight % LCP	General description	Key references
0–15	Processing Aid, Viscosity Reduction	Nobile et al. [62], Cogswell [63–64], Isavey and Modic et al. [65–66] Siegman et al. [67]
15–85	Self-reinforcing resins, In-situ composites	Kiss [68] Baird et al. [69]
85–100	Modified LCP	Jackson [70, 71]

The potential utility of LCP as a processing aid for high viscosity conventional polymers was rigorously pursued by ICI in the early 1980s. While the desired viscosity lowering appears to be dominated by the ratio of the viscosities of the components of the mixture, LCPs are unique in processing both high molecular weight and low viscosity. Two modes of behavior have been observed-blends with viscosity which follow the "rule of mixtures" based on the components and blends with viscosity lower than either component. This latter behavior is not understood. The use of LCPs as processing aids should be the easiest blend application to exploit commercially and, ultimately, may serve to render very difficult to process thermoplastics useful in common processes.

The most alluring blend regime to most researchers is the "in-situ" composite where the LCP phase orients during processing to reinforce the plastic part [72, 73]. The effectiveness of this process is a function of the orientation imparted to the LCP in the chosen process. Published micrographs document morphologies ranging from spheres to fibrils. No quantification of morphology or correlations with process conditions have been published. Adhesion between the LCP and conventional polymer phases is clearly poor. During processing, stress transfer appears to be through the tortuosity of the phases, but this important factor has not been evaluated in depth. Once consistent result in all studies is that the presence of an LCP phase renders the blend brittle. This is probably a consequence of the poor interphase adhesion and requires clarification. Mechanical properties of the blends, especially tensile modulus, follow expectations consistent with simple composite concepts in the absence of adhesion between matrix and reinforcement. For these blends to be useful commercially, the issues of adhesion and morphology control must be resolved. In addition, LCP in this application is in direct competition with glass and other reinforcing fibers, hence, the cost-effectiveness of the LCP approach must be established.

It has been found by Baird and others [74–77] that the presence of LCP may accelerate and presumably direct the crystallization of conventional polymers (PET, etc.). Porter [76] has shown that, by blending biphasic polymers such as the PET-poly HBA copolymers, miscibility may be achieved between the conventional phase of the biphasic polymer with another conventional polymer that component is miscible with, i.e., X7-G/PBT. The latter phenomena may offer direction in the search for useful compatibilizing agents for LCP/conventional polymer systems.

There are examples given in the literature where the presence of small quantities of LCP blended with a conventional polymer results in mechanical properties significantly higher than either component. No explanation for this observation has been advanced. The analogy of "introducing the lignin into a woody morphology consisting only of fibrin" is appealing, but is not consistent with emerging models of LCP structure of LCP/conventional polymer interactions.

As sparse as the dataset describing mainchain nematic LCP blends with conventional polymers is, it is rich compared to the almost non-existent data on the blending of other types of LCPs-side chain polymers, flexible spacer polymers, smectics, etc.

12 LCP/LCP Blends

DeMeuse and Jaffe [79] have studied blends of thermotropic copolyesters with other thermotropic copolyesters, examining both blends of different copolymer ratios with identical chemistry (HBA/HNA type) and blends of different chemistries. Recognizing that each copolymer may be viewed as a blend (chain to chain variations, along chain variations), it was hypothesized that blending offered the means to "engineer the distribution." Initial results indicate that this is the case. Through such blending, the behavior of both the mesophase and the solid state can be systematically modified; for example, transition temperatures can be shifted and power law indices of viscosity can be changed. These results also strongly implied that the "sequence matching" model of the thermotropic copolyesters in the solid state, as extended by Windle, is more likely correct than the other models suggested in the literature. Evidence was accumulated that LCP/LCP miscibility is not universal and, that at least in the mesophase, the concepts of Flory and Karasz are consistent with the observed physical chemistry. The importance of this work is that it offers a direction for achieving sufficient insight into the nature of LCP structure-property relationships to design more appropriate molecules for given end-uses. Related work at the IBM Laboratories in San Jose indicates that transesterification may be responsible for some of the above observations. Careful evaluation of the data cannot rule out transesterification effects, but strongly suggests that transesterification is not causal in the observed behavior.

As in the case of LCP/conventional polymer blending, little data exists on the blending of LCPs of different inherent chain architecture or mesophase symmetry. Publications from the laboratories of Ringsdorf [80] and Finkelmann [81] show phase separation in blends of sidechain nematics with other similar polymers or small molecule analogs. It is now established that, in contrast to the behavior of low molecular weight LCs, LCPs are often immiscible.

Investigations into the blending of LCPs with other polymers, conventional or LC, are in their infancy and little is understood in detail. The literature is sparse, highly observational in nature and very difficult to reproduce. Much of what is published is more an indication of what might be than an accurate database from which to draw conclusions. It is clear, however, that until the adhesion of LCP to other polymers and the rules controlling blend morphology are understood, the field will remain highly empirical and unlikely to yield many commercial successes. Conversely, for all the reasons that polymer blending is an attractive route to modified polymer products, blends containing LCP are especially attractive. Systematic research to understand the physical chemistry of LCP containing blends is likely to produce results of both commercial and scientific impact. This is true for LCP/conventional blends and all LCP blend systems. While initial work should focus on the commercial nematic polymers, other symmetries and phases should not be ignored. The area of compatibilizers-

"molecular glue"- is another worthy of support because of the high payback potential.

13 Molecular Composites

The concept of the molecular composites first originated from the work of Flory [82] on the polymer rigid rods. He studied ternary blends of rigid rods (PBX type, aramids, and copolyesters) with conventional polymers in solvents. The treatment predicts that a critical region will exist where a single isotropic phase consisting of rods randomly dispersed in the coils will exist. This region is very narrow in its stability boundaries and the preservation of this structure into the solids state depends on "beating the kinetics." If this could be accomplished, given the extraordinarily high mechanical properties of the rods, fibres with excellent tensile and compressive properties should result (a significant percentage of rods is always in tension). If the composite material could be fabricated into three-dimensional parts, these parts would likely possess the high level of specific mechanical properties currently achievable and be extremely attractive for aerospace applications. To date, it is unclear whether a true molecular composite has been demonstrated although materials possessing small agglomerations of rods (diameter of structure less than 5 nm) have been produced with very high tensile properties. It is not yet established, however, if these materials offer an advantage in tension. The concepts underlying molecular composite physics are consistent with the concepts of miscible blends; the materials being produced, even at the very small sizes of rod structures observed, fit the definitions associated with immiscible LCP/conventional polymer blends. Molecular composites can be treated with the already established framework of polymer/polymer mixing and do not require new concepts for accurate description. Success of molecular composites will be strongly linked to the economics of the processes and materials employed.

The combination of LCPs with other materials to control the balance of properties and improve cost-effectiveness is clearly an important technology area for increasing the overall utility of LCPs. The problems inhibiting the rapid development of this technology are the same as those slowing LCP acceptance in other areas, namely:

highly property anisotropy in finished parts
poor compressive strength,
poor adhesion to conventional materials, and
high cost.

Acknowledgement. The authors thank Drs. S. H. Jacobson and H. N. Sung for helpful comments and discussion.

14 References

1. Marvel CS (1975) Trends in high temperature polymer synthesis. Macromol Chem C13: 219
2. Overberger C, Moore JA (1970) Ladder polymers. Adv Polymer Sci 7: 113
3. Hergenrother PM (1987) Heat resistant polymers. Encyclopedia of Polymer Science and Engr vol. 7. Wiley, New York, p 639
4. Olabisi O, Robenson LM, Shaw MT (1979) Polymer-Polymer Miscibility, Academic, New York
5. Solc K (ed) (1982) Polymer compatibility and incompatibility: principles and practices. Harwood Academic
6. Paul DR, Newman S (eds) (1978) Polymer blends. Academic, New York
7. Robeson LM In: Culberston BL (ed) Contemporary topics in polymer science vol. 6. Plenum, New York, p 177
8. Huggins ML (1941) J Chem. Phys 9: 440
9. Flory PJ (1941) J Chem Phys 9: 660
10. Paul DR, Newman S (1978) Polymer blends. Academic, New York
11. Chen PN Sr, et al. (1990) High Performance Polymers. 2: 39
12. Makhija S, Pearce E, Kwei TK (1992) J Appl Polym Sci 44: 917
13. St. Clair AK, St. Clair TL, Slemp WS, Ezzell KS (1985) NASA Technical Memoranda, p 87650
14. St. Clair AK, St. Clair TL (1984) In: K.L. Mittal (ed) Polyimides vol 2. Plenum, New York, p 997
15. Johnston NJ (1990) Proc of the interdisciplinary symp on recent advances in polyimides and other high performance polymers. ACS Polym Division P I, Jan
16. Arai M, Cassidy PE, Farley JM (1989) Macromolecules 22: 989
17. Leung L, Williams DJ, Karasz FE, MacKnight WJ (1986) Polym Bulletin 16: 457
18. Guerra G, Williams DJ, Karasz FE, MacKnight WJ (1988) J Polym Sci P 26: 301
19. Guerra G, Choe S, Williams DJ, Karasz FE, MacKnight WE (1988)
    Macromolecules 21: 231
20. Jaffe M (1990) Proc of the interdisciplinary symp on recent advances in polyimides and other high performance polymers. ACS Polym. Div, San Diego, Jan
21. Yokota R, Horiuchi R, Kochi M, Soma H, Mita I (1988) J Polym Sci Letters 26: 215
22. Yoon DY, Ree M, Volksen W, Hofer D, Depero L, Parrish W (1988) Proc of the third international conf on polyimides. Sponsored by SPE, p. 1
23. Mueller WH, Vora RH, Khanna DN (1990) European patent application no: 354, 361
24. Vora RH (1990) European patent application no: 354, 360
25. Chung TS, Vora R, Jaffe M (1991) J. Polym. Sci Part A: Polym. Chem 29: 1207
26. Krause S (1976), In: Paul DR, Newman S (eds) Polymer blends vol 1. Academic, New York, p 15
27. MacKnight WJ, Karasz FE, Fried JR, In: Paul DR, Newman S (eds) Polymer blends vol 1. Academic, New York, p. 185
28. Hoy KL (1966) J Paint Technology 38: 43
29. Koros WJ Personal communication. University of Texas
30. Makhija S, Pearce E, Kwei TK, Liu F (1990) Polym Engr & Sci 30: 798
31. Sperling LH (1981) Interpenetrating polymer networks and related materials. Plenum, New York
32. Klempner D, Frisch KC (eds) (1980) Polymer networks in polymer science and technology vol 10
33. Lipatov YS, Sergeeva LM (1979) Interpenetrating polymeric networks. Naukova Kumka, Kiev
34. Hanky AO, St. Clair TL (1985) SAMPE J 21: 40
35. Egli AH, King LL, St. Clair TL (1986) Proc 18th Nat SAMPE Tech Conference 18, p 440
36. St. Clair TL (1987) Proc of the interdisciplinary symposium on recent advances in polyimides and other high temperature polymers. Sponsored by ACS Polymer Div, Reno, NV, July
37. Zeng H, Mai K (1986) Macromol. Chem 187: 1787
38. Yamamoto Y, Satoh S, Etoh S (1985) SAMPE J 21: 6
39. Steiner PA, Browne JM, Blair MT, McKillen JM (1987) SAMPE J 23, p 8
40. Pascal T, Mercier R, Sillion B (1988) Proc third int conference on polyimides. Ellenville, NY, Nov. 2–4
41. Raghava RS (1988) J Polym Sci Part B: Polym Phys Ed 26: 65
42. Egli AO, St. Clair TL (1987) US patent 4695610 assigned to NASA, September
43. Pater RH (1991) Polym Eng Sci 31: 28
44. Pater RH (1991) Polymer Eng Sci 31: 20

45. Serafini TT, Delvigs P, Lightsey GR (1973) US patent 3745149 assigned to NASA, July
46. Pater RH (1992) US patent 5098961 assigned to NASA, March 24
47. Pater RH (1990) Proc of the Interdisciplinary Symposium on Recent Advances in Polyimides and Other High Temperature Polymers, sponsored by ACS Polymer Div, San Diego, January
48. Rogers FE (1976) US patent 3959350 assigned to DuPont, May
49. Pascal T, Mercier R, Sillion B (1989) Polymer 30: 739
50. Pascal T, Mercier R, Sillion B (1990) Polymer 31: 79
51. Pater RH, Hsiung HJ, Hansen MG (1992) In: Finlayson K (ed) Advances in polymer blends and alloys technology 3, p 59
52. Hsiung HJ, Hansen MG, Pater RH (1991) 36th International SAMPE Symposium, April 15
53. Blizard KG, Baird DG (1987) Polym Eng Sci 27: 653
54. Amano M, Nakagawa K (1987) Polymer 28: 263
55. Fujino K, Ogawa Y, Kawai K (1964) J Appl Polym Sci 8: 2147
56. Utracki LA (1983) Polym Engr Sci 23 (11), p 602
57. Takayangi M, Kajiyama T (1978) U.K. patent application 2,008,598
58. Takayangi M, Ogata T, Morikawa M, Kai T (1978) J Macromol Sci Phys B17: 4
59. Takayangi M, Ogata T, Morikawa M, Kai T (1978) J Macromol Sci Phys B22, 231
60. Takayangi M (1983) Pure Appl Chem 55: 819
61. Takayanagi M, Gotto K (1985) Polym Bull 13: 35
62. Nobile MR, Acierno D, Incarnato L, Amendola E, Nicolais L (1990) J of Appl Polym Sci 41: 2723
63. Cogswell FN, Griffin BP, Rose JB (1983) U.S. patent 4,386,174
64. Cogswell FN, Griffin BP, Rose JB (1984) U.S. patent 4,438,236
65. Isayev AI, Modic M (1987) Polym Compos 8: 158
66. Isayev AI, M. Modic M (1987) Polym Compos 8: 158
67. Siegman A, Dagan A, Kenig S (1985) Polym 26: 1325
68. Kiss G (1987) Polym Engr Sci 27: 410
69. Sukhadia AM, Done D, Baird D (1990) Polym Engr Sci 30: 519
70. Jackson WJ (1976) J of Polym Sci Polym Chem Ed 14: 2043
71. Jackson WJ (1980) J of Appl Polym Sci 25: 1685
72. Sukhadia AM, Datta A, Baird D (1992) Int Polym Process Soc VII 3
73. Baird D, Ramanathan R (1989) In: Culbertson BM (ed) Multiphase macromolecular systems. Plenum, New York
74. Josepha EG, Wilkes GL, Baird DG (1984) In: Blumstein A (ed) Polymer liquid crystals. Plenum, New York
75. Bhattacharya SK, Tandolkar A, Misra A (1987) Molec Cryst Liq Cryst 153: 501
76. Paci M, Barone C, Magagnini PL (1987) J Polym Sci Polym Phys Ed 25: 1595
77. Pracella M, Chiellini E, Dainelli D (1989) Makromol Chem 190: 175
78. Kimura M, Porter RS, Salee G (1983) J Polym Sci Polym Phys Ed 21: 367
79. De Meuse MT, Jaffe M (1988) Mol Cryst Liq Nonlin Opt 157: 535
80. Ringsdorf H, Schmidt W, Schneller A (1982) Makromol Chem Rapid Commun 3: 745
81. Finkelmann H, Kock HJ, Rehage G (1982) Mol Cryst Liq Cryst 89: 23
82. Flory PJ (1978) Macromolecules 11: 1138

Received March 1994

Author Index Volumes 101-117

Author Index Vols. 1-100 see Vol. 100

Subject Index

Printing: Saladruck, Berlin
Binding: Buchbinderei Lüderitz & Bauer, Berlin